CONTEMPORARY MATHEMATICS

Trends in Harmonic Analysis and Its Applications

AMS Special Session on
Harmonic Analysis and Its Applications
March 29–30, 2014
University of Maryland, Baltimore County,
Baltimore, MD

T0293604

Jens G. Christel
Susanna Dar
Azita Mayeli
Gestur Ólafsson
Editors

American Mathematical Society
Providence, Rhode Island

2010 *Mathematics Subject Classification.* Primary 22E27, 32M15, 42C15, 46E22, 41A05, 43A07, 46L40, 43A32.

Library of Congress Cataloging-in-Publication Data

Trends in harmonic analysis and its applications : AMS special session on harmonic analysis and its applications : March 29–30, 2014, University of Maryland, Baltimore County, Baltimore, MD / Jens G. Christensen [and three others], editor.
    pages cm. — (Contemporary mathematics ; volume 650)
    Includes bibliographical references.
    ISBN 978-1-4704-1879-3 (alk. paper)
    1. Harmonic analysis—Congresses.   I. Christensen, Jens Gerlach, 1975– editor.

QA403.T744   2015
515′.2433—dc23

2015011079

Contemporary Mathematics ISSN: 0271-4132 (print); ISSN: 1098-3627 (online)

DOI: http://dx.doi.org/DOI

Contents

Preface

The present volume provides a snapshot of current trends within the field of harmonic analysis, both commutative and non-commutative, and its applications. We hope this volume will instigate collaboration, and motivate upcoming researchers in the field, as well as provide new ideas for established experts. For this reason the articles included in this volume are quite diverse.

The subjects fall into three categories: frame theory, functional analysis and C^*-algebras, and harmonic analysis on manifolds. Frame theory has emerged as an active subject in connection with non-harmonic Fourier series, wavelets and Gabor theory. Recent years have seen an added focus on the analysis of finite dimensional vector spaces using frames (rank one projections) and fusion frames (weighted orthogonal projections). This volume includes new developments on phase retrieval for finite dimensional frames (Bahmanpour, Cahill, Casazza, Jasper, Woodland and Balan, Zou), tight and random fusion frames (Cahill, Casazza, Ehler,Li) and Gabor frames (Oussa). The topics of functional analysis and C^*-algebras are deeply intertwined with harmonic analysis, and advances in one area often directly benefit the other. Included here are articles on non-commutative solenoids (Latrémolière, Packer), Cuntz-Krieger algebras (Bezuglyi, Jorgensen), and amenability of direct limit groups (Dawson, Ólafsson). In the area of harmonic analysis on manifolds there are contributions on spherical Fourier transforms on compact symmetric spaces (Ho, Ólafsson), Besov spaces on compact manifolds (Pesenson) and reproducing kernel spaces on Lie groups (Jorgensen, Pedersen, Tian).

The volume is based on an AMS special session *Harmonic Analysis and Its Applications* organized by Susanna Dann, Azita Mayeli and Gestur Ólafsson at the Fall Eastern Sectional Meeting in Baltimore 2014. We wish to sincerely thank all the participants who made the workshop a success, and in particular the authors who contributed to this volume.

We also thank the American Mathematical Society for making the special session happen, and a special thank you goes to Christine M. Thivierge, Associate Editor for Proceedings, for her effort in making this volume come to fruition.

Frame Theory

Contemporary Mathematics
Volume **650**, 2015
http://dx.doi.org/10.1090/conm/650/13047

Phase retrieval and norm retrieval

Saeid Bahmanpour, Jameson Cahill, Peter G. Casazza, John Jasper,
and Lindsey M. Woodland

ABSTRACT. Phase retrieval has become a very active area of research. We
will classify when phase retrieval by Parseval frames passes to the Naimark
complement and when phase retrieval by projections passes to the orthogonal
complements. We introduce a new concept we call norm retrieval and show
that this is what is necessary for passing phase retrieval to complements. This
leads to a detailed study of norm retrieval and its relationship to phase retrieval. One fundamental result: a frame $\{\varphi_i\}_{i=1}^M$ yields phase retrieval if and
only if $\{T\varphi_i\}_{i=1}^M$ yields norm retrieval for every invertible operator T.

1. Introduction

Phase retrieval is the problem of recovering a signal from the absolute value of
linear measurement coefficients called intensity measurements. Since its mathematical introduction in [**3**], phase retrieval has become a very active area of research.
Often times in engineering problems, during processing the phase of a signal is lost
and hence a method for recovering this signal with this lack of information is necessary. In particular, this problem occurs in speech recognition [**5, 13, 14**], X-ray
crystallography and electron microscopy [**2, 11, 12**] and a number of other areas.
In some scenarios, such as crystal twinning [**10**], the signal is projected onto higher
dimensional subspaces and one has to recover this signal from the norms of these
projections. This is called *phase retrieval by projections* and a deep study of this
area was made in [**6**] (see also [**1**]). In [**6**], they analyze and compare phase retrieval by vectors to phase retrieval by projections. Although they solve numerous
problems in this area, there are still many open fundamental questions about phase
retrieval in both cases.

In [**6**] it was shown that if a family of projections $\{P_i\}_{i=1}^M$ yields phase retrieval,
it need not occur that $\{(I - P_i)\}_{i=1}^M$ yields phase retrieval. In the present paper
we will classify when this occurs and solve the related problem: If a Parseval frame
yields phase retrieval, when does its Naimark complement yield phase retrieval?
The fundamental notion which connects phase retrieval to complements is something we call *norm retrieval*. After showing that norm retrieval is central to these
questions, we make a detailed study of norm retrieval and its relationship to phase

2010 *Mathematics Subject Classification*. Primary 32C15 .

Key words and phrases. Phase retrieval, norm retrieval, frames.

The authors were supported by NSF 1307685; and NSF ATD 1042701; AFOSR DGE51-
FA9550-11-1-0245.

retrieval. In particular, we show that a collection of vectors $\{\varphi_i\}_{i=1}^M$ yields phase retrieval if and only if $\{T\varphi_i\}_{i=1}^M$ yields norm retrieval for every invertible operator T. Another fundamental idea which connects these results is the question of when the identity operator is in the span of $\{P_i\}_{i=1}^M$. Often times when a collection of vectors or projections yield phase retrieval then the identity is in their span. Moreover, in [**6**] they show that having the identity in the span of a family of projections doing phase retrieval, will generally yield that the orthogonal complements do phase retrieval. We analyze this question regarding the identity operator and its association to norm retrieval and thus phase retrieval. We will also give a number of examples throughout showing that these results are best possible. For an up to date review of phase retrieval by vectors and projections please see [**9**].

2. Preliminaries

In this section we introduce some necessary definitions and basic theorems from finite frame theory. Throughout, let $\mathcal{H}_N$ denote an N-dimensional Hilbert space.

DEFINITION 2.1. *A family of vectors* $\{\varphi_i\}_{i=1}^M$ *in* $\mathcal{H}_N$ *is a **frame** if there are constants* $0 < A \leq B < \infty$ *so that for all* $x \in \mathcal{H}_N$,

$$A\|x\|^2 \leq \sum_{i=1}^M |\langle x, \varphi_i \rangle|^2 \leq B\|x\|^2,$$

*where A and B are the **lower and upper frame bounds**, respectively. If we can choose $A = B$ then the frame is called **tight**, and if $A = B = 1$ it is called a **Parseval frame**.*

Note that in the finite dimensional setting, a frame is simply a spanning set of vectors in the Hilbert space.

If $\{\varphi_i\}_{i=1}^M$ is a frame for $\mathcal{H}_N$, then the **analysis operator** of the frame is the operator $T : \mathcal{H}_N \to \ell_2(M)$ given by

$$T(x) = \{\langle x, \varphi_i \rangle\}_{i=1}^M$$

and the **synthesis operator** is the adjoint operator, T^*, which satisfies

$$T^* \left(\{a_i\}_{i=1}^M \right) = \sum_{i=1}^M a_i \varphi_i.$$

The **frame operator** is the positive, self-adjoint, invertible operator $S := T^*T$ on $\mathcal{H}_N$ and satisfies:

$$S(x) = T^*T(x) = \sum_{i=1}^M \langle x, \varphi_i \rangle \varphi_i.$$

Moreover, $\{\varphi_i\}_{i=1}^M$ is a frame if there are constants $0 < A \leq B < \infty$ such that its frame operator S satisfies $AI \leq S \leq BI$ where I is the identity on $\mathcal{H}_N$.

In particular, the frame operator of a Parseval frame is the identity operator. This fact makes Parseval frames very helpful in applications because they possess the property of perfect reconstruction. That is, if $\{\varphi_i\}_{i=1}^M$ is a Parseval frame for $\mathcal{H}_N$, then for any $x \in \mathcal{H}_N$ we have

$$x = \sum_{i=1}^M \langle x, \varphi_i \rangle \varphi_i.$$

There is a direct method for constructing Parseval frames. For $M \geq N$, given an $M \times M$ unitary matrix, if we select any N rows from this matrix, then the column vectors from these rows form a Parseval frame for $\mathcal{H}_N$. Moreover, the leftover set of $M - N$ rows, also have the property that its M columns form a Parseval frame for $\mathcal{H}_{M-N}$. The next theorem, known as Naimark's Theorem, says that this is the only way to obtain Parseval frames.

THEOREM 2.2 (Naimark's Theorem; page 36 of [8]). *Let* $\Phi = \{\varphi_i\}_{i=1}^M$ *be a frame for* $\mathcal{H}_N$ *with analysis operator* T, *let* $\{e_i\}_{i=1}^M$ *be the standard basis of* $\ell_2(M)$, *and let* $P : \ell_2(M) \to \ell_2(M)$ *be the orthogonal projection onto* range(T). *Then the following conditions are equivalent:*

  (1) $\{\varphi_i\}_{i=1}^M$ *is a Parseval frame for* $\mathcal{H}_N$.
  (2) *For all* $i = 1, \ldots, M$, *we have* $Pe_i = T\varphi_i$.
  (3) *There exist* $\psi_1, \ldots, \psi_M \in \mathcal{H}_{M-N}$ *such that* $\{\varphi_i \oplus \psi_i\}_{i=1}^M$ *is an orthonormal basis of* $\mathcal{H}_M$.

Moreover, $\{\psi_i\}_{i=1}^M$ *is a Parseval frame for* $\mathcal{H}_{M-N}$.

Explicitly, we call $\{\psi_i\}_{i=1}^M$ the **Naimark complement** of Φ. If $\Phi = \{\varphi_i\}_{i=1}^M$ is a Parseval frame, then the analysis operator T of the frame is an isometry. So we can associate φ_i with $T\varphi_i = Pe_i$, and with a slight abuse of notation we have:

THEOREM 2.3 (Naimark's Theorem). $\Phi = \{\varphi_i\}_{i=1}^M$ *is a Parseval frame for* $\mathcal{H}_N$ *if and only if there is an* M-*dimensional Hilbert space* $\mathcal{K}_M$ *with an orthonormal basis* $\{e_i\}_{i=1}^M$ *such that the orthogonal projection* $P : \mathcal{K}_M \to \mathcal{H}_N$ *satisfies* $Pe_i = \varphi_i$ *for all* $i = 1, \ldots, M$. *Moreover, the Naimark complement of* Φ *is* $\{(I - P)e_i\}_{i=1}^M$.

Note that Naimark complements are only defined for Parseval frames. Furthermore, Naimark complements are only defined up to unitary equivalence, that is, if $\{\varphi_i\}_{i=1}^M \subseteq \mathcal{H}_N$ and $\{\psi_i\}_{i=1}^M \subseteq \mathcal{H}_{M-N}$ are Naimark complements, and U and V are unitary operators, then $\{U\varphi_i\}_{i=1}^M$ and $\{V\psi_i\}_{i=1}^M$ are also Naimark complements.

Given a sequence of vectors $\{\varphi_i\}_{i=1}^M$ and an orthogonal projection P, throughout this paper we will frequently refer to $\{P\varphi_i\}_{i=1}^M$ as possessing certain properties, such as: full spark, complement property, phase retrieval, and so on. By this we mean that it has these properties in the range of P.

This concludes a brief introduction to finite frames and other terms which are necessary throughout the present paper. For a more in depth review of finite frame theory, the interested reader is referred to [8].

3. Phase retrieval by vectors and Naimark complements

Signal reconstruction is an important topic of research with applications to numerous fields. Recovering the phase of a signal through the use of a redundant system of vectors (i.e. a frame) or a spanning collection of subspaces is currently a well studied topic (see [9] for a survey of this field).

DEFINITION 3.1. *A set of vectors* $\{\varphi_i\}_{i=1}^M$ *in* $\mathbb{R}^N$ (*or* $\mathbb{C}^N$) *yields* **phase retrieval** *if for all* $x, y \in \mathbb{R}^N$ (*or* $\mathbb{C}^N$) *satisfying* $|\langle x, \varphi_i \rangle| = |\langle y, \varphi_i \rangle|$ *for all* $i = 1, \ldots, M$, *then* $x = cy$ *where* $c = \pm 1$ *in* $\mathbb{R}^N$ (*and for* $\mathbb{C}^N$, $c \in \mathbb{T}^1$ *where* $\mathbb{T}^1$ *is the complex unit circle*).

A fundamental result from [3] classifies phase retrieval in the real case by way of the **complement property**.

DEFINITION 3.2. *A frame $\{\varphi_i\}_{i=1}^M$ in $\mathcal{H}_N$ satisfies the **complement property** if for all subsets $\mathcal{I} \subset \{1,\ldots,M\}$, either $\operatorname{span}\{\varphi_i\}_{i\in\mathcal{I}} = \mathcal{H}_N$ or $\operatorname{span}\{\varphi_i\}_{i\in\mathcal{I}^c} = \mathcal{H}_N$.*

Similar to the complement property, the notion of **full spark** for a set of vectors is very useful for guaranteeing that large subsets of the vectors actually span the space.

DEFINITION 3.3. *Given a family of vectors $\Phi = \{\varphi_i\}_{i=1}^M$ in $\mathcal{H}_N$, the **spark** of Φ is defined as the cardinality of the smallest linearly dependent subset of Φ. When $\operatorname{spark}(\Phi) = N + 1$, every subset of size N is linearly independent, and Φ is said to be **full spark**.*

THEOREM 3.4. **[3]** *A frame $\{\varphi_i\}_{i=1}^M$ in $\mathbb{R}^N$ yields phase retrieval if and only if it has the complement property. In particular, a full spark frame with $2N - 1$ vectors yields phase retrieval. Moreover, if $\{\varphi_i\}_{i=1}^M$ yields phase retrieval in $\mathbb{R}^N$, then $M \geq 2N - 1$, and no set of $2N - 2$ vectors yields phase retrieval.*

In general, it is not necessary for a frame to be full spark in order to yield phase retrieval. For example, as long as our frame contains a full spark subset of $2N - 1$ vectors, it will do phase retrieval. However, if the frame contains exactly $2N - 1$ vectors, then clearly it does phase retrieval if and only if it is full spark.

From Theorem 3.4, it is clear that full spark and the complement property are important properties for a frame when it comes to classification results regarding phase retrieval. We will now present results regarding frames with these such properties. These results stand alone but they also help in analyzing further phase retrieval and norm retrieval results.

Next we will compare phase retrieval for Parseval frames and their Naimark complements. For this we need a result from **[4]**:

THEOREM 3.5. **[4]** *Let P be a projection on $\mathcal{H}_M$ with orthonormal basis $\{e_i\}_{i=1}^M$ and let $\mathcal{I} \subset \{1,2,\ldots,M\}$. The following are equivalent:*
(1) $\{Pe_i\}_{i\in\mathcal{I}}$ is linearly independent.
(2) $\{(I-P)e_i\}_{i\in\mathcal{I}^c}$ spans $(I-P)\mathcal{H}_M$.

First, we need to see that the full spark property passes from a frame to its Naimark complement.

PROPOSITION 3.6. *A Parseval frame is full spark if and only if its Naimark complement is full spark.*

PROOF. By Theorem 2.2 any Parseval frame can be written as $\{Pe_i\}_{i=1}^M$ where $\{e_i\}_{i=1}^M$ is an orthonormal basis for $\ell_2(M) \simeq \mathcal{H}_M$ and P is an orthogonal projection. Furthermore, the Naimark complement of this Parseval frame is $\{(I-P)e_i\}_{i=1}^M$. We also have that $\{Pe_i\}_{i=1}^M$ is full spark if and only if for any subset $\mathcal{I} \subseteq \{1,\ldots,M\}$ such that $|\mathcal{I}| = N$ we have that $\{Pe_i\}_{i\in\mathcal{I}}$ is linearly independent and spanning (in the image of P). So under this assumption Theorem 3.5 implies that $\{(I-P)e_i\}_{i\in\mathcal{I}}$ is also linearly independent and spanning (in the image of $I-P$), so $\{(I-P)e_i\}_{i=1}^M$ is also full spark. The other direction follows from the same argument by reversing the roles of P and $I-P$. □

In general, if a Parseval frame yields phase retrieval, its Naimark complement may not yield phase retrieval. This follows from the fact that there may not be enough vectors in the Naimark complement to satisfy the complement property as we see in the next proposition.

PROPOSITION 3.7. *Assume a Parseval frame $\{\varphi_i\}_{i=1}^M$ yields phase retrieval for $\mathbb{R}^N$ and its Naimark complement $\{\psi_i\}_{i=1}^M$ yields phase retrieval for $\mathbb{R}^{M-N}$. Then $2N - 1 \le M \le 2N + 1$.*

PROOF. Since $\{\varphi_i\}_{i=1}^M$ yields phase retrieval in $\mathbb{R}^N$, we have by Theorem 3.4 that $2N - 1 \le M$. If the Naimark complement $\{\psi_i\}_{i=1}^M$ yields phase retrieval in $\mathbb{R}^{M-N}$ then again by Theorem 3.4 we have $2(M - N) - 1 \le M$. That is, $M \le 2N + 1$. $\qquad\square$

Unfortunately, even if we restrict the number of vectors in a Parseval frame to $2N - 1 \le M \le 2N + 1$, its Naimark complement still might not yield phase retrieval.

EXAMPLE 3.8. *Let $\{\varphi_i\}_{i=2}^{2N}$ be a set of full spark vectors in $\mathbb{R}^N$, with $N \ge 3$. Define $\varphi_1 = \varphi_2$ and let S be the frame operator for $\{\varphi_i\}_{i=1}^{2N}$. Note that $\{S^{-\frac{1}{2}}\varphi_i\}_{i=2}^{2N}$ is still a full spark set of vectors. Therefore $\{S^{-\frac{1}{2}}\varphi_i\}_{i=1}^{2N}$ yields phase retrieval. That is, for any partition $\mathcal{I}, \mathcal{I}^c \subset \{1, \dots, 2N\}$, either $\mathcal{I}$ or $\mathcal{I}^c$ has at least N elements from the full spark family $\{S^{-\frac{1}{2}}\varphi_i\}_{i=2}^{2N}$ and hence spans $\mathbb{R}^N$.*

Now we will see that the Naimark complement of $\{S^{-\frac{1}{2}}\varphi_i\}_{i=1}^{2N}$ fails phase retrieval. Partition $\{S^{-\frac{1}{2}}\varphi_i\}_{i=1}^{2N}$ into $\{S^{-\frac{1}{2}}\varphi_i\}_{i=1}^{2}$ and $\{S^{-\frac{1}{2}}\varphi_i\}_{i=3}^{2N}$. Observe that neither set is linearly independent since $\varphi_1 = \varphi_2$ and $N \ge 3$. By Theorem 3.5, the Naimark complement of each set does not span $\mathbb{R}^{2N-N} = \mathbb{R}^N$. Hence this is a partition of the Naimark complement of $\{S^{-\frac{1}{2}}\varphi_i\}_{i=1}^{2N}$ which fails complement property and therefore fails phase retrieval.

With the aid of full spark, we are able to pass phase retrieval to Naimark complements as long as we satisfy the restriction on the number of vectors.

PROPOSITION 3.9. *If $\Phi = \{\varphi_i\}_{i=1}^M$ is a full spark Parseval frame in $\mathbb{R}^N$ and $2N - 1 \le M \le 2N + 1$ then Φ yields phase retrieval in $\mathbb{R}^N$ and the Naimark complement of Φ yields phase retrieval in $\mathbb{R}^{M-N}$.*

PROOF. By Proposition 3.6, the Naimark complement of Φ is full spark in $\mathbb{R}^{M-N}$. Since $2N - 1 \le M$ and $2(M - N) - 1 \le M$, by Theorem 3.4 both Φ and its Naimark complement have the complement property in their respective spaces. $\qquad\square$

4. Phase retrieval by projections and norm retrieval

Phase retrieval for the higher dimensional case is similar to the one dimensional case.

DEFINITION 4.1. *Let $\{W_i\}_{i=1}^M$ be a collection of subspaces of $\mathcal{H}_N$ and let $\{P_i\}_{i=1}^M$ be the orthogonal projections onto these subspaces. We say that $\{W_i\}_{i=1}^M$ (or $\{P_i\}_{i=1}^M$) yields **phase retrieval** if for all $x, y \in \mathcal{H}_N$ satisfying $\|P_i x\| = \|P_i y\|$ for all $i = 1, \dots, M$, then $x = cy$ for some scalar c with $|c| = 1$.*

Recently, a detailed study of phase retrieval by projections appeared in [6]. Originally, it was believed that higher ranked projections gave much less information than vectors and hence phase retrieval by projections would require many more projections than phase retrieval by vectors requires. In [6], the surprising result appears that we do not need more projections to yield phase retrieval.

THEOREM 4.2. [6] *Phase retrieval can be done in $\mathbb{R}^N$ with $2N - 1$ arbitrary rank projections ($0 < \text{rank } P_i < N$).*

This result yields the equally surprising problem,

PROBLEM 4.3. *Can phase retrieval be done in $\mathbb{R}^N$ with fewer than $2N - 1$ projections?*

Another question which naturally arises in this context is given subspaces $\{W_i\}_{i=1}^{M}$ of $\mathcal{H}_N$ which yield phase retrieval, do $\{W_i^\perp\}_{i=1}^{M}$ yield phase retrieval? It is shown in [**6**] that this is not true in general. We introduce a new fundamental property, norm retrieval, which is precisely what is needed to pass phase retrieval to orthogonal complements. Next we make precise the definition of norm retrieval and then show its importance to phase retrieval. After this we will develop the basic properties of norm retrieval.

DEFINITION 4.4. *Let $\{W_i\}_{i=1}^{M}$ be a collection of subspaces in $\mathcal{H}_N$ and define $\{P_i\}_{i=1}^{M}$ to be the orthogonal projections onto each of these subspaces. We say that $\{W_i\}_{i=1}^{M}$ (or $\{P_i\}_{i=1}^{M}$) yields **norm retrieval** if for all $x, y \in \mathcal{H}_N$ satisfying $\|P_i x\| = \|P_i y\|$ for all $i = 1, \dots, M$, then $\|x\| = \|y\|$.*

REMARK 4.5. *Although trivial, it is important to point out that any collection of subspaces which yields phase retrieval necessarily yields norm retrieval. However, the converse need not hold since any orthonormal set of vectors does norm retrieval but has too few vectors to do phase retrieval.*

We will start by showing that norm retrieval is precisely the condition needed to pass phase retrieval to orthogonal complements.

PROPOSITION 4.6. *Let $\{W_i\}_{i=1}^{M}$ be a collection of subspaces in $\mathcal{H}_N$ yielding phase retrieval and let $\{P_i\}_{i=1}^{M}$ be the projections onto these subspaces. The following are equivalent:*

(1) *$\{I - P_i\}_{i=1}^{M}$ yields phase retrieval.*
(2) *$\{I - P_i\}_{i=1}^{M}$ yields norm retrieval.*

PROOF. $(1) \Rightarrow (2)$ Since phase retrieval always implies norm retrieval then this is clear.
$(2) \Rightarrow (1)$ Let $x, y \in \mathcal{H}_N$ be such that $\|(I - P_i)x\|^2 = \|(I - P_i)y\|^2$ for all $i = 1, \dots, M$. Since $\{I - P_i\}_{i=1}^{M}$ yields norm retrieval then this implies that $\|x\|^2 = \|y\|^2$. Since P and $(I - P)$ correspond to orthogonal subspaces then $\|x\|^2 = \|P_i x\|^2 + \|(I - P_i)x\|^2$ for all $i = 1, \dots, M$. Thus, for all $i = 1, \dots, M$ we have

$$\|P_i x\|^2 = \|x\|^2 - \|(I - P_i)x\|^2 = \|y\|^2 - \|(I - P_i)y\|^2 = \|P_i y\|^2.$$

Since $\{W_i\}_{i=1}^{M}$ yields phase retrieval then this implies that $x = cy$ for some scalar $|c| = 1$. Therefore $\{I - P_i\}_{i=1}^{M}$ yields phase retrieval. $\square$

We can think of norm retrieval as giving us one free measurement when trying to do phase retrieval. For example, let $\{e_i\}_{i=1}^{3}$ be an orthonormal basis for $\mathbb{R}^3$ and choose $\{\varphi_1, \varphi_2\}$ so that these 5 vectors are full spark. Hence, these vectors yield phase retrieval in $\mathbb{R}^3$. Now consider the family of vectors $\Phi := \{e_1, e_2, \varphi_1, \varphi_2\}$ in $\mathbb{R}^3$. Φ cannot do phase retrieval in general since it takes at least 5 vectors to do phase retrieval in $\mathbb{R}^3$. However, given unit norm $x, y \in \mathbb{R}^3$, Φ will do phase retrieval in this scenario since we have knowledge of the norms of the signals. Assume

$$|\langle x, e_i \rangle| = |\langle y, e_i \rangle| \text{ and } |\langle x, \varphi_i \rangle| = |\langle y, \varphi_i \rangle| \text{ for } i = 1, 2.$$

Then

$$1 = \|x\|^2 = \sum_{i=1}^{3} |\langle x, e_i \rangle|^2,$$

and similarly for y. This implies

$$|\langle x, e_3 \rangle|^2 = 1 - \sum_{i=1}^{2} |\langle x, e_i \rangle|^2 = 1 - \sum_{i=1}^{2} |\langle y, e_i \rangle|^2 = |\langle y, e_3 \rangle|^2.$$

That is, x, y have the same modulus of inner products with all 5 vectors which yield phase retrieval and so $x = \pm y$.

A fundamental idea is to apply operators to vectors and subspaces which yield phase retrieval or norm retrieval. We now consider when operators preserve these concepts.

PROPOSITION 4.7. *If $\{\varphi_i\}_{i=1}^{M}$ is a frame in $\mathcal{H}_N$ which yields phase retrieval (respectively norm retrieval) then $\{P\varphi_i\}_{i=1}^{M}$ yields phase retrieval (respectively norm retrieval) for all orthogonal projections P on $\mathcal{H}_N$.*

PROOF. Let $x, y \in P(\mathcal{H}_N)$ such that $|\langle x, P\varphi_i \rangle|^2 = |\langle y, P\varphi_i \rangle|^2$ for all $i = \{1, \ldots, M\}$. For all $i \in \{1, \ldots, M\}$, we have

$$|\langle x, \varphi_i \rangle|^2 = |\langle Px, \varphi_i \rangle|^2 = |\langle x, P\varphi_i \rangle|^2 = |\langle y, P\varphi_i \rangle|^2 = |\langle Py, \varphi_i \rangle|^2 = |\langle y, \varphi_i \rangle|^2.$$

Since $\{\varphi_i\}_{i=1}^{M}$ gives phase retrieval (respectively norm retrieval) then this implies $x = cy$ for some scalar $|c| = 1$ (respectively $\|x\| = \|y\|$). Therefore, $\{P\varphi_i\}_{i=1}^{M}$ yields phase retrieval (respectively norm retrieval). □

Although norm retrieval is preserved when applying any projection to the vectors, this does not hold when we apply an invertible operator to the vectors. The next theorem classifies when invertible operators maintain norm retrieval.

THEOREM 4.8. *Let $\{\varphi_i\}_{i=1}^{M}$ be vectors in $\mathcal{H}_N$. The following are equivalent:*

(1) *$\{\varphi_i\}_{i=1}^{M}$ yields phase retrieval.*
(2) *$\{T\varphi_i\}_{i=1}^{M}$ yields phase retrieval for all invertible operators T on $\mathcal{H}_N$.*
(3) *$\{T\varphi_i\}_{i=1}^{M}$ yields norm retrieval for all invertible operators T on $\mathcal{H}_N$.*

PROOF. (1) $\Rightarrow$ (2) Let T be any invertible operator on $\mathbb{R}^N$ and let $x, y \in \mathcal{H}_N$ be such that $|\langle x, T\varphi_i \rangle| = |\langle y, T\varphi_i \rangle|$ for all $i \in \{1, \ldots, M\}$. Then $|\langle T^*x, \varphi_i \rangle| = |\langle T^*y, \varphi_i \rangle|$ for all $i \in \{1, \ldots, M\}$. Since $\{\varphi_i\}_{i=1}^{M}$ yields phase retrieval then this implies $T^*x = cT^*y$ for some scalar $|c| = 1$. Since T is invertible and linear then $(T^*)^{-1}T^*x = (T^*)^{-1}cT^*y$ implies $x = cy$ and $|c| = 1$. Therefore, $\{T\varphi_i\}_{i=1}^{M}$ does phase retrieval.

(2) $\Rightarrow$ (3) Since phase retrieval implies norm retrieval then this is clear.

(3) $\Rightarrow$ (1) Choose nonzero $x, y \in \mathcal{H}_N$ such that $|\langle x, \varphi_i \rangle| = |\langle y, \varphi_i \rangle|$ for all $i \in \{1, \ldots, M\}$. By assumption, $\{T\varphi_i\}_{i=1}^{M}$ yields norm retrieval for all invertible operators T on $\mathcal{H}_N$.

Let T be any invertible operator on $\mathcal{H}_N$, then $(T^*)^{-1}$ is an invertible operator and hence $\{(T^*)^{-1}\varphi_i\}_{i=1}^M$ yields norm retrieval. For $Tx, Ty \in \mathcal{H}_N$, we have

$$
\begin{aligned}
|\langle Tx, (T^*)^{-1}\varphi_i \rangle| &= |\langle T^{-1}Tx, \varphi_i \rangle| \\
&= |\langle x, \varphi_i \rangle| \\
&= |\langle y, \varphi_i \rangle| \\
&= |\langle T^{-1}Ty, \varphi_i \rangle| \\
&= |\langle Ty, (T^*)^{-1}\varphi_i \rangle|,
\end{aligned}
$$

for every $i \in \{1, \ldots, M\}$. Hence, $\|Tx\| = \|Ty\|$ for any invertible operator T on $\mathcal{H}_N$.

Now we will be done if we can show that $\|Tx\| = \|Ty\|$ for any invertible T implies $y = cx$ for some scalar c with $|c| = 1$. First note that since the identity operator is invertible we have that $\|x\| = \|y\|$, so if $y = cx$ then it follows that $|c| = 1$. Now choose an orthonormal basis $\{e_j\}_{j=1}^N$ for $\mathcal{H}_N$ with $e_1 = \frac{x}{\|x\|}$ and suppose $y = \sum_{j=1}^N \alpha_j e_j$, so that $\|y\|^2 = \sum_{j=1}^N \alpha_j^2$. Define the operator T by $Te_1 = e_1$ and $Te_j = \frac{1}{2}e_j$ for $j = 2, \ldots, N$. Now we have that $\|x\|^2 = \|Tx\|^2 = \|Ty\|^2 = \alpha_1^2 + \sum_{j=2}^N \frac{1}{4}\alpha_j^2$, which implies that $\sum_{j=2}^N \alpha_j^2 = \frac{1}{4}\sum_{j=2}^N \alpha_j^2$, and so $\alpha_j = 0$ for $j = 2, \ldots, N$. Therefore, $y = \alpha_1 e_1 = \frac{\alpha_1}{\|x\|}x$ which competes the proof. $\square$

Note that the equivalence of (1) and (2) in the above Theorem was shown in [**3**], so the new part is that (3) is equivalent to both of these.

At first glance one would think that retrieving the norm of a signal would be much easier than recovering the actual signal. However, Theorem 4.8 gives a new classification of phase retrieval in terms of norm retrieval and states that if every invertible operator applied to a frame yields norm retrieval then our original frame yields phase retrieval. This illustrates that recovering the norm of a signal may be more similar to recovering the actual signal than originally thought and hence may not be as easily achievable as anticipated.

PROBLEM 4.9. *Does Theorem 4.8 generalize to subspaces?*

Theorem 4.8 shows that if a frame does not yield phase retrieval, then we cannot apply an invertible operator to it in order to get a frame that does yield phase retrieval. In contrast, it is true that there exists at least one invertible operator which when applied to a frame yields norm retrieval.

PROPOSITION 4.10. *Given $\{\varphi_i\}_{i=1}^M$ spanning $\mathcal{H}_N$, there exists an invertible operator T on $\mathcal{H}_N$ so that the collection of orthogonal projections onto the vectors $\{T\varphi_i\}_{i=1}^M$ yields norm retrieval.*

PROOF. Without loss of generality, assume $\{\varphi_i\}_{i=1}^N$ are linearly independent. Choose an invertible operator T so that $T\varphi_i = e_i$ for all $i = 1, \ldots, N$, where $\{e_i\}_{i=1}^N$ is an orthonormal basis for $\mathcal{H}_N$. Thus, for any $x \in \mathcal{H}_N$, $\|x\|^2 = \sum_{i=1}^N |\langle x, e_i \rangle|^2 = \sum_{i=1}^N |\langle x, T\varphi_i \rangle|^2$. Hence, the collection of orthogonal projections onto the vectors $\{T\varphi_i\}_{i=1}^N$ yields norm retrieval. In particular, $\{T\varphi_i\}_{i=1}^M$ yields norm retrieval. $\square$

5. The identity operator, norm retrieval and phase retrieval

Instead of applying operators to a frame and observing properties regarding the image; we now look at what operators lie in the span of the projections onto a collection of subspaces. In numerous results and examples regarding phase retrieval, we have found that often times when a collection of vectors or projections yields phase retrieval or norm retrieval then the identity is in their span. In this section we classify when this occurs and when it fails. In [**6**], they show that having the identity in the span of a family of projections doing phase retrieval, will generally yield that the orthogonal complements do phase retrieval.

THEOREM 5.1. [**6**] *Assume* $\{W_i\}_{i=1}^M$ *are subspaces of* $\mathbb{R}^N$ *yielding phase retrieval with corresponding orthogonal projections* $\{P_i\}_{i=1}^M$. *If* $I = \sum_{i=1}^M a_i P_i$ *and* $\sum_{i=1}^M a_i \neq 1$, *then* $\{W_i^{\perp}\}_{i=1}^M$ *yields phase retrieval.*

We now state one consequence of Theorem 5.1 which did not appear in [**6**].

THEOREM 5.2. *Let* $\{W_i\}_{i=1}^M$ *be a collection subspaces of* $\mathcal{H}_N$ *that yields phase retrieval, and suppose further that* $\dim(W_i) = K$ *for every* $i = 1, 2, \ldots, M$. *Let* P_i *be the orthogonal projection onto* W_i *and suppose that* $I \in \text{span}\{P_i\}_{i=1}^M$, *then* $\{W_i^{\perp}\}$ *yields phase retrieval.*

PROOF. Let $I = \sum_{i=1}^M a_i P_i$. Then

$$N = \text{Tr}(I) = \text{Tr}(\sum_{i=1}^M a_i P_i) = \sum_{i=1}^M a_i \text{Tr}(P_i) = K \sum_{i=1}^M a_i$$

since $\text{Tr}(P_i) = K$ for every i. Therefore, $\sum_{i=1}^M a_i = \frac{N}{K} > 1$ since $K < N$, so the result follows from Theorem 5.1.   □

Given a collection of orthogonal projections $\{P_i\}_{i=1}^M$, we cannot conclude that they yield phase retrieval just because $I \in \text{span}\{P_i\}_{i=1}^M$. For example, given any projection P, then $I = P + (I - P)$ but certainly $\{P, I - P\}$ will not yield phase retrieval. However, we now show that this is enough to conclude that $\{P_i\}_{i=1}^M$ yields norm retrieval.

PROPOSITION 5.3. *Let* $\{W_i\}_{i=1}^M$ *be subspaces of* $\mathcal{H}_N$, *and let* $\{P_i\}_{i=1}^M$ *be the associated projections. If* $I \in \text{span}\{P_i\}_{i=1}^M$, *then* $\{W_i\}_{i=1}^M$ *gives norm retrieval.*

PROOF. Suppose $I = \sum_{i=1}^M a_i P_i$. Notice for $x \in \mathcal{H}_N$ we have

$$\sum_{i=1}^M a_i \|P_i x\|^2 = \sum_{i=1}^M \langle a_i P_i x, x \rangle = \left\langle \sum_{i=1}^M a_i P_i x, x \right\rangle = \langle I x, x \rangle = \|x\|^2$$

Let $x, y \in \mathcal{H}_N$ such that $\|P_i x\| = \|P_i y\|$ for all $i = 1, \ldots, M$. We have

$$\|x\|^2 = \sum_{i=1}^M a_i \|P_i x\|^2 = \sum_{i=1}^M a_i \|P_i y\|^2 = \|y\|^2.$$

Hence $\{W_i\}_{i=1}^M$ yields norm retrieval.   □

The converse of Proposition 5.3 is far from true. One way to see this (at least for the real case) is as follows: For $2N \leq M \leq N(N+1)/2$ choose any full spark frame $\Phi = \{\varphi_i\}_{i=1}^M$ for $\mathbb{R}^N$ such that $\{\varphi_i \varphi_i^*\}_{i=1}^M$ is linearly independent (which happens generically, see e.g., [**7**]). Let S be the frame operator for Φ and define

$$\psi_i = \frac{S^{-1/2}\varphi_i}{\|S^{-1/2}\varphi_i\|}$$

with $P_i = \psi_i \psi_i^*$ (note that P_i is a rank one orthogonal projection). Since $\{S^{-1/2}\varphi_i\}_{i=1}^M$ is a Parseval frame it follows that

$$(5.1) \qquad\qquad I = \sum_{i=1}^M \|S^{-1/2}\varphi_i\|^2 P_i.$$

Also, since $\{\varphi_i\varphi_i^*\}_{i=1}^M$ is linearly independent it follows that $\{P_i\}_{i=1}^M$ is linearly independent (and so (5.1) is the only way to write I as a linear combination of the P_i's). Also since $\{\varphi_i\}_{i=1}^M$ is full spark we know that $\|S^{-1/2}\varphi_i\| \neq 0$ for every $i = 1, 2, \ldots, M$. Therefore it follows that if $\mathcal{I} \subset \{1, 2, \ldots, M\}$ then $I \notin \operatorname{span}\{P_i\}_{i\in\mathcal{I}}$. Furthermore, since $M \geq 2N$ and $\{\varphi_i\}_{i=1}^M$ is full spark it follows that $\{P_i\}_{i\in\mathcal{I}}$ yields phase retrieval (and hence norm retrieval) whenever $|\mathcal{I}| \geq 2N - 1$.

Although the above example proves that the converse of Proposition 5.3 is false, in the special case where $\sum_{i=1}^M \dim(W_i) = N$ it turns out to be true.

PROPOSITION 5.4. *A collection of unit norm vectors $\{\varphi_i\}_{i=1}^N$ in $\mathcal{H}_N$ yield norm retrieval if and only if $\{\varphi_i\}_{i=1}^N$ are orthogonal.*

Proposition 5.4 is a consequence of the following more general theorem about subspaces.

THEOREM 5.5. *Let $\{W_i\}_{i=1}^M$ be a collection of subspaces of $\mathcal{H}_N$ with the property that $\sum_{i=1}^M \dim(W_i) = N$ and let P_i be the orthogonal projection onto subspace W_i for each $i = 1, \ldots, M$. The following are equivalent:*

(1) *$\{W_i\}_{i=1}^M$ yields norm retrieval*
(2) *$\sum_{i=1}^M P_i = I$.*

PROOF. (2) $\Rightarrow$ (1) Follows from Proposition 5.3.

(1) $\Rightarrow$ (2) Pick some W_j and define V_j to be the span of the $M - 1$ subspaces $\{W_i\}_{i\neq j}$, and let Q be the orthogonal projection onto V_j. Without loss of generality we may assume that W_j is not the zero subspace. Note that

$$\dim V_j \leq \sum_{i\neq j} \dim(W_i) = N - \dim(W_j).$$

Claim 1: $W_j \cap V_j = \{0\}$.
Proof of Claim: Assume to the contrary that $W_j \cap V_j$ is nontrivial. Then

$$\dim \operatorname{span}\{W_i\}_{i=1}^M < \dim V_j + \dim W_j \leq N.$$

This implies that there exists a nonzero $x_0 \in (\operatorname{span}\{W_i\}_{i=1}^M)^\perp$ and hence $P_i x_0 = 0$ for all $i = 1, \ldots, M$. However, since $\{W_i\}_{i=1}^M$ gives norm retrieval, we conclude that $x_0 = 0$, a contradiction. Thus $W_j \cap V = \{0\}$.

Claim 2: $P_j Q = Q P_j = 0$.
Proof of Claim 2: Assume toward a contradiction that $P_j Q \neq 0$, and thus $Q P_j \neq 0$. Set $Y = \{x \in W_j \colon Qx = 0\}$. Since $Q P_j \neq 0$ we see that $Y \neq W_j$.

Let Z be the orthogonal complement of Y in W_j. Since $V_j \cap W_j = \{0\}$ and $Z \subset W_j$, we conclude that $V_j \cap Z = \{0\}$.

Let $z \in Z \setminus \{0\}$. Since $z \notin V_j$ we have $Qz \neq z$, and since $z \notin Y$ we have $Qz \neq 0$. Set $x := Qz \neq 0$ (note $x \neq z$) and $y := (I - Q)z \neq 0$.

Note that

$$\begin{aligned}
\langle P_j x, y \rangle &= \langle P_j Q z, (I-Q)z \rangle = \langle P_j Q z, P_j(I-Q)z \rangle \\
&= \langle P_j Q z, P_j z - P_j Q z \rangle = \langle P_j Q z, z - P_j Q z \rangle \\
&= \langle P_j Q z, z \rangle - \|P_j Q z\|^2 = \langle Q z, P_j z \rangle - \|P_j Q z\|^2 \\
&= \langle Q z, z \rangle - \|P_j Q z\|^2 = \langle Q z, Q z \rangle - \|P_j Q z\|^2 \\
&= \|Q z\|^2 - \|P_j Q z\|^2
\end{aligned}$$

Subclaim: $\langle P_j x, y \rangle$ is nonzero and positive.

Proof of Subclaim: Notice that $\|Q z\|^2 = \|P_j Q z\|^2 + \|(I-P_j)Q z\|^2$ and hence $\|P_j Q z\|^2 \le \|Q z\|^2$. Thus $\langle P_j x, y \rangle = \|Q z\|^2 - \|P_j Q z\|^2 \ge 0$. In particular, if $\langle P_j x, y \rangle = 0$ then this forces $P_j Q z = Q z$. Hence $Q z = x \in W_j$, contradicting the fact that $W_j \cap V_j = \{0\}$. Thus, $\langle P_j x, y \rangle$ is nonzero and positive.

Set $v_1 = x$ and $v_2 = x + \alpha y$ for some $\alpha \in \mathcal{H}$ which will be specified later. Since $P_i(I-Q) = 0$ for $i \ne j$, we have

$$\begin{aligned}
\|P_i v_2\| &= \|P_i(x + \alpha y)\| = \|P_i(Q z + \alpha(I-Q)z)\| \\
&= \|P_i Q z + \alpha P_i(I-Q)z\| = \|P_i Q z\| = \|P_i x\| \\
&= \|P_i v_1\|
\end{aligned}$$

for all $i \ne j$.

If $P_j y = 0$ then we take α to be any nonzero scalar in $\mathcal{H}$, and we have

$$\|P_j v_1\| = \|P_j x\| = \|P_j x + \alpha P_j y\| = \|P_j v_2\|.$$

If $P_j y \ne 0$ then we set $\alpha = -\frac{2\langle P_j x, y \rangle}{\|P_j y\|^2}$ and we have

$$\begin{aligned}
\|P_j v_2\|^2 &= \|P_j x + \alpha P_j y\|^2 \\
&= \|P_j x\|^2 + \alpha\bar{\alpha}\|P_j y\|^2 + \bar{\alpha}\langle P_j x, P_j y \rangle + \alpha\langle P_j y, P_j x \rangle \\
&= \|P_j x\|^2 + \frac{4\langle P_j x, y \rangle\langle y, P_j x \rangle}{\|P_j y\|^4}\|P_j y\|^2 - \frac{2\langle y, P_j x \rangle\langle P_j x, y \rangle}{\|P_j y\|^2} - \frac{2\langle P_j x, y \rangle\langle y, P_j x \rangle}{\|P_j y\|^2} \\
&= \|P_j x\|^2 + \frac{4|\langle P_j x, y \rangle|^2}{\|P_j y\|^2} - \frac{2|\langle P_j x, y \rangle|^2}{\|P_j y\|^2} - \frac{2|\langle P_j x, y \rangle|^2}{\|P_j y\|^2} \\
&= \|P_j x\|^2 = \|P_j v_1\|^2.
\end{aligned}$$

Thus $\|P_i v_1\| = \|P_i v_2\|$ for all $i = 1, \ldots, M$. However, for any $\alpha \in \mathcal{H} \setminus \{0\}$ we have

$$\|v_2\|^2 = \|x\|^2 + |\alpha|^2\|y\|^2 > \|x\|^2 = \|v_1\|^2.$$

Hence $\{W_i\}$ does not yield norm retrieval, a contradiction to our assumption, and so $Q P_j = P_j Q = 0$, which finishes the proof of Claim 2.

Therefore, we have that

$$W_j = \mathrm{im}(P_j) = \ker(Q) = V_j^\perp.$$

But if $i \ne j$ then $W_i \subseteq V_j$ and so $W_i \perp W_j$, from which (2) easily follows. $\qquad\square$

References

[1] C. Bachoc and M. Ehler, *Signal reconstruction from the magnitude of subspace components*, Available online: arXiv:1209.5986.

[2] R. H. Bates and D. Mnyama. *The status of practical Fourier phase retrieval*, in W. H. Hawkes, ed., Advances in Electronics and Electron Physics, **67** (1986), 1–64.

[3] Radu Balan, Pete Casazza, and Dan Edidin, *On signal reconstruction without phase*, Appl. Comput. Harmon. Anal. **20** (2006), no. 3, 345–356, DOI 10.1016/j.acha.2005.07.001. MR2224902 (2007b:94054)

[4] Bernhard G. Bodmann, Peter G. Casazza, Vern I. Paulsen, and Darrin Speegle, *Spanning and independence properties of frame partitions*, Proc. Amer. Math. Soc. **140** (2012), no. 7, 2193–2207, DOI 10.1090/S0002-9939-2011-11072-4. MR2898683

[5] C. Becchetti and L. P. Ricotti. *Speech recognition theory and C++ implementation*. Wiley (1999).

[6] J. Cahill, P.G. Casazza, J. Peterson and L.M. Woodland. *Phase retrieval by projections*, Available online: arXiv:1305.6226.

[7] J.Cahill and X. Chen. *A note on scalable frames*, Available online: arXiv:1301.7292.

[8] *Finite frames*, Applied and Numerical Harmonic Analysis, Birkhäuser/Springer, New York, 2013. Theory and applications; Edited by Peter G. Casazza and Gitta Kutyniok. MR2964005

[9] P.G. Casazza and L.M. Woodland, *Phase retrieval by vectors and projections* Contemporary Math. (to appear)

[10] J. Drenth, *Principles of protein x-ray crystallography*, Springer, 2010.

[11] J. R. Fienup. *Reconstruction of an object from the modulus of its fourier transform*, Optics Letters, **3** (1978), 27–29.

[12] J. R. Fienup. *Phase retrieval algorithms: A comparison*, Applied Optics, **21** (15) (1982), 2758–2768.

[13] L. Rabiner and B. H. Juang. *Fundamentals of speech recognition*. Prentice Hall Signal Processing Series (1993).

[14] Joseph M. Renes, Robin Blume-Kohout, A. J. Scott, and Carlton M. Caves, *Symmetric informationally complete quantum measurements*, J. Math. Phys. **45** (2004), no. 6, 2171–2180, DOI 10.1063/1.1737053. MR2059685 (2004m:81043)

DEPARTMENT OF MATHEMATICS, UNIVERSITY OF MISSOURI, COLUMBIA, MISSOURI 65211-4100
E-mail address: sbgxf@math.missouri.edu

MATHEMATICS DEPARTMENT, DUKE UNIVERSITY, BOX 90320, DURHAM, NORTH CAROLINA 27708-0320
E-mail address: jameson.cahill@gmail.com

DEPARTMENT OF MATHEMATICS, UNIVERSITY OF MISSOURI, COLUMBIA, MISSOURI 65211-4100
E-mail address: casazzap@missouri.edu

DEPARTMENT OF MATHEMATICS, UNIVERSITY OF OREGON, EUGENE, OREGON 97403
E-mail address: jjasper30@gmail.com

DEPARTMENT OF MATHEMATICS, UNIVERSITY OF MISSOURI, COLUMBIA, MISSOURI 65211-4100
E-mail address: lmwvh4@mail.missouri.edu

Contemporary Mathematics
Volume **650**, 2015
http://dx.doi.org/10.1090/conm/650/13030

On Lipschitz inversion of nonlinear redundant representations

Radu Balan and Dongmian Zou

ABSTRACT. In this note we show that reconstruction from magnitudes of frame coefficients (the so called "phase retrieval problem") can be performed using Lipschitz continuous maps. Specifically we show that when the nonlinear analysis map $\alpha : H \to \mathbb{R}^m$ is injective, with $(\alpha(x))_k = |\langle x, f_k \rangle|^2$, where $\{f_1, \cdots, f_m\}$ is a frame for the Hilbert space H, then there exists a left inverse map $\omega : \mathbb{R}^m \to H$ that is Lipschitz continuous. Additionally we obtain that the Lipschitz constant of this inverse map is at most 12 divided by the lower Lipschitz constant of α.

1. Introduction

Let H be an n-dimensional Hilbert space and $\mathcal{F} = \{f_1, f_2, \cdots, f_m\}$ be a spanning set for H. Since H has finite dimension, $\mathcal{F}$ forms a frame for H, that is, there exist two positive constants A and B such that

$$(1.1) \qquad A \|x\|^2 \leq \sum_{k=1}^{m} |\langle x, f_k \rangle|^2 \leq B \|x\|^2, \quad \forall x \in H$$

In this paper, H can be a real or complex Hilbert space and the result applies to both cases. On H we consider the equivalent replation $x \sim y$ if and only if there is a scalar a of magnitude one, $|a| = 1$, so that $y = ax$. Let $\hat{H} = H/\sim$ denote the set of equivalence classes. Note that $\hat{H} \setminus \{0\}$ is equivalent to the cross-product between a real or complex projective space $\mathcal{P}^{n-1}$ of dimension $n-1$ and the positive semiaxis $\mathbb{R}^+ = (0, \infty)$.

Let α denote the nonlinear map

$$(1.2) \qquad \alpha : H \to \mathbb{R}^m, \ \alpha(x) = \left(|\langle x, f_k \rangle|^2 \right)_{1 \leq k \leq m}$$

Note that α induces a nonlinear map which is well defined on $\hat{H}$. By abuse of notation we also denote it by α. The *phase retrieval problem* (or the *phaseless reconstruction problem*) refers to analyzing when α is an injective map, and in this

2010 *Mathematics Subject Classification.* Primary 15A29, 65H10, 90C26.

Key words and phrases. Frames, Lipschitz maps, stability.

The first author was supported in part by NSF Grant DMS-1109498 and DMS-1413249. He also acknowledges fruitful discussions with Krzysztof Nowak and Hugo Woerdeman (both from Drexel University) who pointed out several references, with Stanislav Minsker (Duke University) for pointing out [**ZB**], and Vern Paulsen (University of Houston), Marcin Bownick (University of Oregon) and Friedrich Philipp (Technical University of Berlin).

case to finding "good" left inverses. The frame $\mathcal{F}$ is said to be *phase retrievable* if the nonlinear map α is injective. In this paper we assume α is injective (hence $\mathcal{F}$ is phase retrievable).

A continuous map $f : (X, d_X) \to (Y, d_Y)$, defined between metric spaces X and Y with distances d_X and d_Y respectively, is said to be Lipschitz continuous with Lipschitz constant $Lip(f)$ if

$$(1.3) \qquad Lip(f) := \sup_{x_1 \neq x_2 \in X} \frac{d_Y(f(x_1), f(x_2))}{d_X(x_1, x_2)} < \infty$$

Existing literature (e.g. [**BW**]) establishes that when α is injective, it is also bi-Lipschitz for metric d_1 (the nuclear norm, which is defined in (2.4)) in $\hat{H}$ and Euclidian norm in $\mathbb{R}^m$. As a consequence of these results we obtain that a left inverse of α is Lipschitz when restricted to the image of $\hat{H}$ through α. In this paper we show that this left inverse admits a Lipschitz continuous extension to the entire $\mathbb{R}^m$. Surprisingly we obtain the Lipschitz constant of this extension is just a small factor larger than the minimal Lipschitz constant, a factor that is independent of the dimension n or the number of frame vectors m.

The Lipschitz properties of α is related to the stability of reconstruction. Consider the noisy model for the reconstruction of a signal x with the measurements

$$(1.4) \qquad y = \alpha(x) + \nu$$

where $\nu \in \mathbb{R}^m$ is the noise. The stability of specific reconstruction methods is studied in, for instance, [**BCMN**], [**BH**] and [**CSV**]. In general, if we can find (guaranteed by the result of this paper) a Lipschitz continuous map defined on the whole $\mathbb{R}^m$, say $\omega : (\mathbb{R}^m, \|\cdot\|) \to (\hat{H}, d_1)$, such that $\omega(\alpha(x)) = x$ for all $x \in \hat{H}$, then we have a stable reconstruction in the following sense: Let $x_0 \in \hat{H}$ be the original signal and y_1 be the measurement from the noisy model (1.4) with noise ν_1. Let $x_1 = \omega(y_1)$. Then

$$(1.5) \quad d_1(x_0, x_1) = d_1(\omega(\alpha(x_0)), \omega(y_1)) \leq Lip(\omega) \cdot \|\alpha(x_0) - y_1\| = Lip(\omega) \cdot \|\nu_1\|.$$

Moreover, let y_1 and y_2 be two different measurements of $\alpha(x_0)$ from (1.4) with noise ν_1, ν_2, respectively. Then we have

$$(1.6) \quad d_1(x_1, x_2) = d_1(\omega(y_1), \omega(y_2)) \leq Lip(\omega) \cdot \|y_1 - y_2\| = Lip(\omega) \cdot \|\nu_1 - \nu_2\|.$$

Note that in general (1.5) does not imply (1.6).

2. Notations and Statement of Main Results

The nonlinear map α defined by (1.2) naturally induces a linear map between the space $Sym(H) = \{T : H \to H \, , \, T = T^*\}$ of symmetric operators on H and $\mathbb{R}^m$:

$$(2.1) \qquad \mathcal{A} : Sym(H) \to \mathbb{R}^m \, , \, \mathcal{A}(T) = (\langle T f_k, f_k \rangle)_{1 \leq k \leq m}$$

Note that $\alpha(x) = \mathcal{A}([\![x, x]\!])$ where

$$(2.2) \qquad [\![x, y]\!] = \frac{1}{2}(\langle \cdot, x \rangle y + \langle \cdot, y \rangle x)$$

denotes the symmetric outer product between vectors x and y.

The linear map $\mathcal{A}$ has first been observed in [**BBCE**] and it has been exploited successfully in various papers e.g. [**Ba2, CSV, Ba3**].

Let $S^{p,q}(H)$ denote the set of symmetric operators that have at most p strictly positive eigenvalues and q strictly negative eigenvalues.

In particular $S^{1,0}(H)$ denotes the set of non-negative symmetric operators of rank at most one:

$$(2.3) \qquad S^{1,0}(H) = \{T \in Sym(H) \ s.t. \ \exists x \in H, \forall y \in H \ , \ T(y) = \langle y, x \rangle x\}$$

In [**Ba4**] we studied in more depth geometric and analytic properties of this set. The map α is injective if and only if $\mathcal{A}$ restricted to $S^{1,0}(H)$ is injective. On the space $\hat{H}$ we define the *matrix norm induced metrics* as follows: For every $1 \leq p \leq \infty$ and $x, y \in H$,

$$(2.4) \qquad d_p(\hat{x}, \hat{y}) = \|[\![x, x]\!] - [\![y, y]\!]\|_p = \begin{cases} \left(\sum_{k=1}^n (\sigma_k)^p\right)^{1/p} & for \quad 1 \leq p < \infty \\ \max_{1 \leq k \leq n} \sigma_k & for \quad p = \infty \end{cases}$$

where $(\sigma_k)_{1 \leq k \leq n}$ are the singular values of the matrix $[\![x, x]\!] - [\![y, y]\!]$, which is of rank at most 2. In particular, for $p = 1$, d_1 corresponds to the nuclear norm $\|\cdot\|_1$ in $Sym(H)$ (the sum of singular values); for $p = \infty$, d_∞ corresponds to the operator norm $\|\cdot\|_\infty$ in $Sym(H)$ (the largest singular value). In the following parts, when no subscript is used, $\|\cdot\| = \|\cdot\|_2$.

In previous papers [**Ba4, BW**] we showed a result that is equivalent to the following theorem:

THEOREM 2.1. *If $\mathcal{F}$ is phase retrievable, then there exist constants $a_0, b_0 > 0$ such that for every $x, y \in H$,*

$$(2.5) \qquad \sqrt{a_0} d_1(x, y) \leq \|\alpha(x) - \alpha(y)\| \leq \sqrt{b_0} d_1(x, y)$$

i.e. α is bi-Lipschitz between $(\hat{H}, d_1)$ and $(\mathbb{R}^m, \|\cdot\|)$.

Consequently, the inverse map defined on the range of α from metric space $(\alpha(\hat{H}), \|\cdot\|)$ to $(\hat{H}, d_1)$:

$$(2.6) \qquad \tilde{\omega} : \alpha(\hat{H}) \subset \mathbb{R}^m \to \hat{H} \ , \ \tilde{\omega}(c) = x \ if \ \alpha(x) = c$$

is Lipschitz and its Lipschitz constant is bounded by $\frac{1}{\sqrt{a_0}}$.

Now we state the main result of this paper:

THEOREM 2.2. *Let $\mathcal{F} = \{f_1, \cdots, f_m\}$ be a phase retrievable frame for the n-dimensional Hilbert space H, and let $\alpha : \hat{H} \to \mathbb{R}^m$ denote the injective nonlinear analysis map $\alpha(x) = (|\langle x, f_k \rangle|^2)_{1 \leq k \leq m}$. Then there exists a Lipschitz continuous function $\omega : \mathbb{R}^m \to \hat{H}$ such that $\omega(\alpha(x)) = x$ for all $x \in \hat{H}$. ω has a Lipschitz constant $Lip(\omega)$ between $(\mathbb{R}^m, \|\cdot\|_2)$ and $(\hat{H}, d_1)$ bounded by*

$$(2.7) \qquad Lip(\omega) \leq \frac{12}{\sqrt{a_0}}$$

The proof of Theorem 2.2, presented in the next section, requires construction of a special Lipschitz map. We believe this particular result is interesting in itself and may be used in other constructions. Due to its importance we state it here:

LEMMA 2.3. *Consider the spectral decomposition of any self-adjoint operator in $Sym(H)$, $A = \sum_{k=1}^d \lambda_{m(k)} P_k$, where $\lambda_1 \geq \lambda_2 \geq \cdots \geq \lambda_n$ are the n eigenvalues including multiplicities, and $P_1, \ldots, P_d$ are the orthogonal projections associated to* *the d distinct eigenvalues. Additionally, $m(1) = 1$ and $m(k+1) = m(k) + r(k)$, where $r(k) = rank(P_k)$ is the multiplicity of eigenvalue $\lambda_{m(k)}$. Then the map*

$$(2.8) \qquad \pi : Sym(H) \to S^{1,0}(H) \ , \ \pi(A) = (\lambda_1 - \lambda_2)P_1$$

satisfies the following two properties:

(1) *π is Lipschitz continuous from $(Sym(H), \|\cdot\|_\infty)$ to $(S^{1,0}(H), \|\cdot\|_\infty)$ with Lipschitz constant less than or equal to 6;*
(2) *$\pi(A) = A$ for all $A \in S^{1,0}(H)$.*

The estimates of Theorem 2.2 and Lemma 2.3 are not optimal. In a separate publication [**BZ**] we improve it and extend the estimates to other metrics.

3. Proof of Results

The proof of Theorem 2.2 requires the Kirszbraun Theorem (see, e.g. [**WW**], Ch.10-11). Kirszbraun Theorem applies when two metric spaces have the following property:

DEFINITION 3.1 (Kirszbraun Property (K)). Let X and Y be two metric spaces with metric d_x and d_y respectively. (X, Y) is said to have Property (K) if for any pair of families of closed balls $\{B(x_i, r_i) : i \in I\}$, $\{B(y_i, r_i) : i \in I\}$, such that $d_y(y_i, y_j) \le d_x(x_i, x_j)$ for each $i, j \in I$, it holds that $\bigcap_{i \in I} B(x_i, r_i) \ne \emptyset \Rightarrow \bigcap_{i \in I} B(y_i, r_i) \ne \emptyset$.

Kirszbraun Theorem states the following:

THEOREM 3.2 (Kirszbraun Theorem). *Let X and Y be two metric spaces and (X, Y) has Property (K). Suppose U is a subset of X and $f : U \to Y$ is a Lipschitz map. Then there exists a Lipschitz map $F : X \to Y$ which extends f to X and $Lip(F) = Lip(f)$. In particular, (X, Y) has Property (K) if X and Y are Hilbert spaces and d_X, d_Y are the correspondingly induced metrics.*

Note that we cannot use the Kirszbraun Theorem directly to extend $\tilde{\omega}$. Specifically, our pair of spaces $(\mathbb{R}^m, \hat{H})$ does not satisfy the Kirszbraun Property. We give the following counterexample.

EXAMPLE 3.3. Let $X = \mathbb{R}^m$ for any $m \in \mathbb{N}$ and $Y = \hat{H}$ with $H = \mathbb{C}^2$. We want to show that (X, Y) does not have Property (K). Let $\tilde{y}_1 = (1, 0)$ and $\tilde{y}_2 = (0, \sqrt{3})$ be representitives of $y_1, y_2 \in Y$, respectively. Then $d_1(y_1, y_2) = 4$. Pick any two points x_1, x_2 in X with $\|x_1 - x_2\| = 4$. Then $B(x_1, 2)$ and $B(x_2, 2)$ intersect at $x_3 = (x_1 + x_2)/2 \in X$. It suffices to show that the closed balls $B(y_1, 2)$ and $B(y_2, 2)$ have no intersection in H. Assume on the contrary that the two balls intersect at y_3, then pick a representative of y_3, say $\tilde{y}_3 = (a, b)$ where $a, b \in \mathbb{C}$. It can be computed that

$$(3.1) \qquad d_1(y_1, y_3) = |a|^4 + |b|^4 - 2|a|^2 + 2|b|^2 + 2|a|^2|b|^2 + 1$$

and

$$(3.2) \qquad d_1(y_2, y_3) = |a|^4 + |b|^4 + 6|a|^2 - 6|b|^2 + 2|a|^2|b|^2 + 9$$

Set $d_1(y_1, y_3) = d_1(y_2, y_3) = 2$. Take the difference of the right hand side of (3.1) and (3.2), we have $|b|^2 - |a|^2 = 1$ and thus $|b|^2 \ge 1$. However, the right hand side of (3.1) can be rewritten as $(|a|^2 + |b|^2 - 1)^2 + 4|b|^2$, so $d_1(y_1, y_3) = 2$ would imply that $|b|^2 \le 1/2$. This is a contradiction.

We start with the proof of Lemma 2.3.

PROOF OF LEMMA 2.3. We prove (1) only. (2) follows directly from the expression of π.

Let $A, B \in Sym(H)$ where $A = \sum_{k=1}^{d} \lambda_{m(k)} P_k$ is the spectral decomposition as stated in the lemma and $B = \sum_{k'=1}^{d'} \mu_{m(k')} Q_{k'}$ is a decomposition in the same manner. We now show that

(3.3) $$\|\pi(A) - \pi(B)\|_{\infty} \leq 6 \|A - B\|_{\infty}$$

Assume $\lambda_1 - \lambda_2 \leq \mu_1 - \mu_2$. Otherwise switch the notations for A and B. If $\mu_1 - \mu_2 = 0$ then $\pi(A) = \pi(B) = 0$ and the inequality (3.3) is satisfied. Assume now $\mu_1 - \mu_2 > 0$. Thus Q_1 is of rank 1 and therefore $\|Q_1\|_{\infty} = 1$.

First note thats

(3.4)
$$\begin{aligned}
\pi(A) - \pi(B) &= (\lambda_1 - \lambda_2)P_1 - (\mu_1 - \mu_2)Q_1 \\
&= (\lambda_1 - \lambda_2)(P_1 - Q_1) + (\lambda_1 - \mu_1 - (\lambda_2 - \mu_2))Q_1
\end{aligned}$$

Here $\|P_1\|_{\infty} = \|Q_1\|_{\infty} = 1$. Therefore we have $\|P_1 - Q_1\|_{\infty} \leq 1$ since P_1, Q_1 are both positive semidefinite.

Also, by Weyl's inequality (see [**Bh**] III.2) we have $|\lambda_i - \mu_i| \leq \|A - B\|_{\infty}$ for each i. Apply this to $i = 1, 2$ we get $|\lambda_1 - \mu_1 - (\lambda_2 - \mu_2)| \leq |\lambda_1 - \mu_1| + |\lambda_2 - \mu_2| \leq 2 \|A - B\|_{\infty}$. Thus $|\lambda_1 - \mu_1| + |\lambda_2 - \mu_2| \leq 2 \|A - B\|_{\infty}$.

Let $g := \lambda_1 - \lambda_2$, $\delta := \|A - B\|_{\infty}$, then apply the above inequality to (3.4) we get

(3.5) $$\|\pi(A) - \pi(B)\|_{\infty} \leq g \|P_1 - Q_1\|_{\infty} + 2\delta \leq g + 2\delta$$

If $0 \leq g \leq 4\delta$, then $\|\pi(A) - \pi(B)\|_{\infty} \leq 6\delta$ and we are done. Now we consider the case where $g > 4\delta$. In the complex plane, let $\gamma = \gamma(t)$ be the (directed) circle centered at λ_1 with radius $g/2$. Since $\delta < g/4$ we have $|\lambda_1 - \mu_1| < g/4$ and $|\lambda_2 - \mu_2| < g/4$. Therefore the contour encloses μ_1 but not μ_2.

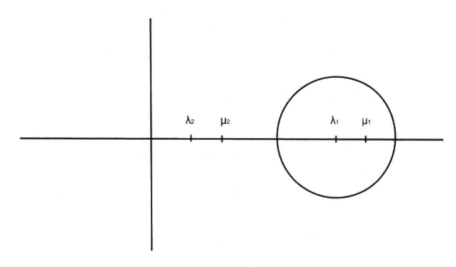

Using holomorphic calculus, we can put

(3.6) $$P_1 = -\frac{1}{2\pi i} \oint_{\gamma} R_A \, dz$$

and

$$(3.7) \qquad Q_1 = -\frac{1}{2\pi i} \oint_\gamma R_B \, dz$$

where $R_A = (A - zI)^{-1}$ and $R_B = (B - zI)^{-1}$.

Now we have

$$(3.8) \qquad \begin{aligned} P_1 - Q_1 &= \frac{1}{2\pi i} \oint_\gamma (R_B - R_A) \, dz \\ &= \frac{1}{2\pi i} \oint_\gamma R_A (B - A) R_B \, dz \end{aligned}$$

Thus

$$(3.9) \qquad \begin{aligned} \|P_1 - Q_1\|_\infty &\leq \frac{1}{2\pi} \cdot 2\pi \cdot \frac{g}{2} \cdot \max_z \|A - zI\|_\infty \|B - A\|_\infty \|B - zI\|_\infty \\ &= \frac{g\delta}{2} \cdot \max_z \max \left\{ \frac{1}{|\lambda_1 - z|}, \frac{1}{|\lambda_2 - z|} \right\} \cdot \\ &\qquad \max_z \max \left\{ \frac{1}{|\mu_1 - z|}, \frac{1}{|\mu_2 - z|} \right\} \\ &= \frac{g\delta}{2} \cdot \frac{2}{g} \cdot \frac{4}{g} \\ &= \frac{4\delta}{g} \end{aligned}$$

Thus by the first inequality in (3.5) we have

$$(3.10) \qquad \|\pi(A) - \pi(B)\|_\infty \leq 4\delta + 2\delta = 6\delta$$

Therefore, we have proved (3.3).

$\square$

REMARK 3.4. Using the integration contour from [**ZB**], one can derive a slightly stronger bound. We plan to present this result in [**BZ**].

REMARK 3.5. Numerical experiments seem to suggest that the optimal Lipschitz constant in Lemma 2.3 is 2.

Now we are ready to prove Theorem 2.2.

PROOF OF THEOREM 2.2. We construct a Lipschitz map $\omega : (\mathbb{R}^m, \|\cdot\|) \to (\hat{H}, d_1)$ such that $\omega(\alpha(x)) = x$ for all $x \in \hat{H}$ and $Lip(\omega) \leq 12/\sqrt{a_0}$.

Let $M = \alpha(\hat{H}) \subset \mathbb{R}^m$. By hypothesis, there is a map $\tilde{\omega}_1 : M \to \hat{H}$ that is Lipschitz continuous and satisfies $\tilde{\omega}_1(\alpha(x)) = x$ for all $x \in \hat{H}$. Additionally, the Lipschitz bound between $(M, \|\cdot\|)$ (that is, M with Euclidian distance) and $(\hat{H}, d_1)$ is given by $1/\sqrt{a_0}$.

First we change the metric on $\hat{H}$ from d_1 to d_2 and embed isometrically $\hat{H}$ into $Sym(H)$ with Frobenius norm (i.e. Euclidian metric):

$$(3.11) \qquad (M, \|\cdot\|) \xrightarrow{\tilde{\omega}_1} (\hat{H}, d_1) \xrightarrow{i_{1,2}} (\hat{H}, d_2) \xrightarrow{\kappa} (Sym(H), \|\cdot\|_{Fr})$$

where $i_{1,2}(x) = x$ is the identity of $\hat{H}$ and κ is the isometry given by

$$(3.12) \qquad \kappa : \hat{H} \to S^{1,0}(H), \quad x \mapsto [\![x, x]\!]$$

Obviously we have $Lip(i_{1,2}) = 1$ and $Lip(\kappa) = 1$. Thus we obtain a map $\tilde{\omega}_2 : (M, \|\cdot\|) \to (Sym(H), \|\cdot\|_{Fr})$ of Lipschitz constant

$$(3.13) \qquad Lip(\tilde{\omega}_2) \leq Lip(\tilde{\omega}_1)Lip(i_{1,2})Lip(\kappa) = \frac{1}{\sqrt{a_0}}$$

Kirszbraun Theorem (Theorem 3.2) extends isometrically $\tilde{\omega}_2$ from M to the entire $\mathbb{R}^m$ with Euclidian metric $\|\cdot\|$. Thus we obtain a Lipschitz map $\omega_2 : (\mathbb{R}^m, \|\cdot\|) \to (Sym(H), \|\cdot\|_{Fr})$ of Lipschitz constant $Lip(\omega_2) = Lip(\tilde{\omega}_2) \leq 1/\sqrt{a_0}$ such that $\omega_2(\alpha(x)) = [\![x, x]\!]$ for all $x \in \hat{H}$.

Now we consider the following maps:

$$(\mathbb{R}^m, \|\cdot\|) \xrightarrow{\omega_2} (Sym(H), \|\cdot\|_{Fr})$$
$$\xrightarrow{I_{2,\infty}} (Sym(H), \|\cdot\|_\infty)$$
$$(3.14) \qquad \xrightarrow{\pi} (S^{1,0}(H), \|\cdot\|_\infty)$$
$$\xrightarrow{\kappa^{-1}} (\hat{H}, d_\infty)$$
$$\xrightarrow{i_{\infty,1}} (\hat{H}, d_1)$$

where $I_{2,\infty}$ and $i_{\infty,1}$ are identity maps that change the metrics. The map ω is defined by

$$(3.15) \qquad \omega : (\mathbb{R}^m, \|\cdot\|) \to (\hat{H}, d_1), \quad \omega = i_{\infty,1} \cdot \kappa^{-1} \cdot \pi \cdot I_{2,\infty} \cdot \omega_2$$

The Lipschitz constant is bounded by

$$Lip(\omega) \leq Lip(\omega_2)Lip(I_{2,\infty})Lip(\pi)Lip(\kappa^{-1})Lip(i_{2,1})$$
$$(3.16) \qquad \leq \frac{1}{\sqrt{a_0}} \cdot 1 \cdot 6 \cdot 1 \cdot 2$$
$$= \frac{12}{\sqrt{a_0}}$$

Hence we obtained (2.7).

$\square$

References

[ABFM] B. Alexeev, A. S. Bandeira, M. Fickus, D. G. Mixon, *Phase Retrieval with Polarization*, SIAM J. Imaging Sci., **7** (1) (2014), 35–66.

[Ba1] Radu Balan, *Equivalence relations and distances between Hilbert frames*, Proc. Amer. Math. Soc. **127** (1999), no. 8, 2353–2366, DOI 10.1090/S0002-9939-99-04826-1. MR1600096 (99j:46025)

[Ba2] R. Balan, *On Signal Reconstruction from Its Spectrogram*, Proceedings of the CISS Conference, Princeton NJ, May 2010.

[Ba3] R. Balan, *Reconstruction of Signals from Magnitudes of Redundant Representations*, available online arXiv:1207.1134v1 [math.FA] 4 July 2012.

[Ba4] R. Balan, *Reconstruction of Signals from Magnitudes of Redundant Representations: The Complex Case*, available online arXiv:1304.1839v1 [math.FA] 6 April 2013.

[BCE1] Radu Balan, Pete Casazza, and Dan Edidin, *On signal reconstruction without phase*, Appl. Comput. Harmon. Anal. **20** (2006), no. 3, 345–356, DOI 10.1016/j.acha.2005.07.001. MR2224902 (2007b:94054)

[BCE2] R. Balan, P. Casazza, D. Edidin, *Equivalence of Reconstruction from the Absolute Value of the Frame Coefficients to a Sparse Representation Problem*, IEEE Signal.Proc.Letters, **14** (5) (2007), 341–343.

[BBCE] Radu Balan, Bernhard G. Bodmann, Peter G. Casazza, and Dan Edidin, *Painless reconstruction from magnitudes of frame coefficients*, J. Fourier Anal. Appl. **15** (2009), no. 4, 488–501, DOI 10.1007/s00041-009-9065-1. MR2549940 (2010m:42066)

[BW] R. Balan and Y. Wang, *Invertibility and Robustness of Phaseless Reconstruction*, available online arXiv:1308.4718v1. Appl. Comp. Harm. Anal., to appear 2014.

[BZ] R. Balan and D. Zou, *Phase Retrieval using Lipschitz Continuous Maps*, available online arXiv:1403.2304v1.

[BCMN] Afonso S. Bandeira, Jameson Cahill, Dustin G. Mixon, and Aaron A. Nelson, *Saving phase: injectivity and stability for phase retrieval*, Appl. Comput. Harmon. Anal. **37** (2014), no. 1, 106–125, DOI 10.1016/j.acha.2013.10.002. MR3202304

[Be] Yoav Benyamini and Joram Lindenstrauss, *Geometric nonlinear functional analysis. Vol. 1*, American Mathematical Society Colloquium Publications, vol. 48, American Mathematical Society, Providence, RI, 2000. MR1727673 (2001b:46001)

[Bh] Rajendra Bhatia, *Matrix analysis*, Graduate Texts in Mathematics, vol. 169, Springer-Verlag, New York, 1997. MR1477662 (98i:15003)

[BH] B. G. Bodmann and N. Hammen, *Stable Phase Retrieval with Low-Redundancy Frames*, available online arXiv:1302.5487v1. Adv. Comput. Math., accepted 10 April 2014.

[CCPW] J. Cahill, P.G. Casazza, J. Peterson, L. Woodland, *Phase retrieval by projections*, available online arXiv: 1305.6226v3

[CSV] Emmanuel J. Candès, Thomas Strohmer, and Vladislav Voroninski, *PhaseLift: exact and stable signal recovery from magnitude measurements via convex programming*, Comm. Pure Appl. Math. **66** (2013), no. 8, 1241–1274, DOI 10.1002/cpa.21432. MR3069958

[CESV] Emmanuel J. Candès, Yonina C. Eldar, Thomas Strohmer, and Vladislav Voroninski, *Phase retrieval via matrix completion*, SIAM J. Imaging Sci. **6** (2013), no. 1, 199–225, DOI 10.1137/110848074. MR3032952

[Ca] Peter G. Casazza, *The art of frame theory*, Taiwanese J. Math. **4** (2000), no. 2, 129–201. MR1757401 (2001f:42046)

[Cah] J. Cahill, personal communication, October 2012.

[FMNW] Matthew Fickus, Dustin G. Mixon, Aaron A. Nelson, and Yang Wang, *Phase retrieval from very few measurements*, Linear Algebra Appl. **449** (2014), 475–499, DOI 10.1016/j.laa.2014.02.011. MR3191879

[HG] Matthew J. Hirn and Erwan Y. Le Gruyer, *A general theorem of existence of quasi absolutely minimal Lipschitz extensions*, Math. Ann. **359** (2014), no. 3-4, 595–628, DOI 10.1007/s00208-013-1003-5. MR3231008

[Ph] F. Philipp, SPIE 2013 Conference Presentation, August 16, 2013, San Diego, CA.

[WAM] I. Waldspurger, A. d'Aspremont, S. Mallat, *Phase recovery, MaxCut and complex semidefinite programming*, Available online: arXiv:1206.0102

[WW] J. H. Wells and L. R. Williams, *Embeddings and extensions in analysis*, Springer-Verlag, New York-Heidelberg, 1975. Ergebnisse der Mathematik und ihrer Grenzgebiete, Band 84. MR0461107 (57 #1092)

[ZB] L. Zwald and G. Blanchard, *On the convergence of eigenspaces in kernel Principal Component Analysis*, Proc. NIPS 05, vol. 18, 1649-1656, MIT Press, 2006.

DEPARTMENT OF MATHEMATICS AND CENTER FOR SCIENTIFIC COMPUTATION AND MATHEMATICAL MODELING, UNIVERSITY OF MARYLAND, COLLEGE PARK, MARYLAND 20742
E-mail address: rvbalan@math.umd.edu

APPLIED MATHEMATICS AND STATISTICS, AND SCIENTIFIC COMPUTATION PROGRAM, UNIVERSITY OF MARYLAND, COLLEGE PARK, MARYLAND 20742
E-mail address: zou@math.umd.edu

Contemporary Mathematics
Volume **650**, 2015
http://dx.doi.org/10.1090/conm/650/13042

Tight and random nonorthogonal fusion frames

Jameson Cahill, Peter G. Casazza, Martin Ehler, and Shidong Li

ABSTRACT. Nonorthogonal fusion frames were introduced in **Cahill, Casazza, and Li** [**4**] to provide a general method for constructing sparse and/or tight fusion frames. In this paper we will resolve some of the fundamental questions left open in **Cahill, Casazza, and Li** [**4**]. First we show that tight nonorthogonal fusion frames are surprisingly easy to construct. In order to do this we will establish classifications of when and how to write certain self adjoint operators as a product (or sum) of (nonorthogonal) projection operators. We also discuss random versions of nonorthogonal fusion frames, derive constructions based on orbits of irreducible finite subgroups of the unitary group, and study the fusion frame potential in the nonorthogonal setting.

1. Introduction

Fusion frames were introduced in [**7**] and further developed in [**8**]. Recently there has been much activity around the idea of fusion frames, see [**6**] and references therein or go to the *Fusion Frame* website: www.fusionframes.org. Loosely speaking, a fusion frame is a collection of subspaces $\{W_i\}_{i=1}^m$ all contained in some bigger Hilbert space $\mathcal{H}$ such that any signal $f \in \mathcal{H}$ can be stably reconstructed from the set of orthogonal projections $\{\pi_i f\}_{i=1}^m$, where π_i denotes the orthogonal projection from $\mathcal{H}$ onto W_i. Typically we think of the dimension of each subspace W_i as being much smaller than the dimension of $\mathcal{H}$ so that a high dimensional signal f can be reconstructed from several low dimensional measurements $\{\pi_i f\}_{i=1}^m$.

In [**4**], the authors introduced *nonorthogonal fusion frames* in order to achieve sparsity of the fusion frame operator. The basic observation in [**4**] is that replacing orthogonal projections π_i in the original definition of fusion frames [**8**] by nonorthogonal projections P_i onto the same subspaces W_i can result in a fusion frame operator which is much sparser. This is because, for example, one can always choose the null space of the projection P_i to contain some basis elements $\{e_{i_j}\}_j$ that are complementary to the subspace W_i, and thereby nullify some columns of P_i, which in turn results in sparsity of the (new) fusion frame operator. One further observation which was made in [**4**] but was not explored very thoroughly there is that tight nonorthogonal fusion frames are much more abundant than tight (orthogonal) fusion frames. This is important since it was shown in [**9**] that there

---

2010 *Mathematics Subject Classification.* Primary 15:42C15.

Key words and phrases. Fusion frame, projection.

The first and seconde authors were supported by NSF 1307685; and NSF ATD 1042701; AFOSR DGE51: FA9550-11-1-0245.

are very few tight fusion frames in general. In this paper, constructions of tight nonorthogonal fusion frames and nonorthogonal fusion frames of a prescribed fusion frame operator are provided.

We now give a formal definition of nonorthogonal fusion frames. Throughout this paper, let $\mathcal{H}_n$ denote an n-dimensional Hilbert space.

DEFINITION 1.1. *An operator* $P : \mathcal{H}_n \to \mathcal{H}_n$ *is called a projection if* $P^2 = P$. *If in addition we have* $P^* = P$ *then* P *is called an orthogonal projection.*

DEFINITION 1.2. *Let* $\{P_i\}_{i=1}^m$ *be a collection of projections on* $\mathcal{H}$ *and* $\{v_i\}_{i=1}^m$ *a collection of positive real numbers. We say* $\{(P_i, v_i)\}_{i=1}^m$ *is a nonorthogonal fusion frame for* $\mathcal{H}$ *if there exist constants* $0 < A \le B < \infty$ *such that*

$$A\|f\|^2 \le \sum_{i=1}^m v_i^2 \|P_i f\|^2 \le B\|f\|^2$$

for every $f \in \mathcal{H}$. *We say it is tight if* $A = B$.

Given a nonorthogonal fusion frame we define the nonorthogonal fusion frame operator $S : \mathcal{H}_n \to \mathcal{H}_n$ by

$$(1) \qquad\qquad Sf = \sum_{i=1}^m v_i^2 P_i^* P_i f.$$

We observe that $\{(v_i, P_i)\}$ is tight if and only if $S = \lambda I$ (where $\lambda = A = B$). Therefore, much of this paper is devoted to studying ways of writing multiples of the identity in the form of the right hand side of equation (1). We will also usually assume that $v_i = 1$ for every $i = 1, \ldots, m$

Before leaving the introduction we collect some basic facts about projections and fix some notation that will be used throughout this paper.

PROPOSITION 1.3. *Let* P *be a projection on* $\mathcal{H}$, $W = \mathrm{im} P$, $W^* \equiv (\ker P)^\perp = [(I - P)(\mathcal{H})]^\perp$. *Denote by* P^* *the adjoint of* P *in* $\mathcal{H}$. *Then:*
(1) $(P^*)^2 = P^*$, *and* $\mathrm{im} P^* = W^*$, $\ker P^* = W^\perp$.
(2) $W^* = \mathrm{im} P^* = \mathrm{im} P^* P$.
(3) P *is an invertible operator mapping* W^* *onto* W.
(4) $\dim(W) = \dim(W^*)$.

COROLLARY 1.4. *Given subspaces* W, W^* *of* $\mathcal{H}$ *with* $\dim W = \dim W^*$, *there is a projection* P *onto* W *with* $P^* P(\mathcal{H}) = W^*$ *if and only if* $(W^*)^\perp \cap W = \{0\}$.

NOTATION 1.5. Throughout this paper we will always use the notation of Proposition 1.3; *i.e.*, P will always stand for a (nonorthogonal) projection, W will always be the image of P, and W^* will always be the image of P^*. Furthermore, we will always use the symbol π_W to denote the *orthogonal* projection onto the subspace $W \subseteq \mathcal{H}_n$.

REMARK 1.6. We will be working with finite dimensional Hilbert spaces. Some of the results here can be generalized to infinite dimensions but we have chosen not to cover this case.

2. Classification of self adjoint operators via projections

Let $T : \mathcal{H}_n \to \mathcal{H}_n$ be a positive, self adjoint, linear operator. The main result of this section is to classify the set

$$\Omega(T) = \{P : P^2 = P, P^*P = T\}.$$

Armed with this result and its consequences, in the next section we will show how to construct large families of sparse and tight non-orthogonal fusion frames. The spectral theorem tells us that $T = \sum_{j=1}^n \lambda_j \pi_j$ where the λ_j's are the eigenvalues of T and π_j is the orthogonal projection onto the one dimensional span of the jth eigenvector of T. Therefore $P \in \Omega(T)$ if and only if P^*P has the same eigenvalues and eigenvectors as T. Also note that if $P \in \Omega(T)$ then $\ker(P) = \operatorname{im}(T)^\perp$, and since a projection is uniquely determined by its kernel and its image we have a natural bijection between $\Omega(T)$ and the set

$$\tilde{\Omega}(T) := \{W \subseteq \mathcal{H}_n : \operatorname{im}(P) = W \text{ for some } P \in \Omega(T)\}.$$

given by

$$\Omega(T) \ni P \mapsto \operatorname{im}(P) \in \tilde{\Omega}(T).$$

We start with two elementary lemmas.

LEMMA 2.1. *Let P be a projection and let $\{e_j\}_{j=1}^k$ be an orthonormal basis of W^* consisting of eigenvectors of P^*P with corresponding nonzero eigenvalues $\{\lambda_j\}$. Then $\{Pe_j\}_{j=1}^k$ is an orthogonal basis for W and $\|Pe_j\| = \sqrt{\lambda_j}$.*

PROOF. Just observe that $\langle Pe_j, Pe_\ell \rangle = \langle P^*Pe_j, e_\ell \rangle = \lambda_j \langle e_j, e_\ell \rangle$. □

LEMMA 2.2. *Let P be a projection and suppose λ is an eigenvalue of P^*P, $\lambda \neq 0$. Then $\lambda \geq 1$. Moreover, $\lambda = 1$ if and only if the corresponding eigenvector is in $W \cap W^*$.*

PROOF. Note that $W^* = \operatorname{im} P^*P$, so all eigenvectors of P^*P corresponding to nonzero eigenvalues are in W^*. Let $x \in W^*$ and write $Px = x + (P-I)x$. Since $x \perp (I-P)x$,

$$(2) \qquad \|Px\|^2 = \|x\|^2 + \|(P-I)x\|^2 \geq \|x\|^2.$$

By the same argument on P^* we get $\|P^*Px\| \geq \|Px\| \geq \|x\|$ for all $x \in W^*$. Therefore, if $P^*Px = \lambda x$ we have that $\lambda \geq 1$.

Finally, by equation (2), $\lambda = 1$ if and only if $(I-P)x = 0$, or $x = Px \in W$. Hence $x \in W \cap W^*$. □

The next proposition allows us reduce our problem to the case when $\operatorname{rank}(T) \leq n/2$.

PROPOSITION 2.3. *Let P be a projection, then we can write*

$$P = P' + \pi_{W \cap W^*}$$

where $\pi_{W \cap W^*}$ is the orthogonal projection onto $W \cap W^*$, and P' is a projection such that all nonzero eigenvalues of P'^*P' are strictly greater than 1.

PROOF. First note that Lemma 2.2 says that $W \cap W^* = \{x : P^*Px = x\}$. Now let W' be the orthogonal complement of $W \cap W^*$ in W and let P' be the projection onto W' along $\ker(P) + W \cap W^*$. Then $P'\pi_{W \cap W^*} = \pi_{W \cap W^*}P' = 0$, so $(P' + \pi_{W \cap W^*})^2 = P'^2 + \pi_{W \cap W^*}^2 = P' + \pi_{W \cap W^*}$. It is clear that $\operatorname{im}(P' + \pi_{W \cap W^*}) = $

W. Since $\ker P = W^{*\perp} \subseteq (W \cap W^*)^\perp$ it follows that $\ker(P) \subseteq \ker(P' + \pi_{W \cap W^*})$ so we must have $\ker(P) = \ker(P' + \pi_{W \cap W^*})$. Therefore $P = P' + \pi_{W \cap W^*}$, and the nonzero eigenvalues of $P'^* P'$ are precisely the nonzero eigenvalues of $P^* P$ which are greater than 1. $\qquad\square$

We can now state the main theorem of this section:

THEOREM 2.4. *Let $T : \mathcal{H}_n \to \mathcal{H}_n$ be a positive, self-adjoint operator of rank $k \leq \frac{n}{2}$. Let $\{\lambda_j\}_{j=1}^k$ be the nonzero eigenvalues of T and suppose $\lambda_j \geq 1$ for $i = 1, \ldots, k$ and let $\{e_j\}_{j=1}^k$ be an orthonormal basis of $\mathrm{im}(T)$ consisting of eigenvectors of T. Then*

$$\tilde{\Omega}(T) = \{ span\{ \frac{1}{\sqrt{\lambda_j}} e_j + \sqrt{\frac{\lambda_j - 1}{\lambda_j}} e_{j+k} \} : \{e_j\}_{j=1}^{2k} \ is \ orthonormal \}.$$

PROOF. First suppose $W \in \tilde{\Omega}(T)$ and let P be the projection onto W along $\mathrm{im}(T)^\perp$. By Lemma 2.1 we know that $\{ \frac{Pe_j}{\|Pe_j\|} \}_{j=1}^k$ is an orthonormal basis for W. We also know that $\|Pe_j\| = \sqrt{\lambda_j}$ for $j = 1, 2, \ldots, k$ and that e_i is in the range of P^* which is orthogonal to $\ker P = Im\ (I - P)$. So

$$\langle e_i, (P - I)e_i \rangle = \langle (P - I)^* e_i, e_i \rangle \langle 0, e_i \rangle = 0.$$

It follows that

$$\lambda_j = \|e_j\|^2 + \|(P - I)e_j\|^2 = 1 + \|(P - I)e_j\|^2$$

which means

$$\|(P - I)e_j\| = \sqrt{\lambda_j - 1}$$

so if we set for $1 \leq j \leq k$,

$$e_{j+k} = \frac{(P - I)e_j}{\sqrt{1 - \lambda_j}},$$

then $\{e_j\}_{j=1}^{2k}$ is an orthonormal set and

$$\frac{Pe_j}{\|Pe_j\|} = \frac{1}{\sqrt{\lambda_j}} e_j + \sqrt{\frac{\lambda_j - 1}{\lambda_j}} e_{j+k}.$$

Conversely suppose $W = span\{ \frac{1}{\sqrt{\lambda_j}} e_j + \sqrt{\frac{\lambda_j - 1}{\lambda_j}} e_{j+k} \}_{j=1}^k$ with $\{e_j\}_{j=1}^{2k}$ orthonormal. Let P be the projection onto W along $\mathrm{im}(T)^\perp$. Notice that for $1 \leq j \leq k$, $e_j = e_j + \sqrt{\lambda_j - 1} e_{j+k} - \sqrt{\lambda_j - 1} e_{j+k}$ with $e_j + \sqrt{\lambda_j - 1} e_{j+k} \in W$ and $-\sqrt{\lambda_j - 1} e_{j+k} \in \mathrm{im}(T)^\perp$, so $Pe_j = e_j + \sqrt{\lambda_j - 1} e_{j+k}$ for $j = 1, \ldots, k$. Similarly $e_j + \sqrt{\lambda_j - 1} e_{j+k} = \lambda_j e_j + (1 - \lambda_j) e_j + \sqrt{\lambda_j - 1} e_{j+k}$ with $\lambda_j e_j \in W^* = imP^*$ and $(1 - \lambda_j)e_j + \sqrt{\lambda_j - 1} e_{j+k} \in W^\perp = \ker(P^*)$, so $P^* Pe_j = \lambda_j e_j$ for $j = 1, \ldots, k$. Therefore, $P^* P$ has the same eigenvectors and corresponding eigenvalues as T, so $P^* P = T$, and $W \in \tilde{\Omega}(T)$. $\qquad\square$

Before proceeding we remark that Theorem 2.4 is independent of our choice of eigenbasis for T. To see this let $\{e_j'\}_{j=1}^k$ be any other eigenbasis for T and let $W = span\{ \frac{1}{\sqrt{\lambda_j}} e_j' + \sqrt{\frac{\lambda_j - 1}{\lambda_j}} e_{j+k}' \}_{j=1}^k$ with $\{e_j'\}_{j=1}^{2k}$ orthonormal. By the second part of the proof of Theorem 2.4 we have that $W \in \tilde{\Omega}(T)$, and so by the first part of the proof we have that in fact $W = \operatorname{span}\{\frac{1}{\sqrt{\lambda_j}}e_j + \sqrt{\frac{\lambda_j-1}{\lambda_j}}e_{j+k}\}_{j=1}^k$ with $\{e_j\}_{j=1}^{2k}$ orthonormal.

We now state several consequences of Theorem 2.4. The first corollary appeared first in [**12**].

COROLLARY 2.5. *If T is a positive self-adjoint operator of rank $\leq \frac{n}{2}$ with all nonzero eigenvalues ≥ 1, then there is a projection P so that $T = P^*P$.*

COROLLARY 2.6. *If T is a positive self-adjoint operator of rank $\leq \frac{n}{2}$, then there is a projection P and a weight $v > 0$ so that $T = v^2 P^*P$.*

PROOF. Let λ_k be the smallest non-zero eigenvalue of T. So all nonzero eigenvalues of $\frac{1}{\lambda_k}T$ are greater than or equal to 1 and by Corollary 2.5 there is a projection P so that $P^*P = \frac{1}{\lambda_k}T$. Let $v = \sqrt{\lambda_k}$ to finish the proof. □

In the rest of this section we will analyze the case where $\operatorname{rank}(T) > n/2$. Our first proposition can be found in [**12**]. But we feel our proof is much more instructive and visual, so we include it here.

PROPOSITION 2.7. *Let T be a positive self-adjoint operator of rank $k > \frac{n}{2}$ with eigenvectors $\{e_j\}_{j=1}^n$ and respective eigenvalues $\{\lambda_j\}_{j=1}^n$. The following are equivalent:*

(1) There is a projection P so that $T = P^*P$.

(2) The nonzero eigenvalues of T are greater than or equal to 1 and we have

$$|\{j : \lambda_j > 1\}| \leq |\{j : \lambda_j = 0\}|.$$

In particular,

$$|\{j : \lambda_j = 1\}| \geq k - \left\lfloor \frac{n}{2} \right\rfloor.$$

PROOF. Let $A_1 = \{j : \lambda_j > 1\}$, $A_2 = \{j : \lambda_j = 0\}$, and $A_3 = \{j : \lambda_j = 1\}$, and let π_i be the orthogonal projection onto $\operatorname{span}\{e_j : j \in A_i\}$ for $i = 1, 2, 3$.

(1) $\Rightarrow$ (2): By Proposition 2.3, we can write

$$P = P' + \pi_{W \cap W^*},$$

where $\pi_{W \cap W^*}$ is the orthogonal projection onto $W \cap W^*$, and P' is the projection onto the orthogonal complement W' of $W \cap W^*$ in W along $ker\ P + W \cap W^*$. Define $W'^* \equiv \operatorname{im} P'^*$. Then P' is an invertible operator from W'^* onto W', $W'^* \perp W \cap W^*$ and $W' \perp W \cap W^*$, and $W' \cap W'^* = \{0\}$. Hence,

$$
\begin{aligned}
2 \dim W'^* &= \dim W' + \dim W'^* \\
&= \dim(W' + W'^*) \\
&\leq \dim W'^* + \dim \operatorname{span}\{e_j : j \in A_2\}.
\end{aligned}
$$

Since $W'^* = span \{e_j : j \in A_1\}$, it follows that $|A_1| \leq |A_2|$.

(2) $\Rightarrow$ (1): Let $T_1 = T(\pi_1 + \pi_2)$, so $T = T_1 + \pi_3$. By our assumption

$$\operatorname{rank} T_1 \leq \frac{n}{2},$$

and all non-zero eigenvalues of T_1 are strictly greater than 1. By Theorem 2.4 there is a projection P' so that $P'^*P' = T_1$. Let $P = P' + \pi_3$. Then $P'\pi_3 = \pi_3 P' = 0$. Hence, $P = P^2$ is a projection and

$$P^*P = P'^*P' + \pi_3 = T_1 + \pi_3 = T.$$ □

We now will see that in general it may happen that $\Omega(T) = \emptyset$.

COROLLARY 2.8. *If* $rank(T) = k > \frac{n}{2}$ *and* T *does not have 1 as an eigenvalue with multiplicity at least* $k - \lfloor \frac{n}{2} \rfloor$, *then* $\Omega(T) = \emptyset$.

REMARK 2.9. Similar to the proof of Corollary 2.6, if T is a positive self-adjoint operator of rank $> \frac{n}{2}$ with eigenvalues $\{\lambda_1 \geq \cdots \geq \lambda_k > 0 = \lambda_{k+1} = \cdots \lambda_n\}$, then there is a projection P and a weight $v = \sqrt{\lambda_k}$ so that $T = v^2 P^* P$ if and only if

$$|\{j : \lambda_j > \lambda_k\}| \leq |\{j : \lambda_j = 0\}|.$$

PROPOSITION 2.10. *Let* $T : \mathcal{H}_n \to \mathcal{H}_n$ *be a positive, self adjoint operator of rank* $k > \frac{n}{2}$ *with eigenvalues* $\{\lambda_j\}_{j=1}^n$ *and whose nonzero eigenvalues are all greater than or equal to 1. If either*
(1) n is even, or
(2) n is odd and T has at least one eigenvalue in the set $\{0, 1, 2\}$
then there are two projections P_1 and P_2 such that $T = P_1^* P_1 + P_2^* P_2$.

PROOF. Let $\{e_j\}_{j=1}^n$ be an orthonormal basis of $\mathcal{H}_n$ consisting of eigenvectors of T with respective eigenvalues $\{\lambda_j\}_{j=1}^n$, in decreasing order.

Case 1: n is even.

Let $V = \text{span}\{e_j\}_{j \in I}$, $|I| = \frac{n}{2}$. Note that $T = T\pi_V + T\pi_{V^\perp}$. Also, since T, π_V, and $\pi_{V^\perp}$ are all diagonal with respect to $\{e_j\}_{j=1}^n$ it follows that T commutes with both π_V and $\pi_{V^\perp}$. Therefore $(T\pi_V)^* = \pi_V^* T^* = \pi_V T = T\pi_V$, so by Theorem 2.4 there is a projection P_1 such that $T\pi_V = P_1^* P_1$. Similarly we can find a projection P_2 such that $T\pi_{V^\perp} = P_2^* P_2$.

Case 2: n is odd and T has an eigenvalue in the set $\{0, 1, 2\}$.

We will look at the case for each eigenvalue separately.

Subcase 1: $\lambda_n = 0$.

Let $\mathcal{H}_1 = \text{span}\{e_j : 1 \leq j \leq n-1\}$. Then $\dim(\mathcal{H}_1)$ is even so we can apply the same argument as above to $\mathcal{H}_1$.
Subcase 2: $\lambda_n = 1$.

Define T_1, T_2 by

$$T_1 e_j = \begin{cases} Te_j & \text{if } j = 1, 2, \ldots, \frac{n-1}{2} \\ 0 & \text{otherwise} \end{cases}$$

$$T_2 e_j = \begin{cases} Te_j & \text{if } j = \frac{n-1}{2} + 1, \ldots, n-1 \\ 0 & \text{otherwise} \end{cases}$$

Then $rank(T_1) = rank(T_2) = \frac{n-1}{2} < \frac{n}{2}$ so by Corollary 2.5, we can write

$$T_i = P_i^* P_i, \quad i = 1, 2.$$

Let π be the orthogonal projection of $\mathcal{H}_n$ onto $\text{span}\{e_n\}$ and let

$$Q = P_2 + \pi,$$

which is clearly a projection. Then we have

$$T = P_1^* P_1 + Q^* Q.$$

Subcase 3: $\lambda_j = 2$ for some j.

Without loss of generality, re-index $\{\lambda_j\}_{j=1}^n$ so that $\lambda_n = 2$. Define T_1, T_2, and π as above. As in the previous case, define two projections $\{P_i\}_{i=1}^2$ so that

$$T_i = P_i^* P_i.$$

Now let $Q_i = P_i + \pi$, $i = 1, 2$. Then

$$T = Q_1^* Q_1 + Q_2^* Q_2. \qquad \square$$

COROLLARY 2.11. *Let $T : \mathcal{H}_n \to \mathcal{H}_n$ be a positive, self adjoint operator of rank $k > \frac{n}{2}$. There is a weight v and projections P_1 and P_2 so that*

$$T = v^2 P_1^* P_1 + v^2 P_2^* P_2.$$

PROOF. Apply Proposition 2.10 to $\frac{1}{\lambda_k} T$ and set $v = \sqrt{\lambda_k}$. $\qquad \square$

It is important to note that, without weighting, we can always write every positive self-adjoint T as the sum of $P_i^* P_i$ with three projections.

COROLLARY 2.12. *If $T : \mathcal{H}_n \to \mathcal{H}_n$ is a positive, self adjoint operator of rank $k > \frac{n}{2}$ whose nonzero eigenvalues are all greater than or equal to 1, then there are projections $\{P_i\}_{i=1}^3$ so that*

$$T = P_1^* P_1 + P_2^* P_2 + P_3^* P_3.$$

PROOF. If n is even, we can write T as the sum of two projections. Suppose n is odd and let $\{e_j\}_{j=1}^n$ be an eigenbasis of T. Suppose $J_1 \cup J_2 \cup J_3 = \{1, \ldots, n\}$ with $|J_i| < \frac{n}{2}$ and let π_i be the orthogonal projection onto $\mathrm{span}\{e_j : j \in J_i\}$ for $i = 1, 2, 3$. Then $T = T(\pi_1 + \pi_2 + \pi_3)$ and $T\pi_i$ satisfies Corollary 2.5 for each i. $\qquad \square$

3. Tight nonorthogonal fusion frames

As applications of the results of the previous section, in this section we address some issues regarding tight nonorthogonal fusion frames. The first theorem shows which sets of dimensions allow the existence of a tight nonorthogonal fusion frame. The corresponding problem for fusion frames has received considerable attention proven to be quite difficult, see [20], [9], and [3].

THEOREM 3.1. *Suppose $n_1 + \cdots + n_m \geq n$, $n_i \leq \frac{n}{2}$. Then there exists a tight nonorthogonal fusion frame $\{P_i\}_{i=1}^m$ ($v_i = 1$ for every i) for $\mathcal{H}_n$ such that $\mathrm{rank}(P_i) = n_i$ for $i = 1, \ldots, m$.*

PROOF. Choose an orthonormal basis $\{e_j\}_{j=1}^n$ for $\mathcal{H}_n$ and choose a collection of subspaces $\{W_i\}_{i=1}^m$ such that:
1) $W_i = \mathrm{span}\{e_j\}_{j \in J_i}$ with $|J_i| = n_i$ for each $i = 1, \ldots, m$, and
2) $W_1 + \cdots + W_m = \mathcal{H}_n$.
Let π_i be the orthogonal projection onto W_i and let $S = \sum_{i=1}^m \pi_i$. Observe that $I = S^{-1} S = \sum_{i=1}^m S^{-1} \pi_i$. Since each π_i is diagonal with respect to $\{e_j\}_{j=1}^n$ it follows that S^{-1} commutes with π_i, so $S^{-1}\pi_i$ is positive and self adjoint for every $i = 1, \ldots, m$. Let γ be the smallest nonzero eigenvalue of any $S^{-1}\pi_i$, then $\frac{1}{\gamma} S^{-1} \pi_i$

satisfies the hypotheses of Corollary 2.5 so there is a projection P_i so that $P_i^* P_i = \frac{1}{\gamma} S^{-1} \pi_i$, and we have

$$\sum_{i=1}^{m} P_i^* P_i = \frac{1}{\gamma} I.$$

$\square$

Theorem 3.1 should be compared with Theorem 3.2.2 in [20]. Also note that the proof of Theorem 3.1 is constructive, cf. [9]. It was shown in [9] that this theorem fails for orthogonal projections for almost all choices of dimensions for the subspaces. The next theorem deals with adding projections to a given nonorthogonal fusion frame it order to get a tight nonorthogonal fusion frame. Somewhat surprisingly, this can always be achieved with only two projections.

THEOREM 3.2. *Let* $\{P_i\}_{i=1}^{m}$ *be projections on* $\mathcal{H}_n$, $n \geq 2$. *Then there are two projections* $\{P_i\}_{i=m+1}^{m+2}$ *and a* λ *so that*

$$\sum_{i=1}^{m+2} P_i^* P_i = \lambda I.$$

PROOF. Let

$$S = \sum_{i=1}^{m} P_i^* P_i,$$

and let $\lambda_1 \geq \lambda_2 \geq \cdots \geq \lambda_n \geq 0$ be the eigenvalues of S. Let $\lambda = \lambda_1 + 1$ and let

$$T = \lambda I - S.$$

Then T is a positive self-adjoint operator with all of its eigenvalues ≥ 1 and at least one eigenvalue equal to one. By Proposition 2.10, we can find projections $\{P_i\}_{i=m+1}^{m+2}$ so that

$$T = P_{m+1}^* P_{m+1} + P_{m+2}^* P_{m+2}.$$

Thus,

$$\lambda I = S + T = \sum_{i=1}^{m+2} P_i^* P_i.$$

$\square$

No such theorem exists for frames or regular (orthogonal) fusion frames. In general we need to add $n - 1$ vectors to a frame in $\mathcal{H}_n$ in order to get a tight frame (see Proposition 2.1 in [10]). However, in this context Theorem 3.2 may be misleading, as the ranks of the projections we need to add could be quite large. The next result tells us how to deal with the case where we want small rank projections.

PROPOSITION 3.3. *If* $\{P_i\}_{i=1}^{m}$ *are projections on* $\mathcal{H}_n$ *and* $k \leq \frac{n}{2}$, *there are projections* $\{Q_i\}_{i=1}^{L}$ *with* $L = \lceil \frac{n}{k} \rceil$ *and* $rank(Q_i) \leq k$, *and a* λ *so that*

$$\sum_{i=1}^{m} P_i^* P_i + \sum_{j=1}^{L} Q_i^* Q_i = \lambda I.$$

PROOF. Let $S = \sum_{i=1}^{m} P_i^* P_i$ and assume S has eigenvectors $\{e_j\}_{j=1}^{n}$ with respective eigenvalues $\lambda_1 \geq \lambda_2 \geq \cdots \geq \lambda_n$. Partition the set $\{1, \ldots, n\}$ into sets $J_1, \ldots, J_L$ with $|J_\ell| \leq k$ for every $\ell = 1, \ldots, L$. Let π_ℓ denote the orthogonal projection onto span$\{e_j\}_{j \in J_\ell}$. Set $\lambda = \lambda_1 + 1$ and let $T_\ell = (\lambda I - S)\pi_\ell$. Then each T_ℓ satisfies the hypotheses of Corollary 2.5 so choose any projection $Q_\ell \in \Omega(T_\ell)$. Now we have that

$$
\begin{aligned}
\sum_{i=1}^{M} P_i^* P_i + \sum_{\ell=1}^{L} Q_\ell^* Q_\ell &= S + \sum_{\ell=1}^{L} T_\ell \\
&= S + \lambda I - S = \lambda I.
\end{aligned}
$$

$\square$

3.1. 2 projections. As an application of the results of the previous section we will give a complete description of when there are two projections $P_i : \mathcal{H}_n \to \mathcal{H}_n$, $i = 1, 2$ such that

$$(3) \qquad\qquad P_1^* P_1 + P_2^* P_2 = \lambda I.$$

Let $W_1 = \text{im}(P_1), W_1^* = \text{im}(P_1^*), W_2 = \text{im}(P_2), W_2^* = \text{im}(P_2^*)$. We will examine this in several cases but first we make some general remarks. Note that if $x \in W_1^*$ such that $P_1^* P_1 x = \alpha x$ (for $\alpha \in \mathbb{R}$) then $P_2^* P_2 x = (\lambda - \alpha)x$, so there is an orthonormal bases $\{e_j\}_{j=1}^{n}$ consisting of eigenvectors of both $P_1^* P_1$ and $P_2^* P_2$. Furthermore, if $P_1^* P_1 x = 0$ then $P_2^* P_2 x = \lambda x$, so $\ker P_1 = W_1^{*\perp} \subseteq W_2^*$, and similarly $W_2^{*\perp} \subseteq W_1^*$.

It follows from (3) that rank$(P_1) + $rank$(P_2) \geq n$. We will examine the cases of equality and strict inequality separately.

PROPOSITION 3.4. *Suppose P_1 and P_2 are projections on $\mathcal{H}_n$ such that $P_1^* P_1 + P_2^* P_2 = \lambda I$ and that rank$(P_1) + $rank$(P_2) = n$. Then either rank$(P_1) \neq$ rank(P_2) and $\lambda = 1$ or rank$(P_1) = $rank$(P_2) = \frac{n}{2}$ and $\lambda \geq 1$.*

PROOF. First suppose without loss of generality that rank$(P_1) = k > $rank$(P_2)$. In this case we have that $k > \frac{n}{2}$, so $\dim(W_1 \cap W_1^*) \geq 2k - n > 0$. Then by Proposition 2.3 we have that $P_1 = P_1' + \pi_{W_1 \cap W_1^*}$ and $P_1^* P_1 + P_2^* P_2 = P_1'^* P_1' + \pi_{W_1 \cap W_1^*} + P_2^* P_2$. Let $x \in W_1 \cap W_1^*$, then $P'x = 0$, and since $x \notin W_2^*$ it follows that $P_2 x = 0$. Therefore $(P_1^* P_1 + P_2^* P_2)x = x$ which means $\lambda = 1$, both P_1 and P_2 are orthogonal projections, and $W_j^* = W_j$ $j = 1, 2$.

Now suppose that n is even, and $\dim(W_1) = \dim(W_2) = \frac{n}{2}$. In this case we have that $W_1^* = W_2^{*\perp}$, so it follows immediately that $P_1^* P_1 = \lambda \pi_{W_1^*}$ and $P_2^* P_2 = \lambda \pi_{W_2^*}$. $\square$

PROPOSITION 3.5. *Suppose P_1 and P_2 are projections on $\mathcal{H}_n$ such that $P_1^* P_1 + P_2^* P_2 = \lambda I$ and that rank$(P_1) + $rank$(P_2) > n$. Then rank$(P_1) = $rank$(P_2)$, $\lambda = 2$, and $W_1^* \cap W_1 = W_2^* \cap W_2$.*

PROOF. First suppose $\dim(W_1) = k > \ell = \dim(W_2)$. Note that $k > \frac{n}{2}$. By the remarks above we know that 0 must be an eigenvalue of $P_1^* P_1$ with multiplicity $n - k$, λ must be an eigenvalue of $P_1^* P_1$ with multiplicity $n - \ell$, and 1 must be an eigenvalue of $P_1^* P_1$ with multiplicity $\dim(W_1^* \cap W_2^*) \geq 2k - n$. Adding up these multiplicities we get $(n - k) + (n - \ell) + (2k - n) = n + k - \ell = n$ which contradicts the fact that $k > \ell$. Therefore, we may assume that $\dim(W_1) = \dim(W_2)$.

By the remarks above we can choose an orthonormal basis $\{e_j\}_{j=1}^n$ of $\mathcal{H}_n$ so that

$$
\begin{aligned}
P_1^* P_1 e_j &= \lambda e_j \text{ and } P_2^* P_2 e_j = 0 \text{ for } j = 1, \ldots, n - k, \\
P_1^* P_1 e_j &= 0 \text{ and } P_2^* P_2 e_j = \lambda e_i \text{ for } j = k + 1, \ldots, n.
\end{aligned}
$$

Since $\dim(W_1 \cap W_1^*), \dim *(W_2 \cap W_2^*) \geq 2k - n$ it follows that

$$
P_1^* P_1 e_j = e_j = P_2^* P_2 e_j \text{ for } j = n - k + 1, \ldots, k.
$$

Therefore $\lambda = 2$ and $W_1 \cap W_1^* = W_2 \cap W_2^*$.   □

Ideally we would like analogous theorems for any number of projections, but this has proven to be a quite difficult problem.

4. Random nonorthogonal fusion frames

Probabilistic versions of frames have been introduced in [13–15]. Here, we extend the concept to nonorthogonal fusion frames.

Let Ω be a locally compact Hausdorff space and $\mathcal{B}(\Omega)$ be the Borel-sigma algebra on Ω endowed with a probability measure μ. We denote the collection of projections on $\mathcal{H}_n$ by $\mathcal{P}_n$, endowed with the induced Borel sigma algebra. We say that a random projector $P : \Omega \to \mathcal{P}_n$, is a random nonorthogonal fusion frame if there are nonnegative constants A and B such that

$$
A\|x\|^2 \leq \int_\Omega \|P(w)x\|^2 d\mu(\omega) \leq B\|x\|^2, \quad \text{for all } x \in \mathcal{H}_n.
$$

The random projector P is called tight if we can choose $A = B$. The random analysis operator F is defined by

$$
F : \mathcal{H}_n \to L_2(\Omega, \mathcal{H}_n, \mu), \quad x \mapsto (\omega \mapsto P(\omega)x).
$$

Its adjoint operator T^* is the random synthesis operator

$$
F^* : L_2(\Omega, \mathcal{H}_n, \mu) \to \mathcal{H}_n, \quad f \mapsto \int_\Omega P(\omega)^* f(\omega) d\mu(\omega).
$$

The random nonorthogonal fusion frame operator $S = F^* F$ then is

$$
S : \mathcal{H}_n \to \mathcal{H}_n, \quad x \mapsto \int_\Omega P(\omega)^* P(\omega) x d\mu(\omega).
$$

Thus, $S = \int_\Omega P(\omega)^* P\omega d\mu(\omega)$ has a matrix representation $\left(\sum_{i=1}^m \langle P_i^* P_i e_k, e_l \rangle \right)_{i,j}$, where $\{e_j\}_{j=1}^n$ is an orthonormal basis for $\mathcal{H}_n$. Moreover, we obtain

$$
An = \sum_{j=1}^n A\|e_j\|^2 = \sum_{j=1}^n \int_\Omega \|P(\omega)e_j\|^2 d\mu(\omega) = \int_\Omega \sum_{j=1}^n \langle P(\omega)e_j, P(\omega)e_j \rangle d\mu(\omega).
$$

Thus, if μ is a tight random nonorthogonal fusion frame for $\mathcal{H}_n$, then the frame bound A satisfies

$$
A = \frac{1}{n} \int_\Omega \operatorname{trace}(P(\omega)^* P(\omega)) d\mu(\omega).
$$

Next, we present a construction of tight (random) nonorthogonal fusion frames that is based on finite groups. Recall that a finite subgroup G of the unitary operators $O(\mathcal{H}_n)$ is called *irreducible* if each orbit Gx, for $0 \neq x \in \mathcal{H}_n$, spans $\mathcal{H}_n$. In other words, any G-invariant subspace is trivial, i.e., either $\mathcal{H}_n$ or $\{0\}$. The following result is a generalization of Theorem 6.3 in [22], where finite frames were considered:

THEOREM 4.1. *If P is a nontrivial random projection and G is an irreducible finite subgroup of $O(\mathcal{H}_n)$, then $\frac{1}{|G|} \sum_{g \in G} g^* P g$ is a tight random nonorthogonal fusion frame.*

PROOF. One can directly check that the fusion frame operator $S : \mathcal{H}_n \to \mathcal{H}_n$,

$$x \mapsto \frac{1}{|G|} \sum_{g \in G} \int_\Omega g^* P(\omega)^* g g^* P(\omega) g x d\mu(\omega) = \frac{1}{|G|} \sum_{g \in G} \int_\Omega g^* P(\omega)^* P(\omega) g x d\mu(\omega)$$

is self-adjoint and positive semi-definite. Since the identity is an element in G and P is not the trivial random projection, S cannot be the zero mapping, so that it has a positive eigenvalue λ. One checks that each $g \in G$ commutes with S. Thus, the λ-eigenspace is a G-invariant subspace. The irreducibility implies that the eigenspace is the full space $\mathcal{H}_n$, so that S is a multiple of the identity. □

For the sake of completeness, we also formulate Theorem 4.1 in terms of finite nonorthogonal fusion frames:

COROLLARY 4.2. *If P is a projection and G an irreducible finite subgroup of $O(\mathcal{H}_n)$, then $\{g^* P g\}_{g \in G}$ is a tight nonorthogonal fusion frame.*

Next, we shall discuss the fusion frame potential, cf. . [5, 14, 20], in the setting of nonorthogonal random projectors. If P is a random projector, then we call

$$\mathcal{R}(P) = \int_\Omega \int_\Omega \langle P(\omega)^* P(\omega), P(\omega')^* P(\omega') \rangle d\mu(\omega) d\mu(\omega')$$

its *random nonorthogonal fusion frame potential.*

PROPOSITION 4.3. *If P is a nontrivial random projection, then*

(4) $$\mathcal{R}(P) \geq \frac{M^2}{n}, \quad \text{where } M = \int_\Omega \|P(\omega)\|_{HS}^2 d\mu(\omega),$$

and equality holds if and only if P is tight.

Note that $\mathcal{R}(P) = \text{trace}(S^2)$, where S is the random nonorthogonal fusion frame operator of P. This way we see that Proposition 4.3 can be proven by following the lines of the analogues results for orthogonal projectors in [1].

If P is a projection, then $\|P\|_{HS}^2 \geq \text{rank}(P)$, and equality holds if and only if P is an orthogonal projection. We therefore have the following:

COROLLARY 4.4. *If P is a nontrivial random projection, then*

(5) $$\mathcal{R}(P) \geq \frac{M^2}{n}, \quad \text{where } M = \int_\Omega \text{rank}(P(\omega)) d\mu(\omega),$$

and equality holds if and only if P is tight and an orthogonal projection almost everywhere.

We can expect that the sample of a tight random nonorthogonal fusion frame approximates a tight nonorthogonal fusion frame when the sample size increases. The following theorem generalizes results in [13]:

PROPOSITION 4.5. *Let $\{P_i\}_{i=1}^m$ be a collection of independent tight random nonorthogonal fusion frames with frame bounds $\{A_i\}_{i=1}^m$, respectively, such that,*

$$M := \frac{1}{m}\sum_{i=1}^m \int_\Omega \|P_i^*(\omega)P_i(\omega)\|_{HS}^2 d\mu(\omega) < \infty.$$

If $S(\omega) = \sum_{i=1}^m P_i(\omega)^*P_i(\omega)$ denotes the nonorthogonal fusion frame operator associated to $\{P_i(\omega)\}_{i=1}^m$, then

$$\mathbb{E}(\|\frac{1}{m}S - AI\|_{HS}^2) = \frac{1}{m}(M - n\tilde{A}),$$

where $A = \frac{1}{m}\sum_{i=1}^m A_i$ and $\tilde{A} = \frac{1}{m}\sum_{i=1}^m A_i^2$.

PROOF. The (k,l)-th entry of the matrix S is given by $S_{k,l} = \sum_{i=1}^m \langle P_i^*P_i e_k, e_l\rangle$, and we observe that $\mathbb{E}(\langle P_i^*P_i e_k, e_l\rangle)) = A_i \delta_{k,l}$. We derive

$$\mathbb{E}(\|\frac{1}{m}S - AI_d\|_{HS}^2) = \mathbb{E}(\sum_{k,l}\left(\frac{1}{m}S_{k,l} - A\delta_{k,l}\right)^2)$$

$$= \mathbb{E}(\sum_{k,l}\frac{1}{m^2}(S_{k,l})^2) - \mathbb{E}(\sum_k \frac{2A}{m}S_{k,k}) + nA^2$$

$$= \mathbb{E}(\sum_{k,l}\frac{1}{m^2}(S_{k,l})^2) - nA^2$$

since $\mathbb{E}(\sum_k \frac{1}{m}S_{k,k}) = nA$. We split the occurring double sum of $(S_{k,l})^2$ into its diagonal and nondiagonal parts so that the independence of $\{P_i\}_{i=1}^m$ yields

$$\mathbb{E}(\|\frac{1}{m}S - AI\|_{HS}^2) = \frac{1}{m}M + \sum_{k,l}\frac{1}{m^2}\sum_{i\neq j}\mathbb{E}(\langle P_j^*P_j e_k, e_l\rangle)\mathbb{E}(\langle P_i^*P_i e_k, e_l\rangle) - nA^2$$

$$= \frac{1}{m}M + \sum_{k,l}\frac{1}{m^2}\sum_{i\neq j}A_j\delta_{k,l}A_i\delta_{k,l} - nA^2$$

$$= \frac{1}{m}M + \frac{n}{m^2}\sum_{i\neq j}A_jA_i - nA^2 = \frac{1}{m}M - \frac{n}{m}\tilde{A}. \qquad \square$$

The Matrix Rosenthal inequality as stated in [**19**] enables the following variation of Proposition 4.5:

THEOREM 4.6. *Let $\{P_i\}_{i=1}^m$ be a collection of independent random tight nonorthogonal fusion frames with frame bound A, such that, $\mathbb{E}\|P_i^*P_i - AI\|_{4p}^{4p} < \infty$. Let S be as in Proposition 4.5. If $p = 1$ or $p \geq 3/2$, then*

$$\mathbb{E}\|\frac{1}{m}S - AI\|_{4p}^{4p} \leq (\frac{4p-1}{m^2})^{2p}\|(\sum_{i=1}^m \mathbb{E}(P_i^*P_i - AI)^2)^{1/2}\|_{4p}^{4p} + (\frac{4p-1}{m})^{4p}\sum_{i=1}^m \mathbb{E}\|P_i^*P_i - AI\|_{4p}^{4p}.$$

In order to replace the expectation used in Proposition 4.5 and Theorem 4.6 with a proper estimate of the norm of the difference, one can apply large deviation bounds, for which we refer, for instance, to [**23**].

Acknowledgements

M.E. has been funded by the Vienna Science and Technology Fund (WWTF) through project VRG12-009.

References

[1] C. Bachoc and M. Ehler, *Tight p-fusion frames*, Appl. Comput. Harmon. Anal. **35** (2013), no. 1, 1–15, DOI 10.1016/j.acha.2012.07.001. MR3053743

[2] Behrens, Richard T. Signal processing applications of oblique projection operators. *IEEE T. Signal Proces.*, 42:1413-1424, 1994.

[3] Bownik, Marcin, Luoto, Kurt, and Richmond, Edward. A combinatorial characterization of tight fusion frames *Preprint*

[4] Jameson Cahill, Peter G. Casazza, and Shidong Li, *Non-orthogonal fusion frames and the sparsity of fusion frame operators*, J. Fourier Anal. Appl. **18** (2012), no. 2, 287–308, DOI 10.1007/s00041-011-9200-7. MR2898730

[5] Peter G. Casazza and Matthew Fickus, *Minimizing fusion frame potential*, Acta Appl. Math. **107** (2009), no. 1-3, 7–24, DOI 10.1007/s10440-008-9377-1. MR2520007 (2010e:42042)

[6] *Finite frames*, Applied and Numerical Harmonic Analysis, Birkhäuser/Springer, New York, 2013. Theory and applications; Edited by Peter G. Casazza and Gitta Kutyniok. MR2964005

[7] Casazza, Peter G. and Kutyniok, Gitta. Frames of subspaces. 2004.

[8] Peter G. Casazza, Gitta Kutyniok, and Shidong Li, *Fusion frames and distributed processing*, Appl. Comput. Harmon. Anal. **25** (2008), no. 1, 114–132, DOI 10.1016/j.acha.2007.10.001. MR2419707 (2009d:42094)

[9] Peter G. Casazza, Matthew Fickus, Dustin G. Mixon, Yang Wang, and Zhengfang Zhou, *Constructing tight fusion frames*, Appl. Comput. Harmon. Anal. **30** (2011), no. 2, 175–187, DOI 10.1016/j.acha.2010.05.002. MR2754774 (2012c:42069)

[10] Casazza, Peter G. and Tremain, Janet. A brief introduction to Hilbert space frame theory and its applications.

[11] John H. Conway, Ronald H. Hardin, and Neil J. A. Sloane, *Packing lines, planes, etc.: packings in Grassmannian spaces*, Experiment. Math. **5** (1996), no. 2, 139–159. MR1418961 (98a:52029)

[12] G. Corach and A. Maestripieri, *Polar decomposition of oblique projections*, Linear Algebra Appl. **433** (2010), no. 3, 511–519, DOI 10.1016/j.laa.2010.03.016. MR2653816 (2011h:47041)

[13] M. Ehler, *Random tight frames*, J. Fourier Anal. Appl. **18** (2012), no. 1, 1–20, DOI 10.1007/s00041-011-9182-5. MR2885555

[14] M. Ehler and K. A. Okoudjou, *Minimization of the probabilistic p-frame potential*, J. Statist. Plann. Inference **142** (2012), no. 3, 645–659, DOI 10.1016/j.jspi.2011.09.001. MR2853573

[15] Martin Ehler and Kasso A. Okoudjou, *Probabilistic frames: an overview*, Finite frames, Appl. Numer. Harmon. Anal., Birkhäuser/Springer, New York, 2013, pp. 415–436, DOI 10.1007/978-0-8176-8373-3_12. MR2964017

[16] William Fulton, *Eigenvalues, invariant factors, highest weights, and Schubert calculus*, Bull. Amer. Math. Soc. (N.S.) **37** (2000), no. 3, 209–249 (electronic), DOI 10.1090/S0273-0979-00-00865-X. MR1754641 (2001g:15023)

[17] Shidong Li and Dunyan Yan, *Frame fundamental sensor modeling and stability of one-sided frame perturbation*, Acta Appl. Math. **107** (2009), no. 1-3, 91–103, DOI 10.1007/s10440-008-9419-8. MR2520011 (2010h:42061)

[18] Shidong Li and Hidemitsu Ogawa, *Pseudoframes for subspaces with applications*, J. Fourier Anal. Appl. **10** (2004), no. 4, 409–431, DOI 10.1007/s00041-004-3039-0. MR2078265 (2005e:42098)

[19] Lester Mackey, Michael I. Jordan, Richard Y. Chen, Brendan Farrell, and Joel A. Tropp, *Matrix concentration inequalities via the method of exchangeable pairs*, Ann. Probab. **42** (2014), no. 3, 906–945, DOI 10.1214/13-AOP892. MR3189061

[20] Pedro G. Massey, Mariano A. Ruiz, and Demetrio Stojanoff, *The structure of minimizers of the frame potential on fusion frames*, J. Fourier Anal. Appl. **16** (2010), no. 4, 514–543, DOI 10.1007/s00041-009-9098-5. MR2671171 (2011f:42038)

[21] Joseph M. Renes, Robin Blume-Kohout, A. J. Scott, and Carlton M. Caves, *Symmetric informationally complete quantum measurements*, J. Math. Phys. **45** (2004), no. 6, 2171–2180, DOI 10.1063/1.1737053. MR2059685 (2004m:81043)

[22] Richard Vale and Shayne Waldron, *Tight frames and their symmetries*, Constr. Approx. **21** (2005), no. 1, 83–112, DOI 10.1007/s00365-004-0560-y. MR2105392 (2005h:42063)

[23] Roman Vershynin, *Introduction to the non-asymptotic analysis of random matrices*, Compressed sensing, Cambridge Univ. Press, Cambridge, 2012, pp. 210–268. MR2963170

MATHEMATICS DEPARTMENT, DUKE UNIVERSITY, DURHAM, NORTH CAROLINA 27708
E-mail address: jcahill@math.duke.edu

DEPARTMENT OF MATHEMATICS, UNIVERSITY OF MISSOURI, COLUMBIA, MISSOURI 65211
E-mail address: casazzap@missouri.edu

DEPARTMENT OF MATHEMATICS, UNIVERSITY OF VIENNA, VIENNA, AUSTRIA
E-mail address: martin.ehler@univie.ac.at

DEPARTMENT OF MATHEMATICS, SAN FRANCISCO STATE UNIVERSITY, SAN FRANCISCO, CALIFORNIA 94132
E-mail address: shidong@sfsu.edu

Contemporary Mathematics
Volume **650**, 2015
http://dx.doi.org/10.1090/conm/650/13032

Decompositions of rational Gabor representations

Vignon Oussa

ABSTRACT. Let $\Gamma = \langle T_k, M_l : k \in \mathbb{Z}^d, l \in B\mathbb{Z}^d \rangle$ be a group of unitary operators where T_k is a translation operator and M_l is a modulation operator acting on $L^2(\mathbb{R}^d)$. Assuming that B is a non-singular rational matrix of order d with at least one entry which is not an integer, we obtain a direct integral irreducible decomposition of the Gabor representation which is defined by the isomorphism $\pi : (\mathbb{Z}_m \times B\mathbb{Z}^d) \rtimes \mathbb{Z}^d \to \Gamma$ where $\pi(\theta, l, k) = e^{\frac{2\pi i}{m}\theta} M_l T_k$. We also show that the left regular representation of $(\mathbb{Z}_m \times B\mathbb{Z}^d) \rtimes \mathbb{Z}^d$ which is identified with Γ via π is unitarily equivalent to a direct sum of card $([\Gamma, \Gamma])$ many disjoint subrepresentations of the type: $L_0, L_1, \cdots, L_{\text{card}([\Gamma,\Gamma])-1}$ such that for $k \neq 1$ the subrepresentation L_k of the left regular representation is disjoint from the Gabor representation. Additionally, we compute the central decompositions of the representations π and L_1. These decompositions are then exploited to give a new proof of the Density Condition of Gabor systems (for the rational case). More precisely, we prove that π is equivalent to a subrepresentation of L_1 if and only if $|\det B| \leq 1$. We also derive characteristics of vectors f in $L^2(\mathbb{R})^d$ such that $\pi(\Gamma)f$ is a Parseval frame in $L^2(\mathbb{R})^d$.

1. Introduction

The concept of applying tools of abstract harmonic analysis to time-frequency analysis, and wavelet theory is not a new idea [1–3, 7, 11]. For example in [1], Larry Baggett gives a direct integral decomposition of the Stone-von Neumann representation of the discrete Heisenberg group acting in $L^2(\mathbb{R})$. Using his decomposition, he was able to provide specific conditions under which this representation is cyclic. In Section 5.5, [7] the author obtains a characterization of tight Weyl-Heisenberg frames in $L^2(\mathbb{R})$ using the Zak transform and a precise computation of the Plancherel measure of a discrete type I group. In [11], the authors present a thorough study of the left regular representations of various subgroups of the reduced Heisenberg groups. Using well-known results of admissibility of unitary representations of locally compact groups, they were able to offer new insights on Gabor theory.

Let B be a non-singular matrix of order d with real entries. For each $k \in \mathbb{Z}^d$ and $l \in B\mathbb{Z}^d$, we define the corresponding unitary operators T_k, M_l such that $T_k f(t) = f(t - k)$ and $M_l f(t) = e^{-2\pi i \langle l, t \rangle} f(t)$ for $f \in L^2(\mathbb{R}^d)$. The operator T_k is called a shift operator, and the operator M_l is called a modulation operator. Let Γ be a subgroup of the group of unitary operators acting on $L^2(\mathbb{R}^d)$ which is generated by the set $\{T_k, M_l : k \in \mathbb{Z}^d, l \in B\mathbb{Z}^d\}$. We write $\Gamma = \langle T_k, M_l : k \in \mathbb{Z}^d, l \in B\mathbb{Z}^d \rangle$. The commutator subgroup of Γ given by $[\Gamma, \Gamma] = \left\{ e^{2\pi i \langle l, Bk \rangle} : l, k \in \mathbb{Z}^d \right\}$ is a subgroup of the one-dimensional torus $\mathbb{T}$. Since $[\Gamma, \Gamma]$ is always contained in the center of the group, then Γ is a nilpotent group which is generated by $2d$ elements. Moreover, Γ is given the discrete topology, and as such it is a locally compact group. We observe that if B has at least one irrational entry, then it is a non-abelian group with an infinite center. If B only has rational entries with at least one entry which is not an integer, then $[\Gamma, \Gamma]$ is a finite group, and Γ is a non-abelian group which is regarded as a finite extension of an abelian group. If all entries of B are integers, then Γ is abelian, and clearly $[\Gamma, \Gamma]$ is trivial. Finally, it is worth mentioning that Γ is a type I group if and only if B only has rational entries [12].

It is easily derived from the work in Section 4, [11] that if B is an integral matrix, then the Gabor representation $\pi : B\mathbb{Z}^d \times \mathbb{Z}^d \to \Gamma \subset \mathcal{U}\left(L^2\left(\mathbb{R}^d\right)\right)$ defined by $\pi(l, k) = M_l T_k$ is equivalent to a subrepresentation of the left regular representation of Γ if and only if B is a unimodular matrix. The techniques used by the authors of [11] rely on the decompositions of the left regular representation and the Gabor representation into their irreducible components. The group generated by the operators M_l and T_k is a commutative group which is isomorphic to $\mathbb{Z}_d \times B\mathbb{Z}_d$. The unitary dual and the Plancherel measure for discrete abelian groups are well-known and rather easy to write down. Thus, a precise direct integral decomposition of the left regular representation is easily obtained as well. Next, using the Zak transform, the authors decompose the representation π into a direct integral of its irreducible components. They are then able to compare both representations. As a result, one can derive from the work in the fourth section of [11] that the representation π is equivalent to a subrepresentation of the left regular representation if and only if B is a unimodular matrix. The main objective of this paper is to generalize these ideas to the more difficult case where $B \in GL(d, \mathbb{Q})$ and Γ is not a commutative group.

Let us assume that B is an invertible rational matrix with at least one entry which is not an element of $\mathbb{Z}$. Denoting the inverse transpose of a given matrix M by $M^\star$, it is not too hard to see that there exists a matrix $A \in GL(d, \mathbb{Z})$ such that $\Lambda = B^\star \mathbb{Z}^d \cap \mathbb{Z}^d = A\mathbb{Z}^d$. Indeed, a precise algorithm for the construction of A is described on Page 809 of [4]. Put

$$(1) \qquad \Gamma_0 = \left\langle \tau, M_l, T_k : l \in B\mathbb{Z}^d, k \in \Lambda, \tau \in [\Gamma, \Gamma] \right\rangle$$

and define $\Gamma_1 = \left\langle \tau, M_l : l \in B\mathbb{Z}^d, \tau \in [\Gamma, \Gamma] \right\rangle$. Then Γ_0 is a normal abelian subgroup of Γ. Moreover, we observe that Γ_1 is a subgroup of Γ_0 of infinite index. Let m be the number of elements in $[\Gamma, \Gamma]$. Clearly, since B has at least one rational entry which is not an integer, it must be the case that $m > 1$. Furthermore, it is easy to see that there is an isomorphism $\pi : \left(\mathbb{Z}_m \times B\mathbb{Z}^d\right) \rtimes \mathbb{Z}^d \to \Gamma \subset \mathcal{U}\left(L^2\left(\mathbb{R}^d\right)\right)$ defined by $\pi(j, Bl, k) = e^{\frac{2\pi j i}{m}} M_{Bl} T_k$. The multiplication law on the semi-direct product group $\left(\mathbb{Z}_m \times B\mathbb{Z}^d\right) \rtimes \mathbb{Z}^d$ is described as follows. Given arbitrary elements

$$(j, Bl, k), (j_1, Bl_1, k_1) \in \left(\mathbb{Z}_m \times B\mathbb{Z}^d\right) \rtimes \mathbb{Z}^d,$$

we define $(j, Bl, k)(j_1, Bl_1, k_1) = ((j + j_1 + \omega(l_1, k)) \bmod m, B(l + l_1), k + k_1)$ where $\omega(l_1, k) \in \mathbb{Z}_m$, and $\langle Bl_1, k \rangle = \frac{\omega(l_1, k)}{m}$. We call π a rational Gabor representation. It is also worth observing that $\pi^{-1}(\Gamma_0) = \left(\mathbb{Z}_m \times B\mathbb{Z}^d\right) \times A\mathbb{Z}^d$ and $\pi^{-1}(\Gamma_1) = \left(\mathbb{Z}_m \times B\mathbb{Z}^d\right) \times \{0\} \simeq \mathbb{Z}_m \times B\mathbb{Z}^d$. Throughout this work, in order to avoid cluster of notations, we will make no distinction between $\left(\mathbb{Z}_m \times B\mathbb{Z}^d\right) \rtimes \mathbb{Z}^d$ and Γ and their corresponding subgroups.

The main results of this paper are summarized in the following propositions. Let

$$\Gamma = \left\langle T_k, M_l : k \in \mathbb{Z}^d, l \in B\mathbb{Z}^d \right\rangle$$

and assume that B is an invertible rational matrix with at least one entry which is not an integer. Let L be the left regular representation of Γ.

PROPOSITION 1. *The left regular representation of Γ is decomposed as follows:*

$$L \simeq \oplus_{k=0}^{m-1} \int_{\frac{\mathbb{R}^d}{B^\star \mathbb{Z}^d} \times \frac{\mathbb{R}^d}{A^\star \mathbb{Z}^d}}^{\oplus} \mathrm{Ind}_{\Gamma_0}^{\Gamma} \chi_{(k,\sigma)} \, d\sigma. \tag{2}$$

Moreover, the measure $d\sigma$ in (2) is a Lebesgue measure, and (2) is not an irreducible decomposition of L.

PROPOSITION 2. *The Gabor representation π is decomposed as follows:*

$$\pi \simeq \int_{\frac{\mathbb{R}^d}{\mathbb{Z}^d} \times \frac{\mathbb{R}^d}{A^\star \mathbb{Z}^d}}^{\oplus} \mathrm{Ind}_{\Gamma_0}^{\Gamma} \chi_{(1,\sigma)} \, d\sigma. \tag{3}$$

Moreover, $d\sigma$ is a Lebesgue measure defined on the torus $\frac{\mathbb{R}^d}{\mathbb{Z}^d} \times \frac{\mathbb{R}^d}{A^\star \mathbb{Z}^d}$ and (3) is an irreducible decomposition of π.

It is worth pointing out here that the decomposition of π given in Proposition 2 is consistent with the decomposition obtained in Lemma 5.39, [7] for the specific case where $d = 1$ and B is the inverse of a natural number larger than one.

PROPOSITION 3. *Let m be the number of elements in the commutator subgroup $[\Gamma, \Gamma]$. There exists a decomposition of the left regular representation of Γ such that*

$$L \simeq \bigoplus_{k=0}^{m-1} L_k \tag{4}$$

and for each $k \in \{0, 1, \cdots, m-1\}$, the representation L_k is disjoint from π whenever $k \neq 1$. Moreover, the Gabor representation π is equivalent to a subrepresentation of the subrepresentation L_1 of L if and only if $|\det B| \leq 1$.

Although this problem of decomposing the representations π and L into their irreducible components is interesting in its own right, we shall also address how these decompositions can be exploited to derive interesting and relevant results in time-frequency analysis. The proof of Proposition 3 allows us to state the following:

(1) There exists a measurable set $\mathbf{E} \subset \mathbb{R}^d$ which is a subset of a fundamental domain for the lattice $B^\star \mathbb{Z}^d \times A^\star \mathbb{Z}^d$, satisfying

$$\mu\left(\mathbf{E}\right) = \frac{1}{|\det\left(B\right)\det\left(A\right)|\dim\left(l^2\left(\Gamma/\Gamma_0\right)\right)}$$

where μ is the Lebesgue measure on $\mathbb{R}^d \times \mathbb{R}^d$.

(2) There exists a unitary map

$$\mathfrak{A} : \int_{\mathbf{E}}^{\oplus} \left(\oplus_{k=1}^{\ell(\sigma)} l^2\left(\frac{\Gamma}{\Gamma_0}\right)\right) d\sigma \to L^2\left(\mathbb{R}^d\right) \tag{5}$$

which intertwines $\int_{\mathbf{E}}^{\oplus} \left(\oplus_{k=1}^{\ell(\sigma)} \operatorname{Ind}_{\Gamma_0}^{\Gamma} \left(\chi_{(1,\sigma)} \right) \right) d\sigma$ with π such that, the multiplicity function ℓ is bounded, the representations $\chi_{(1,\sigma)}$ are characters of the abelian subgroup Γ_0 and

(6)
$$\int_{\mathbf{E}}^{\oplus} \left(\oplus_{k=1}^{\ell(\sigma)} \operatorname{Ind}_{\Gamma_0}^{\Gamma} \left(\chi_{(1,\sigma)} \right) \right) d\sigma$$

is the central decomposition of π (Section 3.4.2, [**7**]).

Moreover, for the case where $|\det(B)| \leq 1$, the multiplicity function ℓ is bounded above by the number of cosets in Γ/Γ_0 while if $|\det(B)| > 1$, then the multiplicity function ℓ is bounded but is greater than the number of cosets in Γ/Γ_0 on a set $\mathbf{E}' \subseteq \mathbf{E}$ of positive Lebesgue measure. This observation that the upper-bound of the multiplicity function behaves differently in each situation may mistakenly appear to be of limited importance. However, at the heart of this observation, lies a new justification of the well-known Density Condition of Gabor systems for the rational case (Theorem 1.3, [**8**]). In fact, we shall offer a new proof of a rational version of the Density Condition for Gabor systems in Proposition 8.

It is also worth pointing out that the central decomposition of π as described above has several useful implications. Following the discussion on Pages 74-75, [**7**], the decomposition given in (6) may be used to:

(1) Characterize the commuting algebra of the representation π and its center.
(2) Characterize representations which are either quasi-equivalent or disjoint from π (see [**7**] Theorem 3.17 and Corollary 3.18).

Additionally, using the central decomposition of π, in the case where the absolute value of the determinant of B is less or equal to one, we obtain a complete characterization of vectors f such that $\pi(\Gamma) f$ is a Parseval frame in $L^2(\mathbb{R}^d)$.

PROPOSITION 4. *Let us suppose that $|\det B| \leq 1$. Then*

(7)
$$\pi \simeq \int_{\mathbf{E}}^{\oplus} \left(\oplus_{k=1}^{\ell(\sigma)} \operatorname{Ind}_{\Gamma_0}^{\Gamma} \left(\chi_{(1,\sigma)} \right) \right) d\sigma$$

with $\ell(\sigma) \leq \operatorname{card}(\Gamma/\Gamma_0)$. Moreover, $\pi(\Gamma) f$ is a Parseval frame in $L^2(\mathbb{R}^d)$ if and only if $f = \mathfrak{A}\left(a(\sigma) \right)_{\sigma \in \mathbf{E}}$ such that for $d\sigma$-almost every $\sigma \in \mathbf{E}$, $\| a(\sigma)(k) \|_{l^2\left(\frac{\Gamma}{\Gamma_0}\right)}^2 = 1$ for $1 \leq k \leq \ell(\sigma)$ and for distinct $k, j \in \{1, \cdots, \ell(\sigma)\}$, $\langle a(\sigma)(k), a(\sigma)(j) \rangle = 0$.

This paper is organized around the proofs of the propositions mentioned above. In Section 2, we fix notations and we revisit well-known concepts such as induced representations and direct integrals which are of central importance. The proof of Proposition 1 is obtained in the third section. The proofs of Propositions 2, 3 and examples are given in the fourth section. Finally, the last section contains the proof of Proposition 4.

2. Preliminaries

Let us start by fixing our notations and conventions. All representations in this paper are assumed to be unitary representations. Given two equivalent representations π and ρ, we write that $\pi \simeq \rho$. We use the same notation for isomorphic groups. That is, if G and H are isomorphic groups, we write that $G \simeq H$. All sets considered in this paper will be assumed to be measurable. Given two disjoint sets

A and B, the disjoint union of the sets is denoted $A \dot\cup B$. Let $\mathbf{H}$ be a Hilbert space. The identity operator acting on $\mathbf{H}$ is denoted $1_{\mathbf{H}}$. The unitary equivalence classes of irreducible unitary representations of G is called the unitary dual of G and is denoted $\widehat{G}$.

Several of the proofs presented in this work rely on basic properties of induced representations and direct integrals. The following discussion is mainly taken from Chapter 6, [6]. Let G be a locally compact group, and let K be a closed subgroup of G. Let us define $q : G \to G/K$ to be the canonical quotient map and let φ be a unitary representation of the group K acting in some Hilbert space which we call $\mathbf{H}$. Next, let $\mathbf{K}_1$ be the set of continuous $\mathbf{H}$-valued functions f defined over G satisfying the following properties:

(1) $q \left(\text{support } (f)\right)$ is compact,
(2) $f (gk) = \varphi (k)^{-1} f (g)$ for $g \in G$ and $k \in K$.

Clearly, G acts on the set $\mathbf{K}_1$ by left translation. Now, to simplify the presentation, let us suppose that G/K admits a G-invariant measure. We remark that in general, this is not always the case. However, the assumption that G/K admits a G-invariant measure holds for the class of groups considered in this paper. We construct a unitary representation of G by endowing $\mathbf{K}_1$ with the following inner product:

$$\langle f, f' \rangle = \int_{G/K} \langle f (g), f' (g) \rangle_{\mathbf{H}} \; d (gK) \text{ for } f, f' \in \mathbf{K}_1.$$

Now, let $\mathbf{K}$ be the Hilbert completion of the space $\mathbf{K}_1$ with respect to this inner product. The translation operators extend to unitary operators on $\mathbf{K}$ inducing a unitary representation $\text{Ind}_K^G (\varphi)$ which is defined as follows:

$$\text{Ind}_K^G (\varphi) (x) f (g) = f \left(x^{-1} g\right) \text{ for } f \in \mathbf{K}.$$

We notice that if φ is a character, then the Hilbert space $\mathbf{K}$ can be naturally identified with $L^2 (G/H)$. Induced representations are natural ways to construct unitary representations. For example, it is easy to prove that if e is the identity element of G and if 1 is the trivial representation of $\{e\}$ then the representation $\text{Ind}_{\{e\}}^G (1)$ is equivalent to the left regular representation of G. Other properties of induction such as induction in stages will be very useful for us. The reader who is not familiar with these notions is invited to refer to Chapter 6 of the book of Folland [6] for a thorough presentation.

We will now present a short introduction to direct integrals; which are heavily used in this paper. For a complete presentation, the reader is referred to Section 7.4, [6]. Let $\{H_\alpha\}_{\alpha \in A}$ be a family of separable Hilbert spaces indexed by a set A. Let μ be a measure defined on A. We define the direct integral of this family of Hilbert spaces with respect to μ as the space which consists of vectors φ defined on the parameter space A such that $\varphi (\alpha)$ is an element of H_α for each $\alpha \in A$ and $\int_A \|\varphi (\alpha)\|_{H_\alpha}^2 \, d\mu (\alpha) < \infty$ with some additional measurability conditions which we will clarify. A family of separable Hilbert spaces $\{H_\alpha\}_{\alpha \in A}$ indexed by a Borel set A is called a field of Hilbert spaces over A. A map $\varphi : A \to \bigcup_{\alpha \in A} H_\alpha$ such that $\varphi (\alpha) \in H_\alpha$ is called a vector field on A. A measurable field of Hilbert spaces over the indexing set A is a field of Hilbert spaces $\{H_\alpha\}_{\alpha \in A}$ together with a countable set $\{e_j\}_j$ of vector fields such that

(1) the functions $\alpha \mapsto \langle e_j(\alpha), e_k(\alpha) \rangle_{H_\alpha}$ are measurable for all j, k,

(2) the linear span of $\{e_k(\alpha)\}_k$ is dense in H_α for each $\alpha \in A$.

The direct integral of the spaces H_α with respect to the measure μ is denoted by $\int_A^\oplus H_\alpha d\mu(\alpha)$ and is the space of measurable vector fields φ on A such that $\int_A \|\varphi(\alpha)\|_{H_\alpha}^2 d\mu(\alpha) < \infty$. The inner product for this Hilbert space is: $\langle \varphi_1, \varphi_2 \rangle = \int_A \langle \varphi_1(\alpha), \varphi_2(\alpha) \rangle_{H_\alpha} d\mu(\alpha)$ for $\varphi_1, \varphi_2 \in \int_A^\oplus H_\alpha d\mu(\alpha)$.

3. The regular representation and its decompositions

In this section, we will discuss the Plancherel theory for Γ. For this purpose, we will need a complete description of the unitary dual of Γ. This will allow us to obtain a central decomposition of the left regular representation of Γ. Also, a proof of Proposition 1 will be given in this section.

Let L be the left regular representation of Γ. Suppose that Γ is not commutative and that B is a rational matrix. We have shown that Γ has an abelian normal subgroup of finite index which we call Γ_0. Moreover, there is a canonical action (74-79, [10]) of Γ on the group $\widehat{\Gamma}_0$ (the unitary dual of Γ_0) such that for each $P \in \Gamma$ and $\chi \in \widehat{\Gamma}_0$, $P \cdot \chi(Q) = \chi(P^{-1}QP)$. Let us suppose that $\chi = \chi_{(\lambda_1, \lambda_2, \lambda_3)}$ is a character in the unitary dual $\widehat{\Gamma}_0$ where $(\lambda_1, \lambda_2, \lambda_3) \in \{0, 1, \cdots, m-1\} \times \frac{\mathbb{R}^d}{B^\star \mathbb{Z}^d} \times \frac{\mathbb{R}^d}{A^\star \mathbb{Z}^d} \simeq \widehat{\Gamma}_0$, and $\chi_{(\lambda_1, \lambda_2, \lambda_3)}\left(e^{\frac{2\pi i k}{m}} M_{Bl} T_{Aj}\right) = e^{\frac{2\pi i \lambda_1 k}{m}} e^{2\pi i \langle \lambda_2, Bl \rangle} e^{2\pi i \langle \lambda_3, Aj \rangle}$. We observe that $\mathbb{R}^d$ is identified with its dual $\widehat{\mathbb{R}^d}$ and $\{0, 1, \cdots, m-1\}$ parametrizes the unitary dual of the commutator subgroup $[\Gamma, \Gamma]$ which is isomorphic to the cyclic group $\mathbb{Z}_m$. For any $\tau \in [\Gamma, \Gamma]$, we have $\tau \cdot \chi_{(\lambda_1, \lambda_2, \lambda_3)} = \chi_{(\lambda_1, \lambda_2, \lambda_3)}$. Moreover, given $M_l, T_k \in \Gamma$,

$$M_l \cdot \chi_{(\lambda_1, \lambda_2, \lambda_3)} = \chi_{(\lambda_1, \lambda_2, \lambda_3)}, \text{ and } T_k \cdot \chi_{(\lambda_1, \lambda_2, \lambda_3)} = \chi_{(\lambda_1, \lambda_2 - k\lambda_1, \lambda_3)}.$$

Next, let $\Gamma_\chi = \{P \in \Gamma : P \cdot \chi = \chi\}$. It is easy to see that the stability subgroup of the character $\chi_{(\lambda_1, \lambda_2, \lambda_3)}$ is described as follows:

$$(8) \qquad \Gamma_{\chi_{(\lambda_1, \lambda_2, \lambda_3)}} = \left\{ \tau M_l T_k \in \Gamma : \tau \in [\Gamma, \Gamma], l \in B\mathbb{Z}^d, k\lambda_1 \in B^\star \mathbb{Z}^d \right\}.$$

It follows from (8) that the stability group $\Gamma_{\chi_{(\lambda_1, \lambda_2, \lambda_3)}}$ contains the normal subgroup Γ_0. Indeed, if $\lambda_1 = 0$ then $\Gamma_{\chi_{(\lambda_1, \lambda_2, \lambda_3)}} = \Gamma$. Otherwise,

$$(9) \qquad \Gamma_{\chi_{(\lambda_1, \lambda_2, \lambda_3)}} = \left\{ \tau M_l T_k \in \Gamma : \tau \in [\Gamma, \Gamma], \, l \in B\mathbb{Z}^d, k \in \left(\frac{1}{\lambda_1} B^\star \mathbb{Z}^d\right) \cap \mathbb{Z}^d \right\}.$$

According to Mackey theory (see Page 76, [10]) and well-known results of Kleppner and Lipsman (Page 460, [10]), every irreducible representation of Γ is of the type $\text{Ind}_{\Gamma_\chi}^\Gamma (\widetilde{\chi} \otimes \widetilde{\sigma})$ where $\widetilde{\chi}$ is an extension of a character χ of Γ_0 to Γ_χ, and $\widetilde{\sigma}$ is the lift of an irreducible representation σ of Γ_χ / Γ_0 to Γ_χ. Also two irreducible representations $\text{Ind}_{\Gamma_\chi}^\Gamma (\widetilde{\chi} \otimes \widetilde{\sigma})$ and $\text{Ind}_{\Gamma_{\chi_1}}^\Gamma (\widetilde{\chi_1} \otimes \widetilde{\sigma})$ are equivalent if and only if the characters χ and χ_1 of Γ_0 belong to the same Γ-orbit. Since Γ is a finite extension of its subgroup Γ_0, then it is well known that there is a measurable set which is a cross-section for the Γ-orbits in $\widehat{\Gamma}_0$. Now, let Σ be a measurable subset of $\widehat{\Gamma}_0$ which is a cross-section for the Γ-orbits in $\widehat{\Gamma}_0$. The unitary dual of Γ is a fiber space which is described as follows:

$$\widehat{\Gamma} = \bigcup_{\chi \in \Sigma} \left\{ \pi_{\chi, \sigma} = \text{Ind}_{\Gamma_\chi}^\Gamma (\widetilde{\chi} \otimes \widetilde{\sigma}) \; : \; \sigma \in \frac{\widehat{\Gamma_\chi}}{\Gamma_0} \right\}.$$

Finally, since Γ is a type I group, there exists a unique standard Borel measure on $\widehat{\Gamma}$ such that the left regular representation of the group Γ is equivalent to a direct integral of all elements in the unitary dual of Γ, and the multiplicity of each irreducible representation occurring is equal to the dimension of the corresponding Hilbert space. So, we obtain a decomposition of the representation L into a direct integral decomposition of its irreducible representations as follows (see Theorem 3.31, [7] and Theorem 5.12, [11])

$$(10) \qquad L \simeq \int_{\Sigma}^{\oplus} \int_{\widehat{\frac{\Gamma_\chi}{\Gamma_0}}}^{\oplus} \bigoplus_{k=1}^{\dim\left(l^2\left(\frac{\Gamma}{\Gamma_\chi}\right)\right)} \pi_{\chi,\sigma} d\omega_\chi(\sigma) d\chi$$

and $\dim\left(l^2\left(\frac{\Gamma}{\Gamma_\chi}\right)\right) \leq \operatorname{card}(\Gamma/\Gamma_0)$. The fact that $\dim\left(l^2\left(\frac{\Gamma}{\Gamma_\chi}\right)\right) \leq \operatorname{card}(\Gamma/\Gamma_0)$ is justified because the number of representative elements of the quotient group $\frac{\Gamma}{\Gamma_\chi}$ is bounded above by the number of elements in $\frac{\Gamma}{\Gamma_0}$. Moreover the direct integral representation in (10) is realized as acting in the Hilbert space

$$(11) \qquad \int_{\Sigma}^{\oplus} \int_{\widehat{\frac{\Gamma_\chi}{\Gamma_0}}}^{\oplus} \bigoplus_{k=1}^{\dim\left(l^2\left(\frac{\Gamma}{\Gamma_\chi}\right)\right)} l^2\left(\frac{\Gamma}{\Gamma_\chi}\right) d\omega_\chi(\sigma) d\chi.$$

Although the decomposition in (10) is canonical, the decomposition provided by Proposition 1 will be more convenient for us.

3.1. Proof of Proposition 1. Let e be the identity element in Γ, and let 1 be the trivial representation of the trivial group $\{e\}$. We observe that $L \simeq \operatorname{Ind}_{\{e\}}^{\Gamma}(1)$. It follows that

$$(12) \qquad L \simeq \operatorname{Ind}_{\Gamma_0}^{\Gamma}\left(\operatorname{Ind}_{\{e\}}^{\Gamma_0}(1)\right)$$

$$\simeq \operatorname{Ind}_{\Gamma_0}^{\Gamma}\left(\int_{\widehat{\Gamma_0}}^{\oplus} \chi_t \, dt\right)$$

$$(13) \qquad \simeq \int_{\widehat{\Gamma_0}}^{\oplus} \operatorname{Ind}_{\Gamma_0}^{\Gamma}(\chi_t) \, dt.$$

The second equivalence given above is coming from the fact that $\operatorname{Ind}_{\{e\}}^{\Gamma_0}(1)$ is equivalent to the left regular representation of the group Γ_0. Since Γ_0 is abelian, its left regular representation admits a direct integral decomposition into elements in the unitary dual of Γ_0, each occurring once. Moreover, the measure dt is a Lebesgue measure (also a Haar measure) supported on the unitary dual of the group, and the Plancherel transform is the unitary operator which is intertwining the representations $\operatorname{Ind}_{\{e\}}^{\Gamma_0}(1)$ and $\int_{\widehat{\Gamma_0}}^{\oplus} \chi_t \, dt$. Based on the discussion above, it is worth mentioning that the representations occurring in (13) are generally reducible since it is not always the case that $\Gamma_0 = \Gamma_{\chi_t}$. We observe that $\widehat{\Gamma_0}$ is parametrized by the group $\mathbb{Z}_m \times \frac{\mathbb{R}^d}{B^\star \mathbb{Z}^d} \times \frac{\mathbb{R}^d}{A^\star \mathbb{Z}^d}$. Thus, identifying $\widehat{\Gamma_0}$ with $\mathbb{Z}_m \times \frac{\mathbb{R}^d}{B^\star \mathbb{Z}^d} \times \frac{\mathbb{R}^d}{A^\star \mathbb{Z}^d}$, we reach the desired result: $L \simeq \bigoplus_{k=0}^{m-1} \int_{\frac{\mathbb{R}^d}{B^\star \mathbb{Z}^d} \times \frac{\mathbb{R}^d}{A^\star \mathbb{Z}^d}}^{\oplus} \operatorname{Ind}_{\Gamma_0}^{\Gamma} \chi_{(k,t)} \, dt$.

REMARK 5. *Referring to (9), we remark that $\Gamma_{\chi_{(1,t_2,t_3)}} = \Gamma_0$, and in this case $\operatorname{Ind}_{\Gamma_0}^{\Gamma}\left(\chi_{(1,t_2,t_3)}\right)$ is an irreducible representation of the group Γ.*

4. Decomposition of π

In this section, we will provide a decomposition of the Gabor representation π. For this purpose, it is convenient to regard the set $\mathbb{R}^d$ as a fiber space, with base space the d-dimensional torus. Next, for any element t in the torus, the corresponding fiber is the set $t + \mathbb{Z}^d$. With this concept in mind, let us define the periodization map $\mathfrak{R} : L^2\left(\mathbb{R}^d\right) \to \int_{\frac{\mathbb{R}^d}{\mathbb{Z}^d}}^{\oplus} l^2\left(\mathbb{Z}^d\right) dt$ such that $\mathfrak{R}f\left(t\right) = \left(f\left(t+k\right)\right)_{k \in \mathbb{Z}^d}$. We remark here that we clearly abuse notation by making no distinction between $\frac{\mathbb{R}^d}{\mathbb{Z}^d}$ and a choice of a measurable subset of $\mathbb{R}^d$ which is a fundamental domain for $\frac{\mathbb{R}^d}{\mathbb{Z}^d}$. Next, the inner product which we endow the direct integral Hilbert space $\int_{\frac{\mathbb{R}^d}{\mathbb{Z}^d}}^{\oplus} l^2\left(\mathbb{Z}^d\right) dt$ with is defined as follows. For any vectors

$$f \text{ and } h \in \int_{\frac{\mathbb{R}^d}{\mathbb{Z}^d}}^{\oplus} l^2\left(\mathbb{Z}^d\right) dt$$

the inner product of f and g is equal to

$$\langle f, h \rangle_{\mathfrak{R}\left(L^2\left(\mathbb{R}^d\right)\right)} = \int_{\frac{\mathbb{R}^d}{\mathbb{Z}^d}} \left(\sum_{k \in \mathbb{Z}^d} f\left(t+k\right) \overline{h\left(t+k\right)} \right) dt,$$

and it is easy to check that $\mathfrak{R}$ is a unitary map.

4.1. Proof of Proposition 2. For $t \in \mathbb{R}^d$, we consider the unitary character $\chi_{(1,-t)} : \Gamma_1 \to \mathbb{T}$ which is defined by $\chi_{(1,-t)}\left(e^{2\pi i z} M_l\right) = e^{2\pi i z} e^{-2\pi i \langle t, l \rangle}$. Next, we compute the action of the unitary representation $\mathrm{Ind}_{\Gamma_1}^{\Gamma} \chi_{(1,-t)}$ of Γ which is acting in the Hilbert space

$$\mathbf{H}_t = \left\{ \begin{array}{c} f : \Gamma \to \mathbb{C} : f\left(PQ\right) = \chi_{(1,-t)}\left(Q\right)^{-1} f\left(P\right), Q \in \Gamma_1 \\ \text{and } \sum_{P\Gamma_1 \in \frac{\Gamma}{\Gamma_1}} \left|f\left(P\right)\right|^2 < \infty \end{array} \right\}.$$

Let Θ be a cross-section for Γ/Γ_1 in Γ. The Hilbert space $\mathbf{H}_t$ is naturally identified with $l^2\left(\Theta\right)$ since for any $Q \in \Gamma_1$, we have $\left|f\left(PQ\right)\right| = \left|f\left(P\right)\right|$. Via this identification, we may realize the representation $\mathrm{Ind}_{\Gamma_1}^{\Gamma} \chi_{(1,-t)}$ as acting in $l^2\left(\Theta\right)$. More precisely, for $a \in l^2\left(\Theta\right)$ and $\rho_t = \mathrm{Ind}_{\Gamma_1}^{\Gamma} \chi_{(1,-t)}$ we have

$$\left(\rho_t\left(X\right) a\right)\left(T_j\right) = \left\{ \begin{array}{ll} a\left(T_{j-k}\right) & \text{if } X = T_k \\ e^{-2\pi i \langle j, l \rangle} e^{-2\pi i \langle t, l \rangle} a\left(T_j\right) & \text{if } X = M_l \\ e^{2\pi i \theta} a\left(T_j\right) & \text{if } X = e^{2\pi i \theta} \end{array} \right. .$$

Defining the unitary operator $\mathfrak{J} : l^2\left(\Theta\right) \to l^2\left(\mathbb{Z}^d\right)$ such that $\left(\mathfrak{J}a\right)\left(j\right) = a\left(T_j\right)$, it is easily checked that:

$$\mathfrak{J}^{-1}\left[\left(\mathfrak{R} X f\right)\left(t\right)\right] = \left\{ \begin{array}{ll} \rho_t\left(T_k\right)\left[\mathfrak{J}^{-1}\left(\mathfrak{R}f\left(t\right)\right)\right] & \text{if } X = T_k \\ \rho_t\left(M_l\right)\left[\mathfrak{J}^{-1}\left(\mathfrak{R}f\left(t\right)\right)\right] & \text{if } X = M_l \\ \rho_t\left(e^{2\pi i \theta}\right)\left[\mathfrak{J}^{-1}\left(\mathfrak{R}f\left(t\right)\right)\right] & \text{if } X = e^{2\pi i \theta} \end{array} \right. .$$

Thus, the representation π is unitarily equivalent to

$$(14) \qquad\qquad \int_{\frac{\mathbb{R}^d}{\mathbb{Z}^d}}^{\oplus} \rho_t \, dt.$$

Now, we remark that ρ_t is not an irreducible representation of the group Γ. Indeed, by inducing in stages (see Page 166, [6]), we obtain that the representation ρ_t is unitarily equivalent to $\mathrm{Ind}_{\Gamma_0}^{\Gamma}\left(\mathrm{Ind}_{\Gamma_1}^{\Gamma_0}\left(\chi_{(1,-t)}\right)\right)$ and $\mathrm{Ind}_{\Gamma_1}^{\Gamma_0}\chi_{(1,-t)}$ acts in the Hilbert space

$$(15) \qquad \mathbf{K}_t = \left\{ \begin{array}{c} f : \Gamma_0 \to \mathbb{C} : f(PQ) = \chi_{(1,-t)}\left(Q^{-1}\right) f(P), Q \in \Gamma_1 \\ \text{and } \sum_{P\Gamma_1 \in \frac{\Gamma_0}{\Gamma_1}} |f(P)|^2 < \infty \end{array} \right\}.$$

Since $\chi_{(1,-t)}$ is a character, for any $f \in \mathbf{K}_t$, we notice that $|f(PQ)| = |f(P)|$ for $Q \in \Gamma_1$. Thus, the Hilbert space $\mathbf{K}_t$ is naturally identified with $l^2\left(\frac{\Gamma_0}{\Gamma_1}\right) \simeq l^2\left(A\mathbb{Z}^d\right)$ where $A\mathbb{Z}^d$ is a parametrizing set for the quotient group $\frac{\Gamma_0}{\Gamma_1}$. Via this identification, we may realize $\mathrm{Ind}_{\Gamma_1}^{\Gamma_0}\chi_{(1,-t)}$ as acting in $l^2\left(A\mathbb{Z}^d\right)$. More precisely, for $T_j, j \in A\mathbb{Z}^d$, we compute the action of $\mathrm{Ind}_{\Gamma_1}^{\Gamma_0}\chi_{(1,-t)}$ as follows:

$$\left[\mathrm{Ind}_{\Gamma_1}^{\Gamma_0}\chi_{(1,-t)}(X) f\right](T_j) = \left\{ \begin{array}{cc} f(T_{j-k}) & \text{if } X = T_k \\ e^{-2\pi i\langle t,l\rangle} f(T_j) & \text{if } X = M_l \\ e^{2\pi i\theta} f(T_j) & \text{if } X = e^{2\pi i\theta} \end{array} \right..$$

Now, let $\mathbf{F}_{A\mathbb{Z}^d}$ be the Fourier transform defined on $l^2\left(A\mathbb{Z}^d\right)$. Given a vector f in $l^2\left(A\mathbb{Z}^d\right)$, it is not too hard to see that

$$\left[\mathbf{F}_{A\mathbb{Z}^d}\left(\mathrm{Ind}_{\Gamma_1}^{\Gamma_0}\chi_{(1,-t)}(X) f\right)\right](\xi) = \left\{ \begin{array}{cc} \chi_\xi(T_k)\mathbf{F}_{A\mathbb{Z}^d} f(\xi) & \text{if } X = T_k \\ e^{-2\pi i\langle t,l\rangle}\mathbf{F}_{A\mathbb{Z}^d} f(\xi) & \text{if } X = M_l \\ e^{2\pi i\theta}\mathbf{F}_{A\mathbb{Z}^d} f(\xi) & \text{if } X = e^{2\pi i\theta} \end{array} \right.$$

where χ_ξ is a character of the discrete group $A\mathbb{Z}^d$. As a result, given $X \in \Gamma$ we obtain:

$$(16) \qquad \mathbf{F}_{A\mathbb{Z}^d} \circ \rho_t(X) \circ \mathbf{F}_{A\mathbb{Z}^d}^{-1} = \int_{\frac{\mathbb{R}^d}{A^\star \mathbb{Z}^d}}^{\oplus} \chi_{(1,-t,\xi)}(X) \, d\xi,$$

where $\chi_{(1,-t,\xi)}$ is a character of Γ_0 defined as follows:

$$\chi_{(1,-t,\xi)}(X) = \left\{ \begin{array}{cc} \chi_\xi(T_k) & \text{if } X = T_k \\ e^{-2\pi i\langle t,l\rangle} & \text{if } X = M_l \\ e^{2\pi i\theta} & \text{if } X = e^{2\pi i\theta} \end{array} \right..$$

Using the fact that induction commutes with direct integral decomposition (see Page 41,[5]) we have

$$(17) \qquad \rho_t \simeq \mathrm{Ind}_{\Gamma_0}^{\Gamma}\left(\int_{\frac{\mathbb{R}^d}{A^\star \mathbb{Z}^d}}^{\oplus} \chi_{(1,-t,\xi)} \, d\xi\right) \simeq \int_{\frac{\mathbb{R}d}{A^\star \mathbb{Z}^d}}^{\oplus} \left(\mathrm{Ind}_{\Gamma_0}^{\Gamma}\chi_{(1,-t,\xi)}\right) \, d\xi.$$

Putting (14) and (17) together, we arrive at: $\pi \simeq \int_{\frac{\mathbb{R}^d}{\mathbb{Z}^d}\times\frac{\mathbb{R}^d}{A^\star\mathbb{Z}^d}}^{\oplus} \mathrm{Ind}_{\Gamma_0}^{\Gamma}\chi_{(1,\sigma)}d\sigma$. Finally, the fact that $\mathrm{Ind}_{\Gamma_0}^{\Gamma}\chi_{(1,\sigma)}$ is an irreducible representation is due to Remark 5. This completes the proof.

4.2. A Second Proof of Proposition 2. We shall offer here a different proof of Proposition 2 by exhibiting an explicit intertwining operator which is a version of the Zak transform for the representation π and the direct integral representation given in (3). Let $C_c\left(\mathbb{R}^d\right)$ be the space of continuous functions on $\mathbb{R}^d$ which are compactly supported. Let $\mathcal{Z}$ be the operator which maps each $f \in C_c\left(\mathbb{R}^d\right)$ to the function

$$(18) \quad (\mathcal{Z}f)\left(x,w,j+A\mathbb{Z}^d\right) = \sum_{k\in\mathbb{Z}^d} f\left(x+Ak+j\right) e^{2\pi i\langle w, Ak\rangle} \equiv \phi\left(x,w,j+A\mathbb{Z}^d\right)$$

where $x, w \in \mathbb{R}^d$ and j is an element of a cross-section for $\frac{\mathbb{Z}^d}{A\mathbb{Z}^d}$ in the lattice $\mathbb{Z}^d$. Given arbitrary $m' \in \mathbb{Z}^d$, we may write $m' = Ak' + j'$ where $k' \in \mathbb{Z}^d$ and j' is an element of a cross-section for $\frac{\mathbb{Z}^d}{A\mathbb{Z}^d}$. Next, it is worth observing that given $m, m' \in \mathbb{Z}^d$, $\phi\left(x, w + A^\star m, j + A\mathbb{Z}^d\right) = \phi\left(x, w, j + A\mathbb{Z}^d\right)$ and $\phi\left(x + m', w, j + A\mathbb{Z}^d\right)$ is equal to $e^{-2\pi i \langle w, Ak' \rangle} \phi\left(x, w, j + j' + A\mathbb{Z}^d\right)$. This observation will later help us explain the meaning of Equality (18). Let Σ_1 and Σ_2 be measurable cross-sections for $\frac{\mathbb{R}^d}{\mathbb{Z}^d}$ and $\frac{\mathbb{R}^d}{A^\star \mathbb{Z}^d}$ respectively in $\mathbb{R}^d$. For example, we may pick $\Sigma_1 = [0, 1)^d$ and $\Sigma_2 = A^\star [0, 1)^d$. Since f is square-integrable, by a periodization argument it is easy to see that

$$(19) \qquad \|f\|_{L^2(\mathbb{R}^d)}^2 = \int_{\Sigma_1} \sum_{m \in \mathbb{Z}^d} |f(x + m)|^2 \, dx.$$

Therefore, the integral on the right of (19) is finite. It immediately follows that

$$\int_{\Sigma_1} \sum_{j + A\mathbb{Z}^d \in \mathbb{Z}^d/A\mathbb{Z}^d} \sum_{k \in \mathbb{Z}^d} |f(x + Ak + j)|^2 \, dt < \infty.$$

Therefore, the sum $\sum_{j + A\mathbb{Z}^d \in \mathbb{Z}^d/A\mathbb{Z}^d} \sum_{k \in \mathbb{Z}^d} |f(x + Ak + j)|^2$ is finite for almost every $x \in \Sigma_1$ and a fixed j which is a cross-section for $\mathbb{Z}^d/A\mathbb{Z}^d$ in $\mathbb{Z}^d$. Next, observe that

$$(20) \qquad \sum_{k \in \mathbb{Z}^d} f(x + Ak + j) \, e^{2\pi i \langle w, Ak \rangle}$$

is a Fourier series of the sequence $(f(x + Ak + j))_{Ak \in A\mathbb{Z}^d} \in l^2\left(A\mathbb{Z}^d\right)$. So, for almost every x and for a fixed j, the function $\phi\left(x, \cdot, j + A\mathbb{Z}^d\right)$ is regarded as a function of $L^2\left(\mathbb{R}^d\right)$ which is $A^\star \mathbb{Z}^d$-periodic (it is an $L^2(\Sigma_2)$ function). In summary, we may regard the function $(\mathscr{Z} f)\left(x, w, j + A\mathbb{Z}^d\right)$ as being defined over the set $\Sigma_1 \times \Sigma_2 \times \frac{\mathbb{Z}^d}{A\mathbb{Z}^d}$. Let us now show that $\mathscr{Z}$ maps $C_c\left(\mathbb{R}^d\right)$ isometrically into the Hilbert space $L^2\left(\Sigma_1 \times \Sigma_2 \times \frac{\mathbb{Z}^d}{A\mathbb{Z}^d}\right)$. Given any square-summable function f in $L^2\left(\mathbb{R}^d\right)$, we have

$$\int_{\mathbb{R}^d} |f(t)|^2 \, dt = \int_{\Sigma_1} \sum_{k \in \mathbb{Z}^d} |f(t + k)|^2 \, dt = \int_{\Sigma_1} \sum_{j + A\mathbb{Z}^d \in \mathbb{Z}^d/A\mathbb{Z}^d} \sum_{k \in \mathbb{Z}^d} |f(t + Ak + j)|^2 \, dt.$$

Regarding $(f(t + Ak + j))_{Ak \in A\mathbb{Z}^d}$ as a square-summable sequence in $l^2\left(A\mathbb{Z}^d\right)$, the function

$$w \mapsto \sum_{k \in \mathbb{Z}^d} f(t + Ak + j) \, e^{2\pi i \langle w, Ak \rangle}$$

is the Fourier transform of the sequence $(f(t + Ak + j))_{Ak \in A\mathbb{Z}^d}$. Appealing to Plancherel theorem,

$$\sum_{k \in \mathbb{Z}^d} |f(t + Ak + j)|^2 = \int_{\Sigma_2} \left| \sum_{k \in \mathbb{Z}^d} f(t + Ak + j) \, e^{2\pi i \langle w, Ak \rangle} \right|^2 \, dw.$$

It follows that

$$\int_{\mathbb{R}^d} |f(t)|^2 \, dt = \int_{\Sigma_1} \sum_{j+A\mathbb{Z}^d \in \mathbb{Z}^d/A\mathbb{Z}^d} \int_{\Sigma_2} \left| \sum_{k\in\mathbb{Z}^d} f(t+Ak+j)\, e^{2\pi i \langle w, Ak \rangle} \right|^2 dw\, dt$$

$$= \int_{\Sigma_1} \int_{\Sigma_2} \sum_{j+A\mathbb{Z}^d \in \mathbb{Z}^d/A\mathbb{Z}^d} \left| \mathcal{Z} f(x,w,j+A\mathbb{Z}^d) \right|^2 dw\, dt$$

$$= \| \mathcal{Z} f \|^2_{L^2 \left(\Sigma_1 \times \Sigma_2 \times \frac{\mathbb{Z}^d}{A\mathbb{Z}^d} \right)}.$$

Now, by density, we may extend the operator $\mathcal{Z}$ to $L^2(\mathbb{R}^d)$, and we shall next show that the extension

$$\mathcal{Z} : L^2(\mathbb{R}^d) \to L^2 \left(\Sigma_1 \times \Sigma_2 \times \frac{\mathbb{Z}^d}{A\mathbb{Z}^d} \right)$$

is unitary. At this point, we only need to show that $\mathcal{Z}$ is surjective. Let φ be any vector in the Hilbert space $L^2 \left(\Sigma_1 \times \Sigma_2 \times \frac{\mathbb{Z}^d}{A\mathbb{Z}^d} \right)$. Clearly for almost every x and given any fixed j, we have $\varphi(x,\cdot,j) \in L^2(\Sigma_2)$. For such x and j, let $(c_\ell(x,j))_{\ell \in A\mathbb{Z}^d}$ be the Fourier transform of $\varphi(x,\cdot,j)$. Next, define $f_\varphi \in L^2(\mathbb{R}^d)$ such that for almost every $x \in \Sigma_1$,

$$f_\varphi(x+A\ell+j) = c_\ell(x,j).$$

Now, for almost every $w \in \Sigma_2$,

$$Z f_\varphi(x,w) = \sum_{\ell\in\mathbb{Z}^d} f_\varphi(x+A\ell+j)\, e^{2\pi i \langle w, A\ell \rangle}$$

$$= \sum_{\ell\in\mathbb{Z}^d} c_\ell(x,j)\, e^{2\pi i \langle w, A\ell \rangle}$$

$$= \varphi(x,w,j+A\mathbb{Z}^d).$$

It remains to show that our version of Zak transform intertwines the representation π with $\int_{\Sigma_1 \times \Sigma_2}^{\oplus} \rho_{(1,x,w)} \, dx\, dw$ such that $\rho_{(1,x,w)}$ is equivalent to the induced representation $\mathrm{Ind}_{\Gamma_0}^{\Gamma} \chi_{(1,-x,w)}$. Let $\mathcal{R} : l^2(\Gamma/\Gamma_0) \to l^2(\mathbb{Z}^d/A\mathbb{Z}^d)$ be a unitary map defined by

$$\mathcal{R} \left(f\left(j+A\mathbb{Z}^d\right)_{j+A\mathbb{Z}^d} \right) = (f(T_j \Gamma_0))_{T_j \Gamma_0}.$$

Put

$$\rho_{(1,x,w)}(X) = \mathcal{R} \circ \mathrm{Ind}_{\Gamma_0}^{\Gamma} \chi_{(1,-x,w)}(X) \circ \mathcal{R}^{-1} \quad \text{for every } X \in \Gamma.$$

It is straightforward to check that

$$(\mathcal{Z} T_j f)(x,w,\cdot) = \sum_{k\in\mathbb{Z}^d} T_j f(x+\cdot)\, e^{2\pi i \langle w, Ak \rangle}$$

$$= \sum_{k\in\mathbb{Z}^d} T_j f(x+(\cdot - j))\, e^{2\pi i \langle w, Ak \rangle}$$

$$= \left[\rho_{(1,x,w)}(T_j) \right] (\mathcal{Z} f)(x,w,\cdot)$$

and

$$
\begin{aligned}
(\mathscr{Z}M_{Bl}f)(x,w,\cdot) &= \sum_{k\in\mathbb{Z}^d} e^{-2\pi i\langle Bl,x+Ak+j\rangle} f(x+\cdot)\, e^{2\pi i\langle w,Ak\rangle}\\
&= e^{-2\pi i\langle Bl,x+j\rangle} \sum_{k\in\mathbb{Z}^d} e^{-2\pi i\langle Bl,Ak\rangle} f(x+\cdot)\, e^{2\pi i\langle w,Ak\rangle}\\
&= e^{-2\pi i\langle Bl,x+j\rangle} f(x+\cdot)\, e^{2\pi i\langle w,Ak\rangle}\\
&= e^{-2\pi i\langle Bl,x+j\rangle} (\mathscr{Z}f)(x,w,\cdot)\\
&= \rho_{(1,x,w)}(M_{Bl})(\mathscr{Z}f)(x,w,\cdot).
\end{aligned}
$$

In summary, given any $X \in \Gamma$,

$$
\mathscr{Z}\circ\pi(X)\circ\mathscr{Z}^{-1} = \int_{\Sigma_1\times\Sigma_2}^{\oplus} \rho_{(1,x,w)}(X)\ dxdw.
$$

LEMMA 6. *Let $\Gamma_1 = A_1\mathbb{Z}^d$ and $\Gamma_2 = A_2\mathbb{Z}^d$ be two lattices of $\mathbb{R}^d$ such that A_1 and A_2 are non-singular matrices and $|\det A_1| \leq |\det A_2|$. Then there exist measurable sets Σ_1, Σ_2 such that Σ_1 is a fundamental domain for $\frac{\mathbb{R}^d}{A_1\mathbb{Z}^d}$ and Σ_2 is a fundamental domain for $\frac{\mathbb{R}^d}{A_2\mathbb{Z}^d}$ and $\Sigma_1 \subseteq \Sigma_2 \subset \mathbb{R}^d$.*

PROOF. According to Theorem 1.2, [8], there exists a measurable set Σ_1 such that Σ_1 tiles $\mathbb{R}^d$ by the lattice $A_1\mathbb{Z}^d$ and packs $\mathbb{R}^d$ by $A_2\mathbb{Z}^d$. By packing, we mean that given any distinct $\gamma, \kappa \in A_2\mathbb{Z}^d$, the set $(\Sigma_1 + \gamma)\cap(\Sigma_1 + \kappa)$ is an empty set and $\sum_{\lambda\in A_2\mathbb{Z}^d} 1_{\Sigma_1}(x+\lambda) \leq 1$ for $x \in \mathbb{R}^d$ where 1_{Σ_1} denotes the characteristic function of the set Σ_1. We would like to construct a set Σ_2 which tiles $\mathbb{R}^d$ by $A_2\mathbb{Z}^d$ such that $\Sigma_1 \subseteq \Sigma_2$. To construct such a set, let Ω be a fundamental domain for $\frac{\mathbb{R}^d}{A_2\mathbb{Z}^d}$.

It follows that, there exists a subset I of $A_2\mathbb{Z}^d$ such that $\Sigma_1 \subseteq \bigcup_{k\in I}(\Omega + k)$ and each $(\Omega + k)\cap\Sigma_1$ is a set of positive Lebesgue measure. Next, for each $k \in I$, we define $\Omega_k = (\Omega + k)\cap\Sigma_1$. We observe that $\Sigma_1 = \bigcup_{k\in I}((\Omega + k)\cap\Sigma_1) = \bigcup_{k\in I}\Omega_k$ where $\Omega_k = (\Omega + k)\cap\Sigma_1$. Put

$$
\Sigma_2 = \left(\Omega - \bigcup_{k\in I}^{\textstyle\cdot}(\Omega_k - k)\right)\dot\cup\Sigma_1.
$$

The disjoint union in the equality above is due to the fact that for distinct $k, j \in I$, the set $(\Omega_k - k)\cap(\Omega_j - j)$ is a null set. This holds because, Σ_1 packs $\mathbb{R}^d$ by $A_2\mathbb{Z}^d$. Finally, we observe that

$$
\Sigma_2 = \left(\Omega - \bigcup_{k\in I}^{\textstyle\cdot}(\Omega_k - k)\right)\dot\cup\left(\bigcup_{k\in I}^{\textstyle\cdot}\Omega_k\right)
$$

and $\Omega = \left(\Omega - \bigcup_{k\in I}^{\textstyle\cdot}(\Omega_k - k)\right)\dot\cup\left(\bigcup_{k\in I}^{\textstyle\cdot}\Omega_k'\right)$ where each Ω_k' is $A_2\mathbb{Z}^d$-congruent with Ω_k. Therefore Σ_2 is a fundamental domain for $\frac{\mathbb{R}^d}{A_2\mathbb{Z}^d}$ which contains Σ_1. This completes the proof. $\square$

Now, we are ready to prove Proposition 3. Part of the proof of Proposition 3 relies on some technical facts related to central decompositions of unitary representations. A good presentation of this theory is found in Section 3.4.2, [7].

4.3. Proof of Proposition 3. From Proposition 2, we know that the representation π is unitarily equivalent to

$$(21) \qquad \int_{\frac{\mathbb{R}^d}{\mathbb{Z}^d} \times \frac{\mathbb{R}^d}{A^\star \mathbb{Z}^d}}^{\oplus} \mathrm{Ind}_{\Gamma_0}^{\Gamma} \chi_{(1,\sigma)} \, d\sigma.$$

We recall that Γ_0 is isomorphic to the discrete group $\mathbb{Z}_m \times B\mathbb{Z}^d \times A\mathbb{Z}^d$ and that Γ_1 is isomorphic to $\mathbb{Z}_m \times B\mathbb{Z}^d$ where m is the number of elements in the commutator group of Γ which is a discrete subgroup of the torus. From Proposition 1, we have

$$(22) \qquad L \simeq \oplus_{k=0}^{m-1} \int_{\frac{\mathbb{R}^d}{B^\star \mathbb{Z}^d} \times \frac{\mathbb{R}^d}{A^\star \mathbb{Z}^d}}^{\oplus} \mathrm{Ind}_{\Gamma_0}^{\Gamma} \left(\chi_{(k,\sigma)} \right) \, d\sigma.$$

Now, put

$$(23) \qquad L_k = \int_{\frac{\mathbb{R}^d}{B^\star \mathbb{Z}^d} \times \frac{\mathbb{R}^d}{A^\star \mathbb{Z}^d}}^{\oplus} \mathrm{Ind}_{\Gamma_0}^{\Gamma} \chi_{(k,\sigma)} \, d\sigma.$$

From (22), it is clear that $L = L_0 \oplus \cdots \oplus L_{m-1}$. Next, for distinct i and j, the representations L_i and L_j described above are disjoint representations. This is due to the fact that if $i \neq j$ then the Γ-orbits of $\chi_{(i,\sigma)}$ and $\chi_{(j,\sigma)}$ are disjoint sets and therefore the induced representations $\mathrm{Ind}_{\Gamma_0}^{\Gamma} \chi_{(i,\sigma)}$ and $\mathrm{Ind}_{\Gamma_0}^{\Gamma} \chi_{(j,\sigma)}$ are disjoint representations. Thus, for $k \neq 1$ the representation L_k must be disjoint from π. Let us assume for now that $|\det B| > 1$ (or $|\det (B^\star)| < 1$.) According to Lemma 6, there exist measurable cross-sections Σ_1, Σ_2 for $\frac{\mathbb{R}^d}{\mathbb{Z}^d} \times \frac{\mathbb{R}^d}{A^\star \mathbb{Z}^d}$ and $\frac{\mathbb{R}^d}{B^\star \mathbb{Z}^d} \times \frac{\mathbb{R}^d}{A^\star \mathbb{Z}^d}$ respectively such that $\Sigma_1, \Sigma_2 \subset \mathbb{R}^2$, $\Sigma_1 \supset \Sigma_2$ and $\Sigma_1 - \Sigma_2$ is a set of positive Lebesgue measure. Therefore,

$$(24) \qquad \pi \simeq \int_{\Sigma_1}^{\oplus} \left(\mathrm{Ind}_{\Gamma_0}^{\Gamma} \left(\chi_{(1,\sigma)} \right) \right) \, d\sigma \text{ and } L_1 \simeq \int_{\Sigma_2}^{\oplus} \left(\mathrm{Ind}_{\Gamma_0}^{\Gamma} \left(\chi_{(1,\sigma)} \right) \right) \, d\sigma$$

and the representations above are realized as acting in the direct integrals of finite dimensional vector spaces: $\int_{\Sigma_1}^{\oplus} l^2 \left(\Gamma/\Gamma_0 \right) \, d\sigma$ and $\int_{\Sigma_2}^{\oplus} l^2 \left(\Gamma/\Gamma_0 \right) \, d\sigma$ respectively. We remark that the direct integrals described in (24) are irreducible decompositions of π and L_1. Now, referring to the central decomposition of the left regular representation which is described in (10) there exists a measurable subset $\mathbf{E}$ of Σ_2 such that the central decomposition of L_1 is given by (see Theorem 3.26, [7])

$$\int_{\mathbf{E}}^{\oplus} \oplus_{k=1}^{\dim \left(l^2 (\Gamma/\Gamma_0) \right)} \mathrm{Ind}_{\Gamma_0}^{\Gamma} \left(\chi_{(1,\sigma)} \right) \, d\sigma.$$

Furthermore, recalling that $L_1 \simeq \int_{\Sigma_2}^{\oplus} \left(\mathrm{Ind}_{\Gamma_0}^{\Gamma} \left(\chi_{(1,\sigma)} \right) \right) \, d\sigma$ and letting μ be the Lebesgue measure on $\mathbb{R}^d \times \mathbb{R}^d$, it is necessarily the case that

$$\mu \left(\mathbf{E} \right) = \frac{\mu \left(\Sigma_2 \right)}{\dim \left(l^2 \left(\Gamma/\Gamma_0 \right) \right)} = \frac{1}{|\det \left(B \right) \det \left(A \right)| \dim \left(l^2 \left(\Gamma/\Gamma_0 \right) \right)}.$$

From the discussion provided at the beginning of the third section, the set $\mathbf{E}$ is obtained by taking a cross-section for the Γ-orbits in $\frac{\mathbb{R}^d}{B^\star \mathbb{Z}^d} \times \frac{\mathbb{R}^d}{A^\star \mathbb{Z}^d}$. Moreover, since $\Sigma_1 \supset \Sigma_2$ and since $\Sigma_1 - \Sigma_2$ is a set of positive Lebesgue measure then $\pi \simeq$

$\int_{\Sigma_1}^\oplus \mathrm{Ind}_{\Gamma_0}^\Gamma \left(\chi_{(1,\sigma)} \right) \, d\sigma \simeq L_1 \oplus \int_{\Sigma_1 - \Sigma_2}^\oplus \mathrm{Ind}_{\Gamma_0}^\Gamma \left(\chi_{(1,\sigma)} \right) \, d\sigma$ and $\int_{\Sigma_1 - \Sigma_2}^\oplus \mathrm{Ind}_{\Gamma_0}^\Gamma \left(\chi_{(1,\sigma)} \right) \, d\sigma$ is a subrepresentation of π. Thus a central decomposition of π is given by

$$\int_{\mathbf{E}}^\oplus \oplus_{k=1}^{u(\sigma)\,\dim\left(l^2(\Gamma/\Gamma_0)\right)} \mathrm{Ind}_{\Gamma_0}^\Gamma \left(\chi_{(1,\sigma)} \right) \, d\sigma$$

and the function $u : \mathbf{E} \to \mathbb{N}$ is greater than one on a subset of positive measure of $\mathbf{E}$. Therefore, according to Theorem 3.26, [7], it is not possible for π to be equivalent to a subrepresentation of the left regular representation of Γ if $|\det B| > 1$. Now, let us suppose that $|\det B| \leq 1$. Then $|\det B^\star| \geq 1$. Appealing to Lemma 6, there exist measurable sets Σ_1 and Σ_2 which are measurable fundamental domains for $\frac{\mathbb{R}^d}{\mathbb{Z}^d} \times \frac{\mathbb{R}^d}{A^\star \mathbb{Z}^d}$, and $\frac{\mathbb{R}^d}{B^\star \mathbb{Z}^d} \times \frac{\mathbb{R}^d}{A^\star \mathbb{Z}^d}$ respectively, such that $\Sigma_1, \Sigma_2 \subset \mathbb{R}^d$ and $\Sigma_1 \subseteq \Sigma_2$. Next,

$$L_1 \simeq \int_{\Sigma_2}^\oplus \mathrm{Ind}_{\Gamma_0}^\Gamma \left(\chi_{(1,\sigma)} \right) \, d\sigma$$

$$\simeq \left(\int_{\Sigma_1}^\oplus \mathrm{Ind}_{\Gamma_0}^\Gamma \left(\chi_{(1,\sigma)} \right) \, d\sigma \right) \oplus \left(\int_{\Sigma_2 - \Sigma_1}^\oplus \mathrm{Ind}_{\Gamma_0}^\Gamma \left(\chi_{(1,\sigma)} \right) \, d\sigma \right)$$

$$\simeq \pi \oplus \left(\int_{\Sigma_2 - \Sigma_1}^\oplus \mathrm{Ind}_{\Gamma_0}^\Gamma \left(\chi_{(1,\sigma)} \right) \, d\sigma \right).$$

Finally, π is equivalent to a subrepresentation of L_1 and is equivalent to a subrepresentation of the left regular representation L.

4.4. Examples. In this subsection, we shall present a few examples to illustrate the results obtained in Propositions 1, 2 and 3.

(1) Let us start with a trivial example. Let $d = 1$ and $B = \frac{2}{3}$. Then $B^\star = \frac{3}{2}, A = 3$, and $A^\star = \frac{1}{3}$. Next, $L \simeq \oplus_{k=0}^2 \int_{[0,\frac{3}{2}) \times [0,\frac{1}{3})}^\oplus \mathrm{Ind}_{\Gamma_0}^\Gamma \chi_{(k,\sigma)} \, d\sigma$ and $\pi \simeq \int_{[0,1) \times [0,\frac{1}{3})}^\oplus \mathrm{Ind}_{\Gamma_0}^\Gamma \chi_{(1,\sigma)} \, d\sigma$. Now, the central decomposition of L_1 is given by

$$\int_{[0,\frac{1}{2}) \times [0,\frac{1}{3})}^\oplus \oplus_{j=1}^3 \mathrm{Ind}_{\Gamma_0}^\Gamma \chi_{(1,\sigma)} \, d\sigma$$

and the central decomposition of the rational Gabor representation π is

$$\int_{[0,\frac{1}{2}) \times [0,\frac{1}{3})}^\oplus \oplus_{j=1}^2 \mathrm{Ind}_{\Gamma_0}^\Gamma \chi_{(1,\sigma)} \, d\sigma.$$

From these decompositions, it is obvious that the rational Gabor representation π is equivalent to a subrepresentation of the left regular representation L.

(2) If we define $B = \begin{bmatrix} \frac{2}{3} & 0 \\ 0 & \frac{3}{2} \end{bmatrix}$, then $B^\star = \begin{bmatrix} \frac{3}{2} & 0 \\ 0 & \frac{2}{3} \end{bmatrix}$, $A = \begin{bmatrix} 3 & 0 \\ 0 & 2 \end{bmatrix}$ and $A^\star = \begin{bmatrix} \frac{1}{3} & 0 \\ 0 & \frac{1}{2} \end{bmatrix}$. Next, the left regular representation of Γ can be decomposed into a direct integral of representations as follows:

$$L \simeq \oplus_{k=0}^5 \int_{\mathbf{S}}^\oplus \mathrm{Ind}_{\Gamma_0}^\Gamma \chi_{(k,\sigma)} \, d\sigma$$

where $\mathbf{S} = \mathbf{S}_1 \times A^\star [0,1)^2$ and

$$\mathbf{S}_1 = \left([0,1) \times \left[0, \frac{2}{3}\right)\right) \cup \left(\left[1, \frac{3}{2}\right) \times \left[\frac{2}{3}, 1\right)\right) \cup \left(\left[-\frac{1}{2}, 0\right) \times \left[-\frac{1}{3}, 0\right)\right)$$

is a common connected fundamental domain for the lattices $B^\star \mathbb{Z}^2$ and $\mathbb{Z}^2$.

FIGURE 1. Illustration of the set $\mathbf{S}_1$.

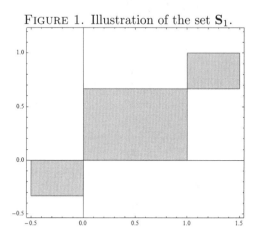

Moreover, we decompose the rational Gabor representation as follows: $\pi \simeq \int_{\mathbf{S}}^{\oplus} \mathrm{Ind}_{\Gamma_0}^{\Gamma} \chi_{(1,\sigma)} \, d\sigma$. One interesting fact to notice here is that: the rational Gabor representation π is actually equivalent to L_1 and

$$L = L_0 \oplus L_1 \oplus L_2 \oplus L_3 \oplus L_4 \oplus L_5.$$

(3) Let $\Gamma = \langle T_k, M_{Bl} : k, l \in \mathbb{Z}^3 \rangle$ where $B = \begin{bmatrix} 1 & 0 & 0 \\ -\frac{1}{5} & \frac{1}{5} & 0 \\ 1 & -1 & 5 \end{bmatrix}$. The inverse

transpose of the matrix B is given by $B^\star = \begin{bmatrix} 1 & 1 & 0 \\ 0 & 5 & 1 \\ 0 & 0 & \frac{1}{5} \end{bmatrix}$. Next, we may

choose the matrix A such that

$$A = \begin{bmatrix} 1 & 1 & 0 \\ 0 & 5 & 5 \\ 0 & 0 & 1 \end{bmatrix} \text{ and } A^\star = \begin{bmatrix} 1 & 0 & 0 \\ -\frac{1}{5} & \frac{1}{5} & 0 \\ 1 & -1 & 1 \end{bmatrix}.$$

Finally, we observe that $\begin{bmatrix} 0 & 0 & 1 \\ 0 & 1 & 5 \\ 1 & \frac{1}{5} & 0 \end{bmatrix} [0,1)^3$ is a common fundamental

domain for both $B^\star \mathbb{Z}^3$ and $\mathbb{Z}^3$. Put

$$\mathbf{S} = \begin{bmatrix} 0 & 0 & 1 \\ 0 & 1 & 5 \\ 1 & \frac{1}{5} & 0 \end{bmatrix} [0,1)^3 \times \begin{bmatrix} 1 & 0 & 0 \\ -\frac{1}{5} & \frac{1}{5} & 0 \\ 1 & -1 & 1 \end{bmatrix} [0,1)^3.$$

Then $L \simeq \oplus_{k=0}^{4} \int_{\mathbf{S}}^{\oplus} \mathrm{Ind}_{\Gamma_0}^{\Gamma} \chi_{(k,t)} dt$ and $\pi \simeq \int_{\mathbf{S}}^{\oplus} \mathrm{Ind}_{\Gamma_0}^{\Gamma} \chi_{(1,t)} dt$.

5. Application to time-frequency analysis

Let π be a unitary representation of a locally compact group X, acting in some Hilbert space $\mathcal{H}$. We say that π is admissible, if and only if there exists some vector $\phi \in \mathcal{H}$ such that the operator W_ϕ^π defined by $W_\phi^\pi : \mathcal{H} \to L^2(X)$, $W_\phi^\pi \psi(x) = \langle \psi, \pi(x)\phi \rangle$ is an isometry of $\mathcal{H}$ into $L^2(X)$.

We continue to assume that B is an invertible rational matrix with at least one entry which is not an integer. Following Proposition 2.14 and Theorem 2.42 of [**7**], the following is immediate.

LEMMA 7. *A representation of Γ is admissible if and only if the representation is equivalent to a subrepresentation of the left regular representation of Γ.*

Given a countable sequence $\{f_i\}_{i\in I}$ of vectors in a Hilbert space $\mathbf{H}$, we say $\{f_i\}_{i\in I}$ forms a frame if and only if there exist strictly positive real numbers A, B such that for any vector $f \in \mathbf{H}$,

$$A\|f\|^2 \leq \sum_{i\in I} |\langle f, f_i \rangle|^2 \leq B\|f\|^2.$$

In the case where $A = B$, the sequence of vectors $\{f_i\}_{i\in I}$ forms a tight frame, and if $A = B = 1$, $\{f_i\}_{i\in I}$ is called a Parseval frame. We remark that an admissible vector for the left regular representation of Γ is a Parseval frame by definition.

The following proposition is well-known for the more general case where B is any invertible matrix (not necessarily a rational matrix.) Although this result is not new, the proof of Proposition 8 is new, and worth presenting in our opinion.

PROPOSITION 8. *Let B be a rational matrix. There exists a vector $g \in L^2(\mathbb{R}^d)$ such that the system $\{M_l T_k g : l \in B\mathbb{Z}^d, k \in \mathbb{Z}^d\}$ is a Parseval frame in $L^2(\mathbb{R}^d)$ if and only if $|\det B| \leq 1$.*

PROOF. The case where B is an element of $GL(d, \mathbb{Z})$ is easily derived from [**11**], Section 4. We shall thus skip this case. So let us assume that B is a rational matrix with at least one entry not in $\mathbb{Z}$. We have shown that the representation π is equivalent to a subrepresentation of the left regular representation of L if and only if $|\det B| \leq 1$. Since Γ is a discrete group, then its left regular representation is admissible if and only if $|\det B| \leq 1$. Thus, the representation π of Γ is admissible if and only if $|\det B| \leq 1$. Suppose that $|\det B| \leq 1$. Then π is admissible and there exists a vector $f \in L^2(\mathbb{R}^d)$ such that the map W_f^π defined by $W_f^\pi h(e^{2\pi i\theta} M_l T_k) = \langle h, e^{2\pi i\theta} M_l T_k f \rangle$ is an isometry. As a result, for any vector $h \in L^2(\mathbb{R}^d)$, we have

$$\left(\sum_{\theta \in [\Gamma,\Gamma]} \sum_{l\in B\mathbb{Z}^d} \sum_{k\in\mathbb{Z}^d} |\langle h, e^{2\pi i\theta} M_l T_k f \rangle|^2 \right)^{1/2} = \|h\|_{L^2(\mathbb{R}^d)}.$$

Next, for $m = \text{card}([\Gamma,\Gamma])$,

$$\sum_{\theta \in [\Gamma,\Gamma]} \sum_{l\in B\mathbb{Z}^d} \sum_{k\in\mathbb{Z}^d} |\langle h, e^{2\pi i\theta} M_l T_k f \rangle|^2 = \sum_{l\in B\mathbb{Z}^d} \sum_{k\in\mathbb{Z}^d} \left|\left\langle h, M_l T_k \left(m^{1/2}f\right)\right\rangle\right|^2.$$

Therefore, if $g = m^{1/2}f$ then

$$\left(\sum_{l\in B\mathbb{Z}^d} \sum_{k\in\mathbb{Z}^d} |\langle h, M_l T_k g \rangle|^2 \right)^{1/2} = \|h\|_{L^2(\mathbb{R}^d)}.$$

For the converse, if we assume that there exists a vector $g \in L^2\left(\mathbb{R}^d\right)$ such that the system

$$\left\{M_l T_k g : l \in B\mathbb{Z}^d, k \in \mathbb{Z}^d\right\}$$

is a Parseval frame in $L^2\left(\mathbb{R}^d\right)$ then it is easy to see that π must be admissible. As a result, it must be the case that $|\det B| \leq 1$. $\qquad\square$

5.1. Proof of Proposition 4.

Let us suppose that $|\det B| \leq 1$. From the proof of Proposition 3, we recall that there exists a unitary map $\mathfrak{A} : \int_{\mathbf{E}}^{\oplus}\left(\oplus_{k=1}^{\ell(\sigma)}l^2\left(\frac{\Gamma}{\Gamma_0}\right)\right)d\sigma \rightarrow L^2\left(\mathbb{R}^d\right)$ which intertwines the representations $\int_{\mathbf{E}}^{\oplus}\left(\oplus_{k=1}^{\ell(\sigma)}\mathrm{Ind}_{\Gamma_0}^{\Gamma}\left(\chi_{(1,\sigma)}\right)\right)d\sigma$ with π such that $\int_{\mathbf{E}}^{\oplus}\left(\oplus_{k=1}^{\ell(\sigma)}\mathrm{Ind}_{\Gamma_0}^{\Gamma}\left(\chi_{(1,\sigma)}\right)\right)d\sigma$ is the central decomposition of π, and $\mathbf{E} \subset \mathbb{R}^d$ is a measurable subset of a fundamental domain for the lattice $B^\star\mathbb{Z}^d \times A^\star\mathbb{Z}^d$ and the multiplicity function ℓ satisfies the condition: $\ell(\sigma) \leq \dim l^2\left(\frac{\Gamma}{\Gamma_0}\right)$. Next, according to the discussion on Page 126, [7] the vector a is an admissible vector for the representation $\tau = \int_{\mathbf{E}}^{\oplus}\left(\oplus_{k=1}^{\ell(\sigma)}\mathrm{Ind}_{\Gamma_0}^{\Gamma}\left(\chi_{(1,\sigma)}\right)\right)d\sigma$ if and only if $a \in \int_{\mathbf{E}}^{\oplus}\left(\oplus_{k=1}^{\ell(\sigma)}l^2\left(\frac{\Gamma}{\Gamma_0}\right)\right)d\sigma$ such that for $d\sigma$-almost every $\sigma \in \mathbf{E}$, $\|a(\sigma)(k)\|_{l^2\left(\frac{\Gamma}{\Gamma_0}\right)}^2 = 1$ for $1 \leq k \leq \ell(\sigma)$ and for distinct $k, j \in \{1, \cdots, \ell(\sigma)\}$ we have $\langle a(\sigma)(k), a(\sigma)(j)\rangle = 0$. Finally, the desired result is obtained by using the fact that $\mathfrak{A}$ intertwines the representations τ with π.

References

[1] Lawrence W. Baggett, *Processing a radar signal and representations of the discrete Heisenberg group*, Colloq. Math. **60/61** (1990), no. 1, 195–203. MR1096368 (92h:22014)

[2] Lawrence W. Baggett and Kathy D. Merrill, *Abstract harmonic analysis and wavelets in* **R**^{n}, The functional and harmonic analysis of wavelets and frames (San Antonio, TX, 1999), Contemp. Math., vol. 247, Amer. Math. Soc., Providence, RI, 1999, pp. 17–27, DOI 10.1090/conm/247/03795. MR1735967 (2001b:42043)

[3] Lawrence W. Baggett, *An abstract interpretation of the wavelet dimension function using group representations*, J. Funct. Anal. **173** (2000), no. 1, 1–20, DOI 10.1006/jfan.1999.3551. MR1760275 (2001j:42028)

[4] Gianfranco Cariolaro, *Unified signal theory*, Springer, London, 2011. With a foreword by Franco P. Preparata. MR3221844

[5] Lawrence J. Corwin and Frederick P. Greenleaf, *Representations of nilpotent Lie groups and their applications. Part I*, Cambridge Studies in Advanced Mathematics, vol. 18, Cambridge University Press, Cambridge, 1990. Basic theory and examples. MR1070979 (92b:22007)

[6] Gerald B. Folland, *A course in abstract harmonic analysis*, Studies in Advanced Mathematics, CRC Press, Boca Raton, FL, 1995. MR1397028 (98c:43001)

[7] Hartmut Führ, *Abstract harmonic analysis of continuous wavelet transforms*, Lecture Notes in Mathematics, vol. 1863, Springer-Verlag, Berlin, 2005. MR2130226 (2006m:43003)

[8] Deguang Han and Yang Wang, *Lattice tiling and the Weyl-Heisenberg frames*, Geom. Funct. Anal. **11** (2001), no. 4, 742–758, DOI 10.1007/PL00001683. MR1866800 (2003j:52021)

[9] Adam Kleppner and Ronald L. Lipsman, *The Plancherel formula for group extensions. I, II*, Ann. Sci. École Norm. Sup. (4) **5** (1972), 459–516; ibid. (4) **6** (1973), 103–132. MR0342641 (49 #7387)

[10] Ronald L. Lipsman, *Group representations*, Lecture Notes in Mathematics, Vol. 388, Springer-Verlag, Berlin-New York, 1974. A survey of some current topics. MR0372116 (51 #8333)

[11] Azita Mayeli and Vignon Oussa, *Regular representations of time-frequency groups*, Math. Nachr. **287** (2014), no. 11-12, 1320–1340, DOI 10.1002/mana.201300019. MR3247020
[12] Elmar Thoma, *Eine Charakterisierung diskreter Gruppen vom Typ I* (German), Invent. Math. **6** (1968), 190–196. MR0248288 (40 #1540)

DEPARTMENT OF MATHEMATICS, BRIDGEWATER STATE UNIVERSITY, BRIDGEWATER, MASSACHUSETTS 02325

Functional Analysis and
C^*-algebras

Contemporary Mathematics
Volume **650**, 2015
http://dx.doi.org/10.1090/conm/650/13008

Representations of Cuntz-Krieger relations, dynamics on Bratteli diagrams, and path-space measures

S. Bezuglyi and Palle E. T. Jorgensen

To the memory of Ola Bratteli

ABSTRACT. We study a new class of representations of the Cuntz-Krieger algebras $\mathcal{O}_A$ constructed by semibranching function systems, naturally related to stationary Bratteli diagrams. The notion of isomorphic semibranching function systems is defined and studied. We show that any isomorphism of such systems implies the equivalence of the corresponding representations of Cuntz-Krieger algebra $\mathcal{O}_A$. In particular, we show that equivalent measures generate equivalent representations of $\mathcal{O}_A$. We use Markov measures which are defined on the path space of stationary Bratteli diagrams to construct isomorphic representations of $\mathcal{O}_A$. To do this, we associate a (strongly) directed graph to a stationary (simple) Bratteli diagram, and show that isomorphic graphs generate isomorphic semibranching function systems. We also consider a class of monic representations of the Cuntz-Krieger algebras, and we classify them up to unitary equivalence. Several examples that illustrate the results are included in the paper.

1. Introduction

Our paper is at the cross-roads of non-commutative and commutative harmonic analysis. The non-commutative component deals with representations of a class of C^*-algebras, called the Cuntz-Krieger algebras, written $\mathcal{O}_A$, and indexed by a matrix A ; it is a matrix over the integers, called the incidence matrix. As the matrix A varies, so does (i) the Bratteli diagram, (ii) the C^*-algebra, and (iii) the associated compact measure space. The representations of $\mathcal{O}_A$ that we study, and find, are motivated by, and have applications to, an important commutative problem; the problem of understanding "dynamics on infinite path-space measures". The latter dynamical systems are called Bratteli diagrams, although this may be a stretch. A Bratteli diagram is a certain discrete graph (see details in the paper), but there is a dual object which is a compact space, from which we get a family of measure-space dynamical systems. It is the latter we study here. We add that the afore mentioned duality, i.e., discrete vs compact, generalizes Pontryagin's duality for Abelian groups: The Pontryagin-dual of a discrete abelian group, is a compact Abelian group, and vice versa. But we study cases here which go beyond the category of groups. Both our non-commutative questions about representations of $\mathcal{O}_A$, and the associated Abelian questions about equivalence of measure spaces, are quite subtle. Our overall plan is to prove theorems on representations of the $\mathcal{O}_A$

C^*-algebras, and then to apply them to classifications of dynamics on this family of compact path-space measure dynamical systems.

1.1. Motivation and earlier papers. Our main theme is the question of deciding equivalence of path-space dynamical systems on Bratteli diagrams X_B. We note that Bratteli diagrams in a variety of guises have found applications to such areas of analysis, both commutative, and non-commutative, as representations of groups, classification of approximately finite dimensional (AF) C^*-algebras, and algebraic K-theory. To give a sample, we mention the following papers [**BrJKR00**], [**BrJKR02a**], [**BrJKR02b**], [**BrJO2004**], [**Ell89**], [**EllMu93**] and the works cited there. In these papers, all three versions of the diagrams are used: (1) graphs (vertices, edges), (2) ordered discrete abelian groups (representing K_0 and K_1 groups), and (3) compact abelian groups (their duals).

In the present paper, we study equivalence of pairs of measures μ arising in this class of dynamical systems (symbolic dynamics). Our answers are presented in terms of the representations of certain Cuntz-Krieger (CK) algebras [**CuKr80**] used in generating the particular system. The CK-algebras are C^*-algebras, but our focus is on their representations in Hilbert spaces of the form $L^2(X_B, \mu)$ where the measures μ are as described. Departing from more traditional dynamical systems, and stochastic processes, we study here systems derived from one-sided endomorphisms, and our measures are typically not Gaussian.

In our setting, we study measures on one-sided infinite paths X_B; i.e., paths represented by infinite words, and with the alphabet in turn represented by vertices in the given Bratteli diagram B. For a given infinite word, the choice of letter at place n is from a set of vertices V_n of B; so with the vertex set depending on the given Bratteli diagram. In the special case of the Cuntz algebra $\mathcal{O}_N$ (N fixed), [**Cu77**, **Cu79**], we may use the same alphabet $\mathbb{Z}_N$ (the cyclic group of order N), at each place in an infinite word. In this case X_B is a Cantor group. The case of $\mathcal{O}_N$ has been studied in recent papers [**DJ14a**, **DJ14b**]. In both cases, Cuntz, and Cuntz-Krieger, by a representation we mean an assignment of isometries in a Hilbert space H, assigning to every letter from the finite alphabet an isometry, and assigned in such a way that distinct isometries have orthogonal ranges, adding up to the identity operator in H, in the case of $\mathcal{O}_N$. By contrast, in the case of the Cuntz-Krieger algebras, the corresponding sum-relation depends on a choice of incidence matrix A, the matrix defining the CK-algebra under consideration, see (2.18) in subsection 2.5 below.

This particular family of representations is motivated in part by quantization of particle systems. Now the "quantized" system must be realized as an L^2-space with respect to a suitable measure on some path-space. Here we study the case of path-space dynamical systems X_B on Bratteli diagrams B. Because of this particular setting, it is not be possible to use of more traditional Gaussian measures (on infinite product spaces); and as an alternative we suggest a family of quasi-stationary Markov measures. In each of our symbolic representations we use one-sided shift to the left; and a system of a finite number of branches (vertices) defining shift to the right, shifting to the right, and filling in a letter from the alphabet. As a result, we get Markov measures on X_B, see e.g., [**Ak12**, **JP12**].

When associated representations of the Cuntz-Krieger algebras are brought to bear, we arrive at useful non-commutative versions of these (commutative) symbolic shift mappings. For earlier related work, see e.g., [**DPS14**, **DJ12**, **Sk97**, **GPS95**].

In more traditional instances of the dynamics problem in its commutative and non-commutative guise, see e.g., [**BrRo81, Hi80**] , there are three typical methods of attacking the question of equivalence or singularity of two measures: (1) Kakutani's theorem for infinite-product measures [**Ka48**], which asserts that two infinite product measures are either equivalent or mutually singular, (2) methods based on entropy considerations, and (3) the method of using the theory of reproducing kernels; see [**Hi80**]. Because of the nature of our setting, one-sided shifts, commutative and non-commutative, we must depart from the setting of Gaussian measures. As a result, in our present study of equivalence or singularity of two measures, vs equivalence (or disjointness) of representations, only ideas from (1) seem to be applicable here.

1.2. Representations of the Cuntz-Krieger algebras. A Cuntz-Krieger algebra $\mathcal{O}_A$ is a C^*-algebra on generators and relations as in (2.18) and depending on a fixed 0-1 matrix A. Recently there has been an increased interest in use of the Cuntz-Krieger algebras and their representations in dynamics (including the study of fractals, and geometric measure theory), in ergodic theory, and in quantization questions from physics. Perhaps this is not surprising since Cuntz, and Cuntz-Krieger algebras are infinite algebras on a finite number of generators, and defined from certain relations. By their nature, these representations reflect intrinsic self-similar inherent in the problem at hand; and thus they serve ideally to encode iterated function systems (IFSs), their dynamics, and their measures. The study of representations of operator algebras related to Bratteli diagrams have found increasing use in pure and applied problems, such as physics, wavelets, fractals, and signals.

Our paper is partitioned in five sections. In Section 2, we collect all necessary definitions and prove some auxiliary results that are used in the paper. More precisely, we first recall the concept of a Bratteli diagram, focusing mostly on the case of a stationary Bratteli diagram. The structure of the path space of a diagram allows considering an analogue of Markov measures. We show in detail how such measures can be constructed. Combinatorial properties of stationary (simple) Bratteli diagrams allow us to define (strongly) directed graphs (we call them coupled graphs) that are naturally associated to such diagrams. It turns out that properties of stationary diagrams can be translated into properties of the corresponding graphs and vice versa. In this section, we also consider one of the main ingredients of our methods, this is the notion of semibranching function systems (s.f.s.) defined on a measure space (X, μ). It turns out that one can define at least two s.f.s. acting on the path space of a stationary Bratteli diagram B, which are indexed by either vertices or edges of B. We find relations between these s.f.s.; they lead to a general concept of refinement of a s.f.s. The importance of s.f.s. follows from the result proved in [**MaPa11**]: any such a system generates a representation of the Cuntz-Krieger algebras on the Hilbert space $L^2(X, \mathcal{B}, \mu)$. This idea is elaborated in the present paper where we study such representations of the Cuntz-Krieger algebras $\mathcal{O}_{\widetilde{A}}$ and $\mathcal{O}_A$. Here $\widetilde{A}$ is the matrix "dual" to A whose non-zero entries are defined by linked edges of B when $s(f) = r(e)$ (see the details below). In order to have a s.f.s., a measure on the path space X_B must be defined.

Amongst the variety of possible measures, we work with quasi-stationary Markov measures and shifts on the path space of a diagram. In Section 3, we define the notion of isomorphic s.f.s. and prove that an isomorphism of two s.f.s. implies the equivalence of the corresponding representations of the Cuntz-Krieger algebra $\mathcal{O}_A$. In particular, equivalent measures will give equivalent representations of $\mathcal{O}_A$. Section 4 is devoted to the study of coupled graphs related to stationary Bratteli diagrams. Isomorphic coupled graphs produce a map (called an admissible map) that, in its turn, generate an isomorphism of s.f.s. This allows us to find new necessary and sufficient conditions for the equivalence of representations of $\mathcal{O}_A$. In the last section we define and study a class of representations of $\mathcal{O}_A$ called *monic* representations. They are characterized by the property that the abelian subalgebra $\mathcal{D}_A$ of $\mathcal{O}_A$ has a cyclic vector. We prove that such representations can arise only when we take a monic system naturally related to the s.f.s defined on the topological Markov chain X_A (see Theorem 5.6).

2. Ingredients of the main results

In this section, we collect the definitions of the main notions and results that are used in the paper. The most important concepts defined below are (stationary) Bratteli diagrams, coupled directed graphs, and semibranching function systems.

2.1. Stationary Bratteli diagrams. Here we give the necessary definitions in the context of general Bratteli diagrams. These definitions are utilized mostly in a particular case of stationary diagrams. We recall that Bratteli diagrams are used in Cantor dynamics to produce convenient models of homeomorphisms of a Cantor set [**GPS95**], [**HPS92**], [**Me06**]. For instance, any substitution dynamical systems is represented as a homeomorphism (Vershik map) acting on the path space of a stationary Bratteli diagram [**F97**], [**DHS99**], [**BKM09**], [**BKMS10**]. Because of the transparent structure of stationary diagrams, many important properties of such dynamical system can be explicitly computed.

DEFINITION 2.1. A *Bratteli diagram* is an infinite graph $B = (V, E)$ such that the vertex set $V = \bigcup_{i \geq 0} V_i$ and the edge set $E = \bigcup_{i \geq 1} E_i$ are partitioned into disjoint subsets V_i and E_i such that
(i) $V_0 = \{v_0\}$ is a single point;
(ii) V_i and E_i are finite sets;
(iii) there exist a range map r and a source map s from E to V such that $r(E_i) = V_i$, $s(E_i) = V_{i-1}$, and $s^{-1}(v) \neq \emptyset$, $r^{-1}(v') \neq \emptyset$ for all $v \in V$ and $v' \in V \setminus V_0$.

The pair (V_i, E_i) or just V_i is called the i-th level of the diagram B. A sequence (finite or infinite) of edges $(e_i : e_i \in E_i)$ such that $r(e_i) = s(e_{i+1})$ is called a *path*. We denote by X_B the set of all infinite paths starting at the vertex v_0. We suppose that X_B is endowed with the clopen topology generated by cylinder sets (finite paths, in other words) such that X_B turns out a Cantor set. This can be done for any simple Bratteli diagram and for a wide class of non-simple diagrams that do not have isolated points.

Given a Bratteli diagram $B = (V, E)$, define a sequence of incidence matrices $F_n = (f_{v,w}^{(n)})$ of B where $f_{v,w}^{(n)} = |\{e \in E_{n+1} : r(e) = v, s(e) = w\}|$ and $v \in V_{n+1}$ and $w \in V_n$. Here and thereafter $|V|$ denotes the cardinality of the set V. The transpose matrix F_n^T will be denoted by A_n.

A Bratteli diagram is called *stationary* if $F_n = F_1 = F$ for every $n \geq 2$. For a stationary diagram B the notation V and E will stand for the sets of vertices of any level and the set of edges between any two consecutive levels below the first one.

With some abuse of terminology, we will also call $A = F^T$ the incidence matrix of the stationary diagram B.

A Bratteli diagram $B' = (V', E')$ is called the *telescoping* of a Bratteli diagram $B = (V, E)$ to a sequence $0 = m_0 < m_1 < \ldots$ if $V'_n = V_{m_n}$ and E'_n is the set of all paths from $V_{m_{n-1}}$ to V_{m_n}, i.e. $E'_n = E_{m_{n-1}} \circ \cdots \circ E_{m_n} = \{(e_{m_{n-1}}, \ldots, e_{m_n}) : e_i \in E_i, r(e_i) = s(e_{i+1})\}$.

Observe that every vertex $v \in V$ is connected to v_0 by a finite path, and the set $E(v_0, v)$ of all such paths is finite. A Bratteli diagram is called *simple* if for any $n > 0$ there exists $m > n$ such that any two vertices $v \in V_n$ and $w \in V_m$ are connected by a finite path. Using the telescoping procedure, we can always assume, without loss of generality, that any pair of vertices from two consecutive levels are connected by at least one edge. In case of a stationary Bratteli diagram B, the simplicity of B is equivalent to the primitivity of the incidence matrix F.

DEFINITION 2.2. Let $B = (V, E)$ be a Bratteli diagram. Two infinite paths $x = (x_i)$ and $y = (y_i)$ from X_B are called *tail equivalent* if there exists i_0 such that $x_i = y_i$ for all $i \geq i_0$. Denote by $\mathcal{R}$ the tail equivalence relation on X_B.

It can be easily seen that a Bratteli diagram is simple if and only if the tail equivalence relation $\mathcal{R}$ is *minimal*; i.e., for arbitrary path $x \in X_B$ the set $\{y \in X_B : y$ is tail equivalent to $x\}$ is dense in X_B.

DEFINITION 2.3. A *cylinder set* in the path space X_B is the set $\{x = (x_i) \in X_B : x_i = e_i, i = 1, \ldots, n\} =: X_w^{(n)}(\overline{e})$, where $\overline{e} = (e_1, \ldots, e_n) \in E(v_0, w)$, $n \geq 1$; we set

$$X_w^{(n)} = \bigcup_{\overline{e} \in E(v_0, w)} X_w^{(n)}(\overline{e}).$$

The cylinder set $X_w^{(n)}(\overline{e})$ will also be denoted as $[\overline{e}] = [(e_0, e_1, \ldots, e_n)]$.

Clearly, $\xi_n = \{X_w^{(n)} : w \in V_n\}$ forms a refining sequence of clopen partition of X_B, $n \in \mathbb{N}$.

We recall the following facts that are widely used for the study of stationary Bratteli diagrams.

Let $\mathcal{A} = \{a_1, \ldots, a_s\}$ be a finite alphabet, $\mathcal{A}^*$ the collection of finite non-empty words over $\mathcal{A}$. Denote by $\Omega = \mathcal{A}^{\mathbb{Z}}$, the set of all two-sided infinite sequences on $\mathcal{A}$. A *substitution* τ is a map $\tau \colon \mathcal{A} \to \mathcal{A}^*$. It extends to maps $\tau \colon \mathcal{A}^* \to \mathcal{A}^*$, and $\tau \colon \Omega \to \Omega$ by concatenation. Denote by T the shift on Ω, that is $T(\ldots x_{-1}.x_0 x_1 \ldots) = (\ldots x_{-1} x_0.x_1 x_2 \ldots)$.

For $x \in \Omega$, let $L_n(x)$ be the set of all words of length n occurring in x; we set $L(x) = \bigcup_{n \in \mathbb{N}} L_n(x)$. The language of τ is the set L_τ of all finite words occurring in $\tau^n(a)$ for some $n \geq 0$ and $a \in \mathcal{A}$. The set $X_\tau ::= \{x \in \Omega : L(x) \subset L_\tau\}$ is T-invariant. The dynamical system (X_τ, T_τ), where T_τ is the restriction of T to the T-invariant set X_τ, is called *the substitution dynamical system* associated to τ.

Depending on properties of τ, the system (X_τ, T_τ) can be minimal or, more generally, aperiodic. As proved in [**DHS99**] (for a minimal homeomorphism T_τ) and [**BKM09**] (for aperiodic T_τ), there is a one-to-one correspondence between substitution dynamical systems and stationary Bratteli diagrams where the dynamics is generated by the so called Vershik map.

The following result simplifies, in general, the study of stationary Bratteli diagrams.

LEMMA 2.4 ([**DHS99**], [**GPS95**]). *Given a stationary Bratteli diagram B, there exists a stationary Bratteli diagram B' such that:*

(1) *$|E(v_0, v)| = 1, \forall v \in V$,*
(2) *the incidence matrix F' is a 0-1 matrix,*
(3) *B and B' are isomorphic Bratteli diagrams.*

A Bratteli diagram B satisfying conditions (1) and (2) of Lemma 2.4 will be called (with some abuse of terminology) a *0-1 Bratteli diagram.*

In fact, an obvious modification of Lemma 2.4 remains true for arbitrary Bratteli diagram.

REMARK 2.5. We observe that the path space of a stationary Bratteli diagram can be endowed with a group structure.

Let F be a $d \times d$ matrix over $\mathbb{Z}$ with transposed A; then A (and F) acts on $\mathbb{R}^d$ by matrix-multiplication: $x \mapsto Ax$, $x \in \mathbb{R}^d$; and this action passes to the quotient $\mathbb{T}^d := \mathbb{R}^d / \mathbb{Z}^d$, $x \pmod{\mathbb{Z}^d} \mapsto Ax \pmod{\mathbb{Z}^d}$. Setting $[x] := x \mod \mathbb{Z}^d$, we write $A[x] := [Ax]$, $x \in \mathbb{T}^d$.

The system of mappings

$$D_F := \mathbb{Z}^d \xrightarrow{F} \mathbb{Z}^d \xrightarrow{F} \cdots \xrightarrow{F} \mathbb{Z}^d \xrightarrow{F} \cdots ,$$

defines the *inductive limit group* D_F. We recall that the discrete abelian group D_F is formed by equivalence classes of elements $(i, x) \in \mathbb{N}_0 \times \mathbb{Z}^d$ with respect to the equivalence relation $((i, x) \sim (j, y)) \Longleftrightarrow (\exists n, m \in \mathbb{N}_0, (F)^n x = (F)^m y)$. If F is invertible, then $(F)^{-i}\mathbb{Z}^d \hookrightarrow (F)^{-i-1}\mathbb{Z}^d$. In this case, the equivalence relation $\sim$ is generated by $(i, x) \sim (i + 1, Fx)$, so that

$$D_F = \left(\bigcup_{k \in \mathbb{Z}} (F)^{-k} \right) \Big/ \mathbb{Z}^d \subset \mathbb{Q}^d.$$

Using Pontryagin duality for locally compact abelian groups, we get the compact dual group $(D_F)^*$ which is realized as a *projective limit group*; it is also called a compact solenoid,

$$Z_A := (D_F)^* = \{(x_k) \in \prod_{k \in \mathbb{N}_0} \mathbb{Z}_d : A[x_{k+1}] = [x_k], \ \forall k \in \mathbb{N}_0\},$$

i.e., $Ax_{k+1} \pmod{\mathbb{Z}^d} = x_k \pmod{\mathbb{Z}^d}$. Here $\mathbb{Z}_d$ denotes the cyclic group of order d, and $\prod \mathbb{Z}_d$ is embedded into $\prod \mathbb{T}^d$ via A.

LEMMA 2.6. *As a compact Cantor space, the path space X_B of a stationary Bratteli diagram B with incidence matrix F is homeomorphic to the compact abelian group $Z_A = (D_F)^*$.*

PROOF. (Sketch) As mentioned above, we can assume that B is a 0-1 Bratteli diagram. For the 0-1 incidence matrix F, there is an alphabet Σ (see [**CuKr80**]) such that

$$X_B = \{(s_k) \in \prod_{k \in \mathbb{N}_0} V : f_{s_k, s_{k+1}} = 1\}.$$

Picking a set of elements V in the finite quotient $\mathbb{Z}^d / A\mathbb{Z}^d$ and using the Pontryagin duality $Z_A = (D_F)^*$, we see that there is a homeomorphism $X_B \longleftrightarrow Z_A$ via $A[x_{s_{k+1}}^{(k+1)}] = x_{s_k}^{(k)}$. $\qquad \square$

It follows from the proved result that the path space X_B has the structure of a compact abelian group Z_A with the probability Haar measure.

2.2. Measures on Bratteli diagrams. Here we give a few definitions and facts related to a class of probability Borel measures on the path space X_B of a Bratteli diagram B, stationary and non-stationary ones. More details can be found in [**BKMS10**] and [**BJ14**]. In order to avoid some unnecessary complications, we make the following assumption: *all measures considered in this paper are assumed to be non-atomic and Borel.*

We will now describe a procedure that would allow us to extend a measures m, which is initially defined on cylinder sets of X_B, to the sigma-algebra $\mathcal{B}(X_B)$ of all Borel sets.

Let X be a compact metric space and let $\mathcal{B} = \mathcal{B}(X)$ be the sigma-algebra of all Borel sets. Suppose that $\mathcal{F}$ and $\mathcal{G}$ are two finitely generated sigma-subalgebras such that $\mathcal{F} \subset \mathcal{G}$. Denote by $\mathcal{M}(\mathcal{F})$ and $\mathcal{M}(\mathcal{G})$ the corresponding algebras of $\mathcal{F}$-measurable and $\mathcal{G}$-measurable functions on X; let $\mathbb{I}$ denote the constant function "one" on X.

A positive operator $\mathcal{E} = \mathcal{E}_{\mathcal{F}\mathcal{G}} : \mathcal{M}(\mathcal{G}) \to \mathcal{M}(\mathcal{F})$ is said to be a *conditional expectation* if
  (i) $\mathcal{E}(\mathbb{I}) = \mathbb{I}$;
  (ii) $\mathcal{E}$ is positive, that is $\mathcal{E}$ maps positive functions in $\mathcal{M}(\mathcal{G})$ onto positive functions in $\mathcal{M}(\mathcal{F})$;
  (iii) $\mathcal{E}(fg) = f\mathcal{E}(g)$ hold for all $g \in \mathcal{M}(\mathcal{G})$ and $f \in \mathcal{M}(\mathcal{F})$.

LEMMA 2.7. *Let $(X, \mathcal{B})$ be as above, and let $(\mathcal{F}_n)_{n \in \mathbb{N}}$ be a sequence of finitely generated sigma-subalgebras such that $\mathcal{F}_n \subset \mathcal{F}_{n+1}$, and let $\mathcal{E}_n : \mathcal{M}(\mathcal{F}_{n+1}) \to \mathcal{M}(\mathcal{F}_n)$ be an associated sequence of conditional expectations such that the following property holds for all $k < l < n$: if $\mathcal{E}_{j,i} = \mathcal{E}_i \circ \cdots \circ \mathcal{E}_{j-1}$, then*

$$\mathcal{E}_{n,l}\mathcal{E}_{l,k} = \mathcal{E}_{n,k}$$

(in the above notation, $\mathcal{E}_n = \mathcal{E}_{n+1,n}$), or, in other words, the following diagram is commutative

$$
\begin{array}{ccc}
\mathcal{M}(\mathcal{F}_n) & \xrightarrow{\mathcal{E}_{n,k}} & \mathcal{M}(\mathcal{F}_k) \\
\searrow{\scriptstyle \mathcal{E}_{n,l}} & & \nearrow{\scriptstyle \mathcal{E}_{l,k}} \\
& \mathcal{M}(\mathcal{F}_l) &
\end{array}
$$

Assume further that

$$\bigcup_n \mathcal{F}_n = \mathcal{B}.$$

Let (μ_n) be a sequence of measures, μ_n defined on $\mathcal{F}_n$ for all $n \in \mathbb{N}$, and assume that the conditional expectations satisfy

(2.1) $\mathcal{E}_n(\mu_{n+1}) = \mu_n, \quad n \in \mathbb{N}.$

Then there is a unique measure $\widetilde{\mu}$ on $\mathcal{B}$ such that

(2.2) $\mathcal{E}_{\mathcal{F}_n, \mathcal{B}}(\widetilde{\mu}) = \mu_n, \quad n \in \mathbb{N}.$

PROOF. We give a sketch of the proof. Let $C(X)$ denote the space of continuous functions. Define $\mathcal{A} = \bigcup_n C(X) \cap \mathcal{M}(\mathcal{F}_n)$. Then $\mathcal{A}$ is closed under the complex conjugacy and separates points in X. Hence, it is uniformly dense in $C(X)$ by the Stone-Weierstrass theorem.

For $f \in \mathcal{A}$, pick $n \in \mathbb{N}$ such that $f \in \mathcal{M}(\mathcal{F}_n)$, and set

$$(2.3) \qquad\qquad L(f) = \int_X f d\mu_n$$

Using (iii) of the definition of the conditional expectation and relation (2.1), we note that L is a well defined linear functional on $\mathcal{A}$. Indeed, for any $f \in \mathcal{M}(\mathcal{F}_n)$,

$$\int_X f d\mu_{n+1} = \int_X \mathcal{E}_n(f) d\mu_{n+1} = \int_X f d(\mathcal{E}_n(\mu_{n+1})) = \int_X f d\mu_n.$$

Now using the Stone-Weierstrass theorem on $\mathcal{A}$, we note that L in (2.3) extends by closure to $C(X)$. Denote this uniquely defined extension by $\widetilde{L}$. The Riesz' theorem applied to $\widetilde{L}$ yields a unique probability measure $\widetilde{\mu}$, defined on $\mathcal{B}$, such that

$$\widetilde{L}(f) = \int_X f d\widetilde{\mu}$$

holds for all $f \in C(X)$. By standard arguments, we can show that $\mathcal{E}(\widetilde{\mu}) = \mu_n$, $\forall n \in \mathbb{N}$. The result then follows. $\square$

REMARK 2.8. In order to illustrate the described above method, one can consider, for instance, the case of the infinite Cartesian product $(X, \nu) = (\prod_i X_i, \prod_i \nu_i)$ of compact measure spaces with $\mathcal{F}_n = \mathcal{B}(X_1) \times \cdots \times \mathcal{B}(X_n)$. Then $\mathcal{M}(\mathcal{F}_n)$ is formed by functions $f(x)$ depending on the first n coordinates, i.e. $f(x) = f_n(x_1, \ldots, x_n)$. The conditional expectation $\mathcal{E}_n : \mathcal{M}(\mathcal{F}_{n+1}) \to \mathcal{M}(\mathcal{F}_n)$ is defined by

$$(\mathcal{E}_n(f_{n+1}))(x_1, \ldots, x_n) = \int_{X_{n+1}} f_{n+1}(x_1, \ldots, x_{n+1}) d\nu_{n+1}(x_{n+1}).$$

A direct computation shows that for the measure $\mu_n = \nu_1 \times \cdots \times \nu_n$ on $\mathcal{F}_n$, we have $\mathcal{E}_n(\mu_{n+1}) = \mu_n$ and Lemma 2.7 is applicable.

Another sort of examples that explains the result of the lemma is based on Bratteli diagrams. Given a Bratteli diagram $B = (V, E)$, define $\mathcal{F}_n$ as the sigma-subalgebra generated by cylinder sets $[\bar{e}] = [(e_0, \ldots, e_n)]$ of length n. Then the measure μ_n on $\mathcal{F}_n$ can be computed by formulas given, for example, in (2.5) or (2.7). It is not hard to verify that again one has that the relation $\mathcal{E}_n(\mu_{n+1}) = \mu_n$ holds for Markov measures (they are defined below).

In what follows, we will consider some specific classes of measures on Bratteli diagrams.

$\mathcal{R}$-invariant measures. Let B be a Bratteli diagram with sequence of incidence matrices (F_n). It is said that a Borel measure μ on X_B is $\mathcal{R}$-invariant if for any $n \in \mathbb{N}$, any vertex $w \in V_n$, and any paths $\bar{e}$ and $\bar{e}'$ from $E(v_0, w)$ one has $\mu(X_w^{(n)}(\bar{e})) = \mu(X_w^{(n)}(\bar{e}'))$. Given an invariant measure μ, we set $\mu^{(n)} = (\mu_v^{(n)} : v \in V_n)$ where $\mu_v^{(n)} = \mu(X_v^{(n)})$. Then μ is completely determined by a sequence of positive probability vectors $(\mu^{(n)})$ satisfying the property

$$(2.4) \qquad\qquad A_n \mu^{(n+1)} = \mu^{(n)}, \quad n \in \mathbb{N}.$$

Denote by $\mathcal{M}(\mathcal{R})$ and $\mathcal{M}_1(\mathcal{R})$ the sets of $\mathcal{R}$-invariant measures and of probability $\mathcal{R}$-invariant measures, respectively. If B is a simple Bratteli diagram, then these sets coincide.

In this paper we mostly deal with simple stationary Bratteli diagrams. Let B be a stationary Bratteli diagram defined by a primitive matrix $A = F^T$. Suppose that λ is the Perron-Frobenius eigenvalue of A, and $\mathbf{x} = (\mathbf{x}_1, \ldots, \mathbf{x}_K)^T$ is the corresponding strictly positive eigenvector normalized by the condition $\sum_{i=1}^{K} \mathbf{x}_i = 1$. It is well known that, for a simple stationary Bratteli diagram, there exists a unique ergodic $\mathcal{R}$-invariant measure μ on X_B, that is $\mathcal{M}_1(\mathcal{R}) = \{\mu\}$. This measure μ is completely determined by its values on cylinder sets

$$(2.5) \qquad \mu(X_i^{(n)}(\overline{e})) = \frac{\mathbf{x}_i}{\lambda^{n-1}},$$

where $i \in V_n$, and $\overline{e}$ is a finite path with $r(\overline{e}) = i$.

Markov measures. More generally, we can consider a class of Borel probability measures on the path space X_B called *Markov measures* because of a clear analogue with the case of Markov chains. In particular, this class contains all $\mathcal{R}$-invariant probability measures [**BJ14**].

Let (P_n) be a sequence of non-negative matrices with entries $(p_{v,e}^{(n)})$ where $v \in V_n, e \in E_{n+1}, n = 0, 1, \ldots$ Thus, the size of P_n is $|V_n| \times |E_{n+1}|$. In particular, P_0 is a row vector. To define a Markov measure m, we require that the sequence (P_n) satisfies the following properties:

$$(2.6) \qquad (a) \ p_{v,e}^{(n)} > 0 \iff s(e) = v; \qquad (b) \ \sum_{e:s(e)=v} p_{v,e}^{(n)} = 1.$$

Then we set for any cylinder set $[\overline{e}] = [(e_0, e_1, \ldots, e_n)]$

$$(2.7) \qquad m([\overline{e}]) = p_{v_0,e_0}^{(0)} p_{s(e_1),e_1}^{(1)} \cdots p_{s(e_n),e_n}^{(n)}.$$

To emphasize that m is generated by a sequence of stochastic matrices, we will also write down $m = m(P_n)$.

Now we return to the definition of a Markov measure and show that, for any such a measure $m(P_n)$ on a Bratteli diagram $B = (V, E)$, we can inductively define a sequence of probability vectors $q^{(k)} = (q^{(k)}(v) : v \in V_k)$, $k \geq 1$, by the following formula

$$q^{(k)}(v) = \sum_{e \in E_k : r(e) = v} q^{(k-1)}(s(e)) p_{s(e),e}^{(k)}$$

where $P_k = (p_{s(e),e}^{(k)})$ and $q^{(0)} = 1$.

Let ν be a probability $\mathcal{R}$-invariant measure. Then one can show that in this case the vectors $q^{(k)}$ coincide with $\nu(X_v^{(k)})$ defined in (2.4).

The following result makes a link between the two classes of measures.

LEMMA 2.9 ([**BJ14**]). *Let $\nu \in \mathcal{M}_1(\mathcal{R})$. Then there exists a sequence of stochastic matrices (P_n) such that $\nu = m(P_n)$. In other words, every probability Borel $\mathcal{R}$-invariant measure is a Markov measure.*

For a stationary Bratteli diagram B, it is natural to distinguish and study a special subset of Markov measures $\nu = \nu(P)$, the so called *stationary Markov measures*. They are obtained when all matrices $P_n, n \in \mathbb{N}$, are the same and equal to a fixed matrix P. Formula (2.7) is transformed then as follows:

$$(2.8) \qquad \nu([\overline{e}]) = p_{v_0,e_0}^{(0)} p_{s(e_1),e_1} \cdots p_{s(e_n),e_n}.$$

2.3. Graphs coupled with Bratteli diagrams. In this subsection we show how can one associate a directed graph $G = (T, P)$ to a stationary Bratteli diagram. It will be clear that the suggested construction can be used in more general settings but we are focused here on the case of stationary diagrams only. Moreover, without loss of generality, we assume that a given stationary Bratteli diagram B is a 0-1 simple diagram; that is it satisfies conditions (1) and (2) of Lemma 2.4.

Let E be the edge set between the first and second levels of B and let A be the transpose of the incidence matrix. By the made assumption, the diagram has only single edges between the vertices of consecutive levels.

REMARK 2.10. We note that there is a one-to-one correspondence between non-zero entries of A and edges of E:

$$a_{i,j} \longleftrightarrow e \ \text{ iff } \ s(e) = i, r(e) = j, \ i, j \in V.$$

This simple observation will be regularly exploited below.

DEFINITION 2.11. (1) We say that a *pair of edges* $(e, f) \in E \times E$ *is linked if* $r(e) = s(f)$. Denote by $\mathcal{L}(E)$ the set of linked pairs.

(ii) Let $e \longleftrightarrow a_{i,j}$ and $f \longleftrightarrow a_{k,l}$ be the correspondences defined in Remark 2.10. Then $(e, f) \in \mathcal{L}(E)$ if and only if $k = j$. In this case, we say that $a_{j,l}$ *follows* $a_{i,j}$.

Next, we want to associate a directed graph to a stationary Bratteli diagram with 0-1 matrix A. This graph, G, will be uniquely defined by the matrix A so that we can write down $G = G(A)$.

DEFINITION 2.12. Let A be a 0-1 matrix. Then the set of vertices, T, of the directed graph $G = G(A)$ is formed non-zero entries $a_{i,j}$ of A. To define the set of directed edges, P, of G, we say that there is an arrow (directed edge) from $a_{i,j}$ to $a_{k,l}$ if and only if $a_{k,l}$ follows $a_{i,j}$ (i.e. $j = k$). By definition, $G = (T, P)$ is called a *coupled graph.*

It follows from this definition that for a fixed non-zero entry $a_{i,j}$ of A (or a vertex $t \in T$) the number of incoming edges for t equals the number of non-zero entries of A in the i-th column, and the number of outgoing edges for t equals the number of non-zero entries of A in the j-th row.

Any graph which is isomorphic to $G(A)$ can be treated as a graph coupled to A.

EXAMPLE 2.13. Let

$$A = \begin{pmatrix} 1 & 1 & 0 \\ 0 & 1 & 1 \\ 1 & 0 & 1 \end{pmatrix}.$$

Then $G(A)$ can be represented as follows.

Obviously, such a representation of $G(A)$ is unique up to an isomorphism.

We recall that a graph G is called strongly connected if for any two vertices t_1 and t_2 of G there exist a path from t_1 to t_2 and a path from t_2 to t_1.

PROPOSITION 2.14. *Suppose that B is a stationary 0-1 Bratteli diagram. Then*

(1) the diagram is simple if and only if the coupled graph G is strongly connected;

(2) there is a one-to-one correspondence between the path space X_B of B and the set of infinite paths X_G of the coupled graph G.

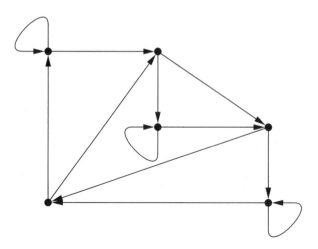

FIGURE 1. Graph defined by the matrix A

PROOF. (1) Suppose that B is simple. In other words, this means that A is primitive. We need to show that for any two vertices $t, t' \in T$ there are paths from t to t' and from t' to t. Let t correspond to $a_{i,j}$ and let t' correspond to $a_{i',j'}$. By simplicity of B, there exists a finite path $\bar{e} = (e_1, \ldots, e_n)$ in the path space X_B such that $s(e_1) = i, r(e_1) = j$, and $s(e_n) = i', r(e_n) = j'$. According to Remark 2.10, the path $\bar{e}$ determines uniquely a sequence of non-zero entries $a_{s(e_1),r(e_1)}, a_{s(e_2),r(e_2)}, \ldots, a_{s(e_n),r(e_n)}$ of A. By Definition 2.12, this sequence corresponds to a path in $G(A)$ that starts at t and ends at t'. Similarly, one can show that there exists a path from t' to t.

(2) This statement follows, in fact, from (1) because the same method can be applied to any infinite path. □

REMARK 2.15. Let Γ be a directed graph with the set of vertices denoted by W. Then one can associate a 0-1 matrix $\overline{A} = (\overline{a}_{v,w})$ of the size $|W| \times |W|$, which is usually called an adjacency matrix. By definition, $\overline{a}_{v,w} = 1$ if and only if there exists a directed edge e in Γ from the vertex v to the vertex w. In the case when $\Gamma = G(A)$ for some 0-1 matrix A (a stationary Bratteli diagram B in other words), we observe that, because of the identification of non-zero entries of A and edges from E, the adjacency matrix has the size $|E| \times |E|$ is determined by the rule $(a_{e,f} = 1) \Longleftrightarrow (r(e) = s(f))$.

Question. It would be interesting to find out what kind of directed strongly connected graphs are isomorphic to graphs obtained from 0-1 stationary simple Bratteli diagrams.

2.4. Semibranching function systems. Here we give the definition of a semibranching function system following [**MaPa11**].

DEFINITION 2.16. (1) Let (X, μ) be a probability measure space with non-atomic measure μ. We consider a finite family $\{\sigma_i : i \in \Lambda\}$ of one-to-one μ-measurable maps σ_i defined on a subset D_i of X and let $R_i = \sigma_i(D_i)$. The family

$\{\sigma_i\}$ is called a *semibranching function system (s.f.s.)* if the following conditions hold:

(i) $\mu(R_i \cap R_j) = 0$ for $i \neq j$ and $\mu(X \setminus \bigcup_{i \in \Lambda} R_i) = 0$;

(ii) $\mu \circ \sigma_i << \mu$ and

$$\rho_\mu(x, \sigma_i) := \frac{d\mu \circ \sigma_i}{d\mu}(x) > 0 \quad \text{for } \mu\text{-a.e. } x \in D_i;$$

(iii) there exists an endomorphism $\sigma : X \to X$ (called a *coding map*) such that $\sigma \circ \sigma_i(x) = x$ for μ-a.e. $x \in D_i$, $i \in \Lambda$.

If, additionally to properties (i) - (iii), we have $\bigcup_{i \in \Lambda} D_i = X$ (μ-a.e.), then the s.f.s. $\{\sigma_i : i \in \Lambda\}$ is called *saturated*.

(2) We also say that a saturated s.f.s. satisfies *condition (C-K)* if for any $i \in \Lambda$ there exists a subset $\Lambda_i \subset \Lambda$ such that up to a set of measure zero

$$D_i = \bigcup_{j \in \Lambda_i} R_j.$$

In this case, condition (C-K) defines a 0-1 matrix $\widetilde{A}$ by the rule:

(2.9) $\qquad\qquad \widetilde{a}_{i,j} = 1 \iff j \in \Lambda_i, \ i \in \Lambda.$

Then the matrix $\widetilde{A}$ is of the size $|\Lambda| \times |\Lambda|$.

The following two examples of s.f.s. (see Examples 2.17 and 2.20), which are generated by a stationary Bratteli diagram, will play the key role in our constructions.

EXAMPLE 2.17. Let B be a stationary simple 0-1 Bratteli diagram. We construct a s.f.s. $\widetilde{\Sigma}$ defined on the path space X_B endowed with a Markov measure m which, as we will see below, must have some additional properties to satisfy Definition 2.16. This s.f.s. is determined by the edge set E which is the set of edges between any two consecutive levels of B. This set plays the role of the index set Λ that was used in Definition 2.16. For any $e \in E$, we denote

(2.10) $\qquad\qquad D_e = \{y = (y_i) \in X_B : s(y_1) = r(e)\},$

(2.11) $\qquad\qquad R_e = \{y = (y_i) \in X_B : y_1 = e\}.$

Then we see that D_e depends on $r(e)$ only, so that $D_e = D_{e'}$ if $r(e) = r(e')$, and $D_e \cap D_{e'} = \emptyset$ if $r(e) \neq r(e')$.

To define a s.f.s. $\{\sigma_e : e \in E\}$ we consider the map

(2.12) $\qquad\qquad \sigma_e(y) := (y_0', e, y_1, y_2, \ldots)$

is a one-to-one continuous map from D_e onto R_e. Here the edge y_0' is uniquely determined by e as the edge connecting v_0 and $s(e)$. Let $\sigma : X_B \to X_B$ be also defined as follows: for any $x = (x_i)_{i \geq 0} \in X_B$

(2.13) $\qquad\qquad \sigma(x) := (z_0', x_2, x_3, \ldots)$

where again z_0' is uniquely determined by the vertex $s(x_2)$. Then it follows from (2.12) and (2.13) that the map σ is onto and satisfies the relation

$$\sigma \circ \sigma_e(x) = x, \quad x \in D_e;$$

hence σ is a coding map.

We immediately deduce from (2.11) that $\{R_e : e \in E\}$ constitutes a partition of X_B into clopen sets. Relation (2.10) implies that the s.f.s. $\{\sigma_e : e \in E\}$ is a saturated s.f.s. Moreover, we claim that it satisfies condition (C-K), that is

$$(2.14) \qquad D_e = \bigcup_{f:s(f)=r(e)} R_f, \quad e \in E.$$

Indeed, $y = (y_i) \in D_e \iff s(y_1) = r(e) \iff \exists f = y_1$ such that $y = (y_0, f, y_2, \dots) \iff y \in \bigcup_{f:s(f)=r(e)} R_f$.

Relation (2.14) shows that the non-zero entries of the 0-1 matrix $\tilde{A}$ from Definition 2.16 are defined by the rule:

$$(\tilde{a}_{e,f} = 1) \iff (s(f) = r(e)) \iff ((e,f) \in \mathcal{L}(E)).$$

In order to clarify the nature of the matrix $\tilde{A}$, we observe that $\tilde{A}$ coincides with the adjacency matrix of the graph $G(A)$ constructed by the initial matrix A (see Remark 2.15).

Next, we observe that $\sigma : X_B \to X_B$ is a finite-to-one continuous map. Indeed, given $x = (x_i) \in X_B$, one can verify that

$$|\sigma^{-1}(x)| = |r^{-1}(r(x_1))| = \sum_{u \in V} f_{v,u}.$$

In other words, we see that $\sigma(x) = \sigma(y)$ if and only if $r(x_1) = r(y_1)$ where $x = (x_i), y = (y_i)$, and if $x \in D_{e_1} = \cdots = D_{e_k}$, then $\sigma^{-1}(x) = \{z_1, \dots, z_k\}$ where $z_i \in R_{e_i}$.

Thus, it remains to find out under what conditions property (ii) of Definition 2.16 holds. In other words, the Radon-Nikodym derivative $\rho_m(x, \sigma_e)$ must be positive on the set D_e for μ-a.e. $x \in D_e$. Since the path space X_B is naturally partitioned into a refining sequence of clopen partitions formed by cylinder sets of fixed length, we can apply de Possel's theorem (see, for instance, [**SG77**]). Let $\mathcal{Q}_n$ be a partition of X_B into cylinder sets $[\bar{e}]$ where each finite path $\bar{e}$ has length n. Suppose that m is a Borel probability measure on X_B and φ is a measurable one-to-one map on X_B. By de Possel's theorem, we have for μ-a.a x,

$$\rho(x, \varphi) = \lim_{n \to \infty} \frac{m(\varphi([\bar{e}(n)]))}{[\bar{e}(n)]}$$

where $\{x\} = \bigcap_n [\bar{e}(n)]$.

We first consider the case when m is the unique $\mathcal{R}$-invariant measure μ. If $\bar{f} = (f_0, f_1, \dots, f_n) \in D_e$, then $\sigma_e(\bar{f}) = (f_0', e, f_1, \dots, f_n)$. By (2.5), we obtain for μ-a.e. $x \in D_e$

$$\mu([\bar{f}]) = \frac{\mathbf{x}_{r(\bar{f})}}{\lambda^{n-1}}, \quad \mu(\sigma_e([\bar{f}])) = \frac{\mathbf{x}_{r(\bar{f})}}{\lambda^n}$$

and therefore

$$(2.15) \qquad \rho_\mu(x, \sigma_e) = \lambda^{-1}.$$

In case of a stationary Markov measure $\nu = \nu(P)$, we see from (2.8) that if $\{x\} = \bigcap_n [f(n)] \in D_e$ then

$$\nu([\bar{f}(n)]) = p_{v_0, f_0}^{(0)} p_{s(f_1), f_1} \cdots p_{s(f_n), f_n},$$

$$\nu(\sigma_e([\bar{f}(n)])) = p_{v_0, f_0'}^{(0)} p_{s(e), e} p_{s(f_1), f_1} \cdots p_{s(f_n), f_n},$$

and finally

$$(2.16) \qquad\qquad \rho_\nu(x, \sigma_e) = \frac{p^{(0)}_{v_0, f_0'} p_{s(e),e}}{p^{(0)}_{v_0, f_0}}$$

(the meaning of f_0' was explained above).

It follows from (2.15) and (2.16) that the Radon-Nikodym derivatives ρ_μ and ρ_ν are positive on D_e if and only if all entries of the vector p_0 are positive. The latter, in particular, means that the support of ν is the whole space X_B.

In case of an arbitrary Markov measure m, we obtain more restrictive conditions under which the Radon-Nikodym derivative $\rho_m(x, \sigma_e)$ is positive on D_e. If $x = (x_i) \in D_e$ is determined by the sequence $[\overline{f}(n)] = [(f_0, f_1, \ldots, f_n)]$ such that $x_i = f_i, i = 0, 1, \ldots, n$, we see that

$$m([f(n)]) = p^{(0)}_{v_0, f_0} p^{(1)}_{s(f_1), f_1} \cdots p^{(n)}_{s(f_n), f_n}$$

and

$$m(\sigma_e([f(n)])) = p^{(0)}_{v_0, f_0'} p^{(1)}_{s(e), e} p^{(2)}_{s(f_1), f_1} \cdots p^{(n+1)}_{s(f_n), f_n}.$$

LEMMA 2.18. *Let m be a Markov measure on the path space of a stationary 0-1 Bratteli diagram, then $\rho_m(x, \sigma_e) > 0$ on D_e if and only if*

$$(2.17) \qquad\qquad 0 < \prod_{i=1}^{\infty} \frac{p^{(i+1)}_{s(f_i), f_i}}{p^{(i)}_{s(f_i), f_i}} < \infty$$

for any $x = \bigcap_n [\overline{f}(n)] \in D_e$

PROOF. This results follows immediately from the relation

$$
\begin{aligned}
\frac{dm \circ \sigma_e}{dm}(x) &= \lim_{n \to \infty} \frac{m(\sigma_e([f(n)]))}{m([f(n)])} \\
&= \lim_{n \to \infty} \frac{p^{(0)}_{v_0, f_0'} p^{(1)}_{s(e), e} p^{(2)}_{s(f_1), f_1} \cdots p^{(n+1)}_{s(f_n), f_n}}{p^{(0)}_{v_0, f_0} p^{(1)}_{s(f_1), f_1} \cdots p^{(n)}_{s(f_n), f_n}} \\
&= \frac{p^{(0)}_{v_0, f_0'} p^{(1)}_{s(e), e}}{p^{(0)}_{v_0, f_0}} \prod_{i=1}^{\infty} \frac{p^{(i+1)}_{s(f_i), f_i}}{p^{(i)}_{s(f_i), f_i}}.
\end{aligned}
$$

$\square$

A Markov measure satisfying (2.17) is called a quasi-stationary measure. We remark that condition (2.17) appeared first in [**DJ14b**] in a different context.

Based on the above discussion, we can summarize the mentioned results in the following theorem.

THEOREM 2.19. *Given a 0-1 stationary simple Bratteli diagram B with edge set E, the collection of maps $\{\sigma_e : D_e \to R_e\}$, $e \in E$, defined as in Example 2.17 on the space (X_B, m), forms a saturated s.f.s. $\widetilde{\Sigma}$ satisfying (C-K) condition where the Markov measure m is either the unique $\mathcal{R}$-invariant measure μ, or a stationary Markov measure ν of full support, or a quasi-stationary measure Markov measure.*

EXAMPLE 2.20. We now give the other example of a s.f.s., denoted by Σ, which is naturally related to a stationary 0-1 Bratteli diagram B with the incidence matrix A. This sort of examples was first considered in [**MaPa11**].

Let $A = (a_{i,j} : i, j = 1, \ldots, n)$ be a 0-1 primitive matrix, and let X_A be the corresponding topological Markov chain where $(x \in X_A) \iff (x = (x_j)_{j \geq 1} : a_{x_j, x_{j+1}} = 1)$. For each $i = 1, \ldots, n$, we define

$$D_i = \{x = (x_k) \in X_A : a_{i,x_1} = 1\}, \quad R_i = \{x = (x_k) \in X_A : x_1 = i\},$$

$$\sigma_i : D_i \to R_i, \quad \sigma_i(x) = ix,$$

and

$$\sigma : X_A \to X_A, \quad \sigma(x_1, x_2, \ldots, x_k, \ldots) = (x_2, x_3, \ldots, x_k, \ldots).$$

Then $D_i = \bigcup_{j:a_{i,j}=1} R_j$ for each i. This means that the matrix associated to the s.f.s. Σ according to (2.9) is exactly A.

In order to finish the definition of the s.f.s. Σ, one needs to specify a measure m on X_A satisfying Definition 2.16. To do this, one can take, for instance, the Hausdorff measure (as was done in [**MaPa11**]) or some Markov measures analogous to those considered in Example 2.17. We will not discuss the details here.

REMARK 2.21. In Examples 2.17 and 2.20 we defined two s.f.s., $\widetilde{\Sigma}$ and Σ, built by a 0-1 stationary Bratteli diagram B with the incidence matrix A. Here we discuss some relations between these two s.f.s. To use more consistent notation, we will add the symbol 'tilde', to any object of $\widetilde{\Sigma}$.

We first observe that these s.f.s. can be thought to be defined on the same space X_B because the sets X_A and X_B are naturally homeomorphic. We have $X_A = (x_1, x_2, \cdots) \mapsto (e_0, e_1, e_2, \cdots) \in X_B$ where the edge e_i is uniquely defined by the properties $s(e_i) = x_i, r(e_i) = x_{i+1}, i \geq 0$, with $x_0 = v_0$. Thus, we identify these spaces of infinite paths and consider $\widetilde{\Sigma}$ and Σ as the s.f.s. defined on the same space X_B.

We recall that the index sets of $\widetilde{\Sigma}$ and Σ are the edge set E and the vertex set V of B, respectively. It follows from the definitions of $\widetilde{\Sigma}$ and Σ, that for every $i \in V$

$$R_i = \bigcup_{e \in s^{-1}(i)} \widetilde{R}_e, \quad D_i = \bigcup_{e:a_{i,s(e)}=1} \widetilde{D}_e.$$

Next, we see that the transformations $\widetilde{\sigma}$, σ, $\widetilde{\sigma}_e$, and σ_i are related as follows:

$$\sigma_i|_{\widetilde{D}_e} = \widetilde{\sigma}_e, \quad s(e) = i \in V,$$

and $\sigma = \widetilde{\sigma}$. Moreover, it follows from the above relations that if μ is a measure on X_B satisfying (ii) of Definition 2.16 for $\widetilde{\Sigma}$, then the measure μ satisfies the same property for the s.f.s. Σ.

The relations between the objects of $\widetilde{\Sigma}$ and Σ show that $\widetilde{\Sigma}$ refines the subsets that constitute Σ, and moreover the restriction of the σ_i on the corresponding subsets $\widetilde{D}_e$ coincides with $\widetilde{\sigma}_e$. In this case we will say that $\widetilde{\Sigma}$ is a *refinement* of Σ.

In order to illustrate this notion, we consider the example of Bratteli diagram with incidence matrix

$$A = \begin{pmatrix} 1 & 1 & 0 \\ 0 & 1 & 1 \\ 1 & 0 & 1 \end{pmatrix}.$$

Let $1, 2, 3$ stand for vertices of B. Then we have the following relations $D_1 = \widetilde{D}_1 \cup \widetilde{D}_2$, $D_2 = \widetilde{D}_2 \cup \widetilde{D}_3$, and $D_3 = \widetilde{D}_1 \cup \widetilde{D}_3$ (we remind that $\widetilde{D}_e = \widetilde{D}_{r(e)}$). Similarly, $R_i = \bigcup_{e:s(e)=i} \widetilde{R}_e$, $i = 1, 2, 3$. Finally, if one computes, for instance, $\sigma_1|_{\widetilde{D}_2}$, then it is the same as σ_e where $r(e) = 2$.

It is clear that the described above situation can happen in more abstract settings when two s.f.s., say $\widetilde{\Gamma} = (\widetilde{\gamma}_{\widetilde{\omega}} : \widetilde{D}_{\widetilde{\omega}} \to \widetilde{R}_{\widetilde{\omega}}, \widetilde{\omega} \in \widetilde{\Omega})$ and $\Gamma = (\gamma_\omega : D_\omega \to R_\omega, \omega \in \Omega)$, are defined on a measure space (X, m). Then we say that $\widetilde{\Gamma}$ *refines* Γ if the partition of X formed by $(\widetilde{R}_{\widetilde{\omega}})$ refines that formed by (R_ω), and for every $\omega \in \Omega$

$$D_\omega = \bigcup_{\widetilde{\omega} \in \Lambda_\omega} \widetilde{D}_{\widetilde{\omega}}, \quad R_\omega = \bigcup_{\widetilde{\omega} \in \Xi_\omega} \widetilde{R}_{\widetilde{\omega}},$$

$$\gamma_\omega|_{\widetilde{D}_{\widetilde{\omega}}} = \widetilde{\gamma}_{\widetilde{\omega}}, \quad \widetilde{\omega} \in \Lambda_\omega.$$

Suppose that the s.f.s. Γ and $\widetilde{\Gamma}$ both satisfy (C-K) condition, and therefore they define two matrices A and $\widetilde{A}$ respectively. We will see below (see Theorem 2.22) that any s.f.s. with (C-K) condition Γ generates a representation of $\mathcal{O}_A$ on $L^2(X, m)$. In the described case, we get representations of $\mathcal{O}_A$ and $\mathcal{O}_{\widetilde{A}}$. It would be interesting to find out how these representations are related each other when one of these s.f.s. refines the other one. We answer this question for s.f.s. $\widetilde{\Sigma}$ and Σ in Section 4.

2.5. The Cuntz-Krieger algebra $\mathcal{O}_A$. Let A be a primitive $n \times n$ matrix with 0-1 entries $a_{i,j}$. The *Cuntz-Krieger algebra* $\mathcal{O}_A$ is generated by partial isometries $S_1, \dots, S_n$ satisfying the relations

$$(2.18) \qquad \sum_{i=1}^n S_i S_i^* = 1, \quad S_i^* S_i = \sum_{j=1}^n a_{i,j} S_j S_j^*.$$

The Cuntz algebra $\mathcal{O}_N$ corresponds to the special case of A when all entries are ones, and $S_i S_i^* = 1$ for all i, i.e., the generators are isometries, as opposed to *partial* isometries.

Let $I = i_1 \cdots i_k$ be a finite word over the alphabet $\{1, \dots, n\}$. Define $S_I = S_{i_1} \cdots S_{i_k}$. Let $\mathcal{D}_A$ be a subalgebra of $\mathcal{O}_A$ generated by $\{S_I S_I^* : I$ is any finite word$\}$. It is well known (see, for instance, [**CuKr80**]) that $\mathcal{D}_A$ is isomorphic to the commutative C^*-algebra $C(X_A)$ of the complex-valued functions defined on the space X_A of infinite path of the topological Markov chain, i.e., $X_A = ((x_j)_{j \geq 0} : a_{x_j, x_{j+1}} = 1)$.

It turns out that any s.f.s. satisfying (C-K) condition is a source for construction of representations of the Cuntz=Krieger algebra. The next theorem shows how such representations of $\mathcal{O}_A$ are arisen.

THEOREM 2.22 ([**MaPa11**]). *Let $\{\sigma_i : i \in \Lambda\}$ be a s.f.s. with coding map σ defined on a probability measure space (X, m). Suppose that it satisfies condition (C-K). Let $\widetilde{A}$ be a 0-1 matrix defined by relation (2.9). Then the operators $T_i = T_i(m)$ and $T_i^* = T_i^*(m)$ acting on $L^2(X, m)$ by formulas*

$$(2.19) \qquad (T_i \xi)(x) = \chi_{R_i}(x) \rho_m(\sigma(x), \sigma_i)^{-1/2} \xi(\sigma(x)), \ i \in \Lambda, \xi \in L^2(X, m)$$

and

$$(2.20) \qquad (T_i^* \xi)(x) = \chi_{D_i}(x) \rho_m(x, \sigma_i)^{1/2} \xi(\sigma_i(x)), \ i \in \Lambda, \ \xi \in L^2(X, m)$$

satisfy (2.18) and generate a representation $\pi = \pi(m)$ of $\mathcal{O}_{\widetilde{A}}$.

It is worth noting that $T_i(m)T_i^*(m)$ is a projection on $L^2(X, m)$ given by the multiplication operator by χ_{R_i}, and similarly $T_i^*(m)T_i(m)$ is a projection realized by multiplication by χ_{D_i}. When the measure m is clearly understood we will write simply T_i and T_i^*, $i \in \Lambda$.

Theorem 2.22 will be applied below in the case when the s.f.s. is taken from Example 2.17.

PROPOSITION 2.23. *Suppose that a s.f.s. $\{\sigma_i : i \in \Lambda\}$ and (X, m) are as in Theorem 2.22. Let m' be another probability measure equivalent to m. Then the representations $\pi(m)$ and $\pi(m')$ defined as in Theorem 2.22 are equivalent.*

The *proof* of the proposition is straightforward and contained in a more general result (see Theorem 3.4) proved in the next section so that we can omit it.

Consider a simple 0-1 stationary Bratteli diagram $B = B(V, E)$. We identify here the set of vertices V with $\{1, \ldots, n\}$. Then we have two s.f.s. Σ and $\widetilde{\Sigma}$ described in Examples 2.20 and 2.17. Let A and $\widetilde{A}$ be the corresponding 0-1 matrices defined as in those examples. Thus, we have two Cuntz–Krieger algebras, $\mathcal{O}_A$ and $\mathcal{O}_{\widetilde{A}}$.

LEMMA 2.24. *The commutative subalgebras $\mathcal{D}_A$ and $\mathcal{D}_{\widetilde{A}}$ of $\mathcal{O}_A$ and $\mathcal{O}_{\widetilde{A}}$ respectively are isomorphic.*

PROOF. We observe that the commutative C^*-algebras $C(X_A)$ and $C(X_B)$ are naturally isomorphic. Indeed, we can identify the characteristic functions of cylinder sets from $C(X_A)$ and $C(X_B)$ by the following rule. Let $i_1, \ldots, i_{n+1}$ be a finite sequence of vertices of B such that $a_{i_k, i_{k+1}} = 1$, $k = 1, \ldots, n$. Then we associate to every pair (i_k, i_{k+1}) the uniquely determined edge e_k in E such that $s(e_k) = i_k, r(e_k) = i_{k+1}$. This defines a finite path in X_B; we can think that this path begins at v_0 since the edge e_0 is completely determined by $i_1 = s(e_1)$. It follows that a one-to-one correspondence

$$\chi_{[i_1, \ldots, i_n]} \longleftrightarrow \chi_{[e_0, e_1, \ldots, e_n]}$$

between the characteristic functions of the cylinder sets is well defined and can be extended by linearity on the algebra generated by characteristic functions. Moreover, this algebra is dense in the space of continuous functions.

Next, we recall that the entries of 0-1 matrix $\widetilde{A}$ is enumerated by edges from E, so that we can consider the topological Markov chain $X_{\widetilde{A}} = \{(e_i) : \widetilde{a}_{e_i, e_{i+1}} = 1\}$. It is obvious that there is one-to-one correspondence between elements of $X_{\widetilde{A}}$ and X_B because $(x_{e_i}) \in X_B$ if and only if $s(e_{i+1}) = r(e_i)$ if and only if $\widetilde{a}_{e_i, e_{i+1}} = 1$. The remaining argument is clear. $\square$

3. Isomorphic semibranching function systems

From now on, we will use the following *convention*: all relations on a measure space between functions, transformations, sets, etc. are understood as mod 0 relations; that is they are true up to a set of measure zero. For instance, a measurable map $F : (X, m) \to (X', m')$ is said to be "onto" if $m'(X' \setminus F(X)) = 0$. Because these properties are obvious as a rule, we usually omit the words "almost everywhere" and mod 0 notation.

DEFINITION 3.1. Let $\{\sigma_i : i \in \Lambda\}$ and $\{\sigma_i' : i \in \Lambda\}$ be two saturated s.f.s. defined on measure spaces (X, m) and (X', m') respectively. Let $\sigma : X \to X$ and

$\sigma' : X' \to X'$ be the corresponding coding maps. Suppose that there is a family of measurable maps $\{\varphi_i : i \in \Lambda\}$ defined on $R_i \subset X$ such that for all $i \in \Lambda$:

(a) $\varphi_i : R_i \to R_i'$ is one-to-one and onto

(b) $(m' \circ \varphi_i)|_{R_i} \sim \mu|_{R_i}$, that is $\dfrac{dm' \circ \varphi_i}{dm}(x) > 0$ if and only if $x \in R_i$;

(c) $\varphi_i \circ \sigma_i(x) = \sigma_i' \circ \Phi_i(x)$, $x \in D_i$.

Here the map $\Phi_i : D_i \to D_i'$ is defined by the following rule:

$$\Phi_i(x) = \varphi_j(x), \quad \forall x \in R_j, \ j \in \Lambda_i,$$

where $D_i = \bigcup_{j \in \Lambda_i} R_j$. Then we say that $\{\sigma_i : i \in \Lambda\}$ and $\{\sigma_i' : i \in \Lambda\}$ are isomorphic s.f.s.

REMARK 3.2. (1) It follows immediately that, in conditions of Definition 3.1, the collection of maps $\{\varphi_i : i \in \Lambda\}$ determines uniquely a one-to-one onto map $F : X \to X'$ such that

$$F(x) = \varphi_i(x), \quad x \in R_i, \ i \in \Lambda.$$

Moreover, $\mu' \circ F \sim \mu$ and for $x \in R_i$

$$\frac{dm' \circ F}{dm}(x) = \frac{dm' \circ \varphi_i}{dm}(x).$$

Since $\{R_i : i \in \Lambda\}$ forms a partition of X, the Radon-Nikodym derivative $\dfrac{dm' \circ F}{dm}(x)$ is positive on X.

(2) We notice that conditions (a) - (c) of Definition 3.1 can be rewritten in terms of F; for example, (c) looks as follows: $F \circ \sigma_i(x) = \sigma_i' \circ F(x)$, $x \in D_i$.

(3) (C-K) condition is invariant with respect to an isomorphism of s.f.s. $\{\sigma_i : i \in \Lambda\}$ and $\{\sigma_i' : i \in \Lambda\}$.

We will use the following lemma below.

LEMMA 3.3. *Given a s.f.s. as in Definition 2.16, the following statements hold: for any $i \in \Lambda$*

$$(1) \quad \sigma(R_i) = D_i, \text{ and } \sigma(y) = \sigma_i^{-1}(y), \quad y \in R_i;$$

$$(2) \quad \Phi_i \circ \sigma(x) = \sigma' \circ \varphi_i(x), \ x \in R_i;$$

$$(3) \quad \frac{dm \circ \sigma}{dm}(z) = \rho_m(\sigma(z), \sigma_i)^{-1}.$$

PROOF. (1) Indeed, if $x \in D_i$, then $y = \sigma_i(x) \in R_i$. Hence the relation $\sigma \circ \sigma_i(x) = x$ implies $\sigma(y) = x$, that is $\sigma_i^{-1}(y) = \sigma(y)$.

(2) It follows from Definition 3.1 (c) that for any $x \in R_i$

$$\varphi_i^{-1} \circ \sigma_i' \circ \Phi_i \circ \sigma(x) = x$$

or $\Phi_i \circ \sigma(x) = (\sigma_i')^{-1} \circ \varphi_i(x)$. By (1), we have $\Phi_i \circ \sigma(x) = \sigma' \circ \varphi_i(x)$.

(3) If $z = \sigma_i(y)$, then by (1) $\sigma(z) = y$ and

$$\frac{dm(\sigma(z))}{dm(z)} = \frac{dm(\sigma \circ \sigma_i(y))}{dm(\sigma_i(y))}$$

$$= \frac{dm(\sigma(z))}{dm(\sigma_i \circ \sigma(z))}$$

$$= \rho_m(\sigma(z), \sigma_i)^{-1}.$$

$\square$

THEOREM 3.4. *Let $\{\sigma_i : i \in \Lambda\}$ and $\{\sigma_i' : i \in \Lambda\}$ be two isomorphic saturated s.f.s. defined on measure spaces (X, m) and (X', m'), respectively. Let also $\sigma : X \to X$ and $\sigma' : X' \to X'$ be the corresponding coding maps. Suppose that $\{T_i = T_i(m) : i \in \Lambda\}$ and $\{T_i' = T_i'(m') : i \in \Lambda\}$ are operators acting respectively on $L^2(X, m)$ and $L^2(X', m')$ according to the formulas:*

$$(3.1) \qquad (T_i\psi)(x) = \chi_{R_i}(x)\rho_m(\sigma(x), \sigma_i)^{-1/2}\psi(\sigma(x)), \ i \in \Lambda, \ \psi \in L^2(X, m)$$

$$(3.2) \qquad (T_i'\xi)(x) = \chi_{R_i'}(x)\rho_{m'}(\sigma'(x), \sigma_i')^{-1/2}\xi(\sigma'(x)), \ i \in \Lambda, \ \xi \in L^2(X', m')$$

If $\widetilde{A}$ is the matrix defined by (2.9), then the representations π and π' of $\mathcal{O}_{\widetilde{A}}$ determined by $\{T_i : i \in \Lambda\}$ and $\{T_i' : i \in \Lambda\}$, respectively, are unitarily equivalent.

PROOF. We will show that there exists an isometry operator $U : L^2(X', m') \to L^2(X, m)$ such that

$$(3.3) \qquad (UT_i'\xi)(x) = (T_iU\xi)(x), \quad \xi \in L^2(X', m'), \quad i \in \Lambda.$$

Define $U = U_F$ by setting

$$(3.4) \qquad (U\xi)(x) = \sqrt{\frac{dm' \circ F}{dm}}(x)\xi(F(x)), \ \xi \in L^2(X', m')$$

where $F : X \to X'$ is a map defined in Remark 3.2. Firstly, we check that U is an isometry. In the respective L^2-norms, we have:

$$
\begin{aligned}
\|U\xi\|^2 &= \int_X \frac{dm' \circ F}{dm}(x)\overline{\xi}(F(x))\xi(Fx)dm(x) \\
&= \int_X |\xi(F(x))|^2 dm'(F(x)) \\
&= \|\xi\|^2.
\end{aligned}
$$

We used here the fact that F is a measurable one-to-one map from X onto X'. Secondly, it is easy to notice that U is onto.

Next, we check that (3.3) holds. In what follows, we will use the relation $F \circ \sigma(x) = \Phi_i \circ \sigma(x) = \sigma' \circ \varphi_i(x)$ when $x \in R_i$, and the relation of Lemma 3.3 (2).

$$
\begin{aligned}
(T_iU\xi)(x) &= \chi_{R_i}(x)\rho_m(\sigma(x), \sigma_i)^{-1/2}(U\xi)(\sigma(x)) \\
&= \chi_{R_i}(x)\rho_m(\sigma(x), \sigma_i)^{-1/2}\left(\frac{dm' \circ F}{dm}(x)\right)^{1/2}\xi(F \circ \sigma(x)) \\
&= \chi_{R_i}(x)\rho_m(\sigma(x), \sigma_i)^{-1/2}\left(\frac{dm'(\sigma' \circ \varphi_i(x))}{dm(\sigma(x))}\right)^{1/2}\xi(\sigma' \circ \varphi_i(x)) \\
&= \chi_{R_i}(x)\left(\frac{dm(\sigma(x))}{dm(x)}\right)^{1/2}\left(\frac{dm'(\sigma' \circ \varphi_i(x))}{dm(\sigma(x))}\right)^{1/2}\xi(\sigma' \circ \varphi_i(x)) \\
&= \chi_{R_i}(x)\left(\frac{dm'(\sigma' \circ \varphi_i(x))}{dm(x)}\right)^{1/2}\xi(\sigma' \circ \varphi_i(x))
\end{aligned}
$$

On the other hand, we have

$$
\begin{aligned}
(UT_i'\xi)(x) &= \left(\frac{dm' \circ F}{dm}(x)\right)^{1/2}(T_i(m')\xi)(F(x)) \\
&= \left(\frac{dm' \circ F}{dm}(x)\right)^{1/2}\chi_{R_i'}(F(x))\rho_{m'}(\sigma' \circ F(x), \sigma_i')^{-1/2}\xi(\sigma' \circ F(x)).
\end{aligned}
$$

We observe that $\chi_{R'_i}(F(x)) = \chi_{R'_i}(\varphi_i(x)) = \chi_{R_i}(x)$ and if $x \in R_i$, then

$$
\begin{aligned}
\rho_{m'}(\sigma' \circ F(x), \sigma'_i)^{-1/2} &= \left(\frac{dm'(\sigma' \circ F(x))}{dm'(\sigma'_i \circ \sigma' \circ F(x))} \right)^{1/2} \\
&= \left(\frac{dm'(\sigma' \circ \varphi_i(x))}{dm'(\varphi_i(x))} \right)^{1/2}
\end{aligned}
$$

Substituting the letter into the expression for $UT_i(m')$, we find

$$
\begin{aligned}
(UT_i(m')\xi)(x) &= \chi_{R_i}(x) \left(\frac{dm' \circ F}{dm}(x) \cdot \frac{dm'(\sigma' \circ \varphi_i(x))}{dm'(\varphi_i(x))} \right)^{1/2} \xi(\sigma' \circ \varphi_i(x)) \\
&= \chi_{R_i}(x) \left(\frac{dm'(\sigma' \circ \varphi_i(x))}{dm(x)} \right)^{1/2} \xi(\sigma' \circ \varphi_i(x)).
\end{aligned}
$$

Comparing the found expressions for UT'_i and $T_i U$, we see that (3.3) holds, and therefore the representations π and π' of $\mathcal{O}_{\tilde{A}}$ are unitarily equivalent. $\qquad\square$

4. Isomorphic s.f.s. on stationary Bratteli diagrams and equivalent representations of $\mathcal{O}_A$

Let B and B' be two stationary simple 0-1 Bratteli diagrams with incidence matrices F and F' whose edge sets are E and E' respectively. Denote as usual $A = F^T$, $A' = (F')^T$. Let now $\{\sigma_e : e \in E\}$ and $\{\sigma'_{e'} : e' \in E'\}$ be two s.f.s. defined on (X_B, m) and $(X_{B'}, m')$ (see Example 2.17) where Borel probability measures are not specified at this stage. Recall also that we denoted by $\mathcal{L}(E)$ ($\mathcal{L}(E')$) the set of linked pairs of edges of E (E').

Our aim is to find out under what conditions these s.f.s. are isomorphic. We observe that the same problem can be considered on a single stationary 0-1 Bratteli diagram B. In this case, the s.f.s. $\{\sigma'_e : e \in E\}$ is obtained from $\{\sigma_e : e \in E\}$ by rearranging edges from E.

DEFINITION 4.1. Let the diagrams B and B' be as above. Suppose that $|E| = |E'|$. Then a one-to-one map $\alpha : E \to E'$ is called *admissible* if for any edges $e, f \in E$

$$
r(e) = s(f) \iff r((\alpha(e)) = s(\alpha(f)),
$$

that is $\alpha \times \alpha(\mathcal{L}(E)) = \mathcal{L}(E')$.

It follows from this definition that α is admissible if and only if $\alpha^{-1} : E' \to E$ is admissible.

EXAMPLE 4.2. We give an example of an admissible map defined on the stationary 0-1 Bratteli diagram B where the matrix A is taken as in Example 2.13. Based on Remark 2.10 we establish the one-to-one correspondence between edges of E and non-zero entries of A as follows: $e_1 \leftrightarrow a_{1,1}, e_2 \leftrightarrow a_{1,2}, e_3 \leftrightarrow a_{2,2}, e_4 \leftrightarrow a_{2,3}, e_5 \leftrightarrow a_{3,1}, e_6 \leftrightarrow a_{3,3}$. If we define

$$
\alpha(e_1) = e_3, \ \alpha(e_3) = e_6, \ \alpha(e_6) = e_1, \ \alpha(e_2) = e_4, \ \alpha(e_4) = e_5, \ \alpha(e_5) = e_2,
$$

then we see that α is an admissible map. This fact is verified directly by considering the set of linked pairs for B.

LEMMA 4.3. *Suppose that $\alpha : E \to E'$ is an admissible map where E and E' are the edge sets of 0-1 stationary simple Bratteli diagrams B an B'. Then α generates a homeomorphism $\overline{\alpha} : X_B \to X_{B'}$ defined as follows: for any $x = (x_i) \in X_B$, set $\overline{\alpha}(x) = y$ where $y = (y_0, \alpha(x_1), \alpha(x_2). \ldots) \in X_{B'}$ and y_0 is uniquely determined by $\alpha(x_1)$ (for consistency, we will denote $y_0 = \alpha(x_0)$).*

PROOF. It easily follows from Definition 4.1 that $\overline{\alpha}$ is one-to-one and onto. The fact that $\overline{\alpha}$ is continuous is deduced from the observation that the preimage of any cylinder set is a cylinder set. □

Assuming that B and B' are 0-1 simple stationary Bratteli diagrams such that an admissible map $\alpha : E \to E'$ exists, we consider possible relations between $\overline{\alpha}$ and measures m and m' on the path spaces X_B and $X_{B'}$. According to subsection 2.2, we will discuss three cases: the $\mathcal{R}$-invariant measure μ, stationary Markov measures ν, and general Markov measures m.

(I) Let μ and μ' be the probability measures on X_B and $X_{B'}$ invariant with respect to the tail equivalence relations $\mathcal{R}$ and $\mathcal{R}'$. Any such a measure is completely determined by its values on cylinder sets (see (2.5). Let $(\mathbf{x}, \lambda)$ and $(\mathbf{x}', \lambda')$ be Perron-Frobenius data for matrices A and A', respectively.

Since $\overline{\alpha} : X_B \to X_{B'}$, the measure μ' defined by $(\mathbf{x}', \lambda')$ is pulled back to X_B and determines a new measure $\mu' \circ \overline{\alpha}$ on X_B.

LEMMA 4.4. *If $\lambda \neq \lambda'$, then the measures $\mu' \circ \overline{\alpha}$ and μ are singular. If $\lambda = \lambda'$, then $\mu' \circ \overline{\alpha} = \mu$.*

PROOF. We recall that $\overline{\alpha}$ maps cylinder sets of X_B onto cylinder sets of $X_{B'}$. Their measures are computed according to (2.5): if $\overline{e} = (e_0, e_1, \ldots, e_n)$, then

$$\mu([\overline{e}]) = \frac{\mathbf{x}_{r(e_n)}}{\lambda^{n-1}}, \quad \mu'(\overline{\alpha}([\overline{e}])) = \frac{\mathbf{x}_{r(\alpha_n(e_n))}}{(\lambda')^{n-1}}.$$

Applying de Possel's theorem, we see that for $\{x\} = \bigcap_n [\overline{e}(n)]$

$$(4.1) \qquad \left(0 < \frac{d\mu' \circ \overline{\alpha}}{d\mu}(x) = \lim_{n \to \infty} \frac{\mu'(\overline{\alpha}([\overline{e}]))}{\mu([\overline{e}])} < \infty \right) \iff \lambda = \lambda'.$$

In fact the above Radon-Nikodym derivative must be equal to one. This observation can be deduced from the following argument. Note that $(x, y) \in \mathcal{R}$ if and only if $(\overline{\alpha}(x), \overline{\alpha}(y)) \in \mathcal{R}'$. If $[\mathcal{R}]$ is the full group of transformations generated by $\mathcal{R}$, then

$$\overline{\alpha}^{-1}[\mathcal{R}']\overline{\alpha} = [\mathcal{R}].$$

This means that $\mu' \circ \overline{\alpha}$ is an $\mathcal{R}$-invariant ergodic measure. Since the set $\mathcal{M}_1(\mathcal{R})$ is a singleton, we conclude that $\mu' \circ \overline{\alpha} = \mu$. □

In particular, if $B = B'$ and α is an admissible map from E onto E, then the homeomorphism $\overline{\alpha} : X_B \to X_B$ preserves the measure μ, and $\overline{\alpha}$ belongs to the normalizer $N[\mathcal{R}]$.

(II) Consider now the case of stationary Markov measures $\nu = \nu(P)$ and $\nu' = \nu'(P')$ defined on X_B and $X_{B'}$ by Markov matrices P and P' according to (2.8).

LEMMA 4.5. *Let* $\overline{\alpha} : X_B \to X_{B'}$ *be defined as above. Then* $\nu' \circ \overline{\alpha}$ *is equivalent to* ν *if and only if for any* $e \in E$

$$(4.2) \qquad\qquad p'_{s(\alpha(e)),\alpha(e)} = p_{s(e),e}.$$

PROOF. Let $\{x\}$ be a point that is uniquely determined by a nested sequence of cylinder sets $([\overline{e}(n)])$. Then $\overline{\alpha}([\overline{e}(n)]) = [\alpha(e_0), \alpha(e_1), \ldots, \alpha(e_n)]$ and we find using relation (2.8) that for any n

$$(4.3) \qquad\qquad \frac{\nu'(\overline{\alpha}([\overline{e}(n)]))}{\nu'[\overline{e}(n)]} = \frac{(p')^{(0)}_{v_0,\alpha(e_0)}}{p^{(0)}_{v_0,e_0}} \prod_{i=1}^{n} \frac{p'_{s(\alpha(e_i)),\alpha(e_i)}}{p_{s(e_i),e_i}}$$

If the condition of the lemma holds, then the Radon-Nikodym derivative $\dfrac{d\nu' \circ \overline{\alpha}}{d\nu}(x)$ is finite and positive.

Conversely, suppose that $\nu' \circ \overline{\alpha} \sim \nu$. Then we see from (4.3) that the product

$$\prod_{i=1}^{\infty} \frac{p'_{s(\alpha(e_i)),\alpha(e_i)}}{p_{s(e_i),e_i}}$$

must converge. Indeed, this follows from the following observation: let T be a finite set containing 1, then $0 < \prod_i t_i < \infty$, $t_i \in T$, if and only if $t_i = 1$ for all sufficiently large i. Since B is simple, then for any $e \in E$ and ν-almost all $x = (x_i) \in X_B$ the $|\{i : x_i = e\}| = \infty$. This proves the result. $\qquad\square$

EXAMPLE 4.6. In order to illustrate the results proved in Lemma 4.5, we consider the same Bratteli diagram B and matrix A as in Examples 2.13 and 4.2. For $\alpha : E \to E$ defined in Example 4.2, define the matrix P (that determines a stationary Markov measure on X_B) as follows

$$P = \begin{pmatrix} p & q & 0 & 0 & 0 & 0 \\ 0 & 0 & p & q & 0 & 0 \\ 0 & 0 & 0 & 0 & q & p \end{pmatrix}$$

where $p, q \in (0,1), p+q = 1$. It is straightforward to verify that for chosen α and P the condition of α-invariance (4.2) holds.

(III) In the case of arbitrary Markov measures m and m' defined on X_B and $X_{B'}$ by sequences of stochastic matrices (P_n) and P'_n, we can proceed as in case (II). By the same method as in Lemma 4.5, we can show that the following statement holds (an easy proof is omitted).

LEMMA 4.7. *Let* B *and* B' *be given as above and let* $m = m(P_n)$ *and* $m'm'(P'_n)$ *be Markov measures defined on* X_B *and* $X_{B'}$. *Then for* $\overline{\alpha} : X_B \to X_{B'}$, *the measure* $m' \circ \overline{\alpha}$ *is equivalent to* m *if and only if for any* $x = (x_i) \in X_B$

$$(4.4) \qquad\qquad 0 < \prod_{i=1}^{\infty} \frac{(p')^{(i)}_{s(\alpha(x_i)),\alpha(x_i)}}{p^{(i)}_{s(x_i),x_i}} < \infty.$$

EXAMPLE 4.8. We use the 0-1 stationary Bratteli diagram and the admissible map α from Example 4.2. To define a Markov measure $m = m(P_n)$, we can set for

$n \in \mathbb{N}$

$$P_n = \begin{pmatrix} p & q & 0 & 0 & 0 & 0 \\ 0 & 0 & p+\varepsilon_n & q-\varepsilon_n & 0 & 0 \\ 0 & 0 & 0 & 0 & q+\delta_n & p-\delta_n \end{pmatrix}$$

where the sequences ε_n and δ_n are chosen so that the product (4.4) converges.

THEOREM 4.9. *Let B and B' be two 0-1 stationary simple Bratteli diagrams and let $\{\sigma_e : e \in E\}$ and $\{\sigma'_{e'} : e' \in E'\}$ be the corresponding s.f.s. defined on (X_B, m) and $(X_{B'}, m')$. Suppose that $\alpha : E \to E'$ is an admissible map and $\overline{\alpha} : X_B \to X_{B'}$ is the one-to-one transformation generated by α. Then $\overline{\alpha}$ implements the isomorphism of $\{\sigma_e : e \in E\}$ and $\{\sigma'_{\alpha(e)} : e \in E\}$ if and only if at least one of the following conditions hold:*

- *$m = \mu$, $m' = \mu'$ where μ and μ' are the measures invariant with respect to $\mathcal{R}$ and $\mathcal{R}'$ respectively satisfying the invariance relation $\mu' \circ \overline{\alpha} = \mu$;*
- *$m = \nu$, $m' = \nu'$ where $\nu = \nu(P)$ and $\nu' = \nu(P')$ are stationary Markov measures satisfying condition (4.2);*
- *$m = m(P_n)$ and $m' = m'(P'_n)$ are Markov measures satisfying condition (4.4).*

PROOF. The proof is based on Lemmas 4.4, 4.5, and 4.7 and the following observations. According to Definition 3.1, we need to check several properties that the s.f.s. $\{\sigma_e : e \in E\}$ and $\{\sigma'_{\alpha(e)} : e \in E\}$ must satisfy.

Firstly, we note that for any $e \in E$, one has $\overline{\alpha} : R_e \to R'_{\alpha(e)}$. Indeed, $x = (x_i) \in R_e \iff x_1 = e$, hence $\alpha(x_1) = \alpha(e)$, and therefore $\overline{\alpha}(x) = (\alpha(x_i)) \in R'_{\alpha(e)}$. Since $\alpha : E \to E'$ is one-to-one and onto, the map $\overline{\alpha} : R_e \to R'_{\alpha(e)}$ is also one-to-one and onto.

Secondly, it follows from the proved fact that $\overline{\alpha}(D_e) = D'_{\alpha(e)}$. Indeed,

$$\overline{\alpha}(D_e) = \bigcup_{f:s(f)=r(e)} \overline{\alpha}(R_f) = \bigcup_{f:s(f)=r(e)} R'_{\alpha(f)} = \bigcup_{f':s(f')=r(\alpha(e))} R'_{f'} = D_{\alpha(e)}.$$

Thirdly, we claim that for any $e \in E$ and $x \in D_e$

$$\overline{\alpha} \circ \sigma_e(x) = \sigma'_{\alpha(e)} \circ \overline{\alpha}(x).$$

Compute

$$\overline{\alpha} \circ \sigma_e(x) = \overline{\alpha}(x'_0, e, x_1, \dots) = (\alpha(x_0), \alpha(e), \alpha(x_1), \dots)$$

and

$$\sigma'_{\alpha(e)} \circ \overline{\alpha}(x) = \sigma'_{\alpha(e)}(\alpha(x_0), \alpha(x_1), \dots) = (y'_0, \alpha(e), \alpha(x_1), \dots).$$

It remains to notice that $\alpha(x_0) = y'_0$ since these edges are determined by $s(\alpha(e))$.

In general, the measure $m' \circ \overline{\alpha}$ is not equivalent to m. The proved lemmas give necessary and sufficient conditions for such an equivalence. They are the same as those given in the theorem. The proof is complete. $\square$

Theorems 3.4 and 4.9 allow us to deduce the following result.

THEOREM 4.10. *Let B, B', $\{\sigma_e : e \in E\}$ and $\{\sigma'_{e'} : e' \in E'\}$ be as in Theorem 4.9. Suppose that for an admissible map $\alpha : E \to E'$ the transformation $\overline{\alpha} : (X_B, m) \to (X_{B'}, m')$ implements an isomorphism of s.f.s. $\{\sigma_e : e \in E\}$ and $\{\sigma'_{\alpha(e)} : e \in E\}$ (this, in particular, means that the measures m and m' satisfy the conditions of Theorem 4.9; and the both s.f.s. have the same matrix $\widetilde{A}$ defined by*

(2.9)). *Let π and π'_α be representations of $\mathcal{O}_{\widetilde{A}}$ generated by operators $\{T_e = T_e(m) : e \in E\}$ and $\{T'_{\alpha(e)} = T'_{\alpha(e)}(m') : e \in E\}$ according to the following formulas:*

$$(4.5) \qquad (T_e\psi)(x) = \chi_{R_e}(x)\rho_m(\sigma(x),\sigma_e)^{-1/2}\psi(\sigma(x)), \ \psi \in L^2(X,m),$$

$$(4.6) \qquad (T'_{\alpha(e)}\xi)(x) = \chi_{R'_{\alpha(e)}}(x)\rho_{m'}(\sigma'(x),\sigma'_{\alpha(e)})^{-1/2}\xi(\sigma'(x)), \ \xi \in L^2(X',m')$$

where σ and σ' are coding maps acting on X_B and $X_{B'}$. Then the representation π and π_α of $\mathcal{O}_{\widetilde{A}}$ are unitarily equivalent.

We remark that Theorem 4.12 (given below) allow us to conclude that not only the representations of $\mathcal{O}_{\widetilde{A}}$ are equivalent but the representations of $\mathcal{O}_A$ constructed by formulas (4.9) are also equivalent.

PROOF. The proof of this theorem is absolutely similar to that of Theorem 3.4 so that we can omit the details. We define an isometric operator $V : L^2(X_{B'},m') \to L^2(X_B,m)$ by

$$(V\xi)(x) = \left(\frac{dm' \circ \overline{\alpha}}{dm}(x)\right)^{1/2} \xi(\overline{\alpha}(x)).$$

Then we can verify by direct computations that for every $e \in E$ and $\xi \in L^2(X_{B'},m')$

$$(T_e V\xi)(x) = (V T_{\alpha(e)}\xi)(x).$$

This means that the representations π and π'_α are equivalent. $\qquad\square$

As was mentioned above, the proved theorem looks simpler when $B = B'$ and $\alpha : E \to E$ is an admissible map of the set E.

We give below an example of two 0-1 simple stationary Bratteli diagrams whose s.f.s. are isomorphic.

EXAMPLE 4.11. Let B and B' be Bratteli diagrams defined by matrices

$$A = \begin{pmatrix} 1 & 0 & 1 & 1 \\ 0 & 1 & 0 & 1 \\ 0 & 1 & 0 & 0 \\ 1 & 0 & 1 & 0 \end{pmatrix}, \quad A' = \begin{pmatrix} 1 & 0 & 1 & 1 \\ 0 & 1 & 1 & 0 \\ 1 & 0 & 0 & 1 \\ 0 & 1 & 0 & 0 \end{pmatrix}$$

respectively. The sets of edges $E = \{e_1,\ldots,e_8\}$ and $E' = \{e'_1,\ldots,e'_8\}$ of B and B' are shown on the following figures:

and

Define an admissible map $\alpha : E \to E'$ as follows.

$$\alpha(e_1) = e'_1, \ \alpha(e_2) = e'_3, \ \alpha(e_3) = e'_2, \ \alpha(e_4) = e'_4,$$
$$\alpha(e_5) = e'_5, \ \alpha(e_6) = e'_8, \ \alpha(e_7) = e'_7, \ \alpha(e_8) = e'_6.$$

It is straightforward to check that $\alpha \times \alpha(\mathcal{L}(E)) = \mathcal{L}(E')$. Therefore, one can apply Theorem 4.9 to obtain that the corresponding s.f.s. $\{\sigma_e : e \in E\}$ and $\{\sigma'_{\overline{\alpha}(e)} : e \in E\}$ are isomorphic.

Suppose that one has a 0-1 simple stationary Bratteli diagram B whose transpose to the incidence matrix is A. Let the other conditions of Theorem 4.10 be satisfied. Then we can construct the unitarily equivalent representations $\widetilde{\pi}$ and $\widetilde{\pi}'$ (we changed the notation) of $\mathcal{O}_{\widetilde{A}}$ generated respectively by the s.f.s. $\{\sigma_e : e \in E\}$ and $\{\sigma_{\alpha(e)} : e \in E\}$. In the next statements we show that these representations define simultaneously some representations π and π' of $\mathcal{O}_A$ which are also unitarily equivalent (undefined symbols are taken from Theorem 4.10).

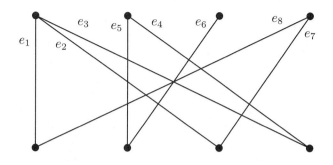

FIGURE 2. Edge set E of B.

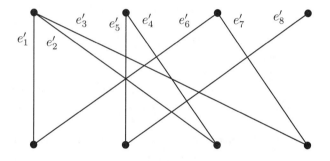

FIGURE 3. Edge set E' of B'.

THEOREM 4.12. *Suppose that $B = B(V, E)$ is a 0-1 simple stationary Bratteli diagram, E is the edge set of B, V is the vertex set, and A is the corresponding 0-1 matrix. For the s.f.s. $\{\sigma_e : e \in E\}$, consider the representation $\widetilde{\pi}$ of $\widetilde{A}$ generated on $L^2(X, m)$ by the operators*

$$(4.7) \qquad (\widetilde{T}_e \psi)(x) = \chi_{R_e}(x)\rho_m(\sigma(x), \sigma_e)^{-1/2}\psi(\sigma(x)), \ e \in E, \ \psi \in L^2(X, m),$$

$$(4.8) \qquad (\widetilde{T}_e^* \xi)(x) = \chi_{D_e}(x)\rho_m(x, \sigma_e)^{1/2}\xi(\sigma_e(x)), \ e \in E, \ \xi \in L^2(X, m).$$

For each $i \in V$, define

$$(4.9) \qquad\qquad T_i = \sum_{e:s(e)=i} \widetilde{T}_e.$$

Then the collection of operators $\{T_i : i \in V\}$ generates a representation of $\mathcal{O}_A$ on $L^2(X, m)$.

PROOF. To prove the result it suffices to check that

$$(4.10) \qquad \sum_{i \in V} T_i T_i^* = 1, \qquad T_i^* T_i = \sum_{j \in V} a_{i,j} T_j T_j^*$$

assuming that the operators $\{\widetilde{T}_e : e \in E\}$ satisfy (4.10) when the matrix A is replaced by $\widetilde{A}$.

To show that the first relation in (4.10) holds, we compute

$$\sum_{i \in V} T_i T_i^* = \sum_{i \in V} (\sum_{e:s(e)=i} \widetilde{T}_e)(\sum_{e':s(e')=i} \widetilde{T}_{e'}^*)$$

$$= \sum_{i \in V} (\sum_{e:s(e)=i} \widetilde{T}_e \widetilde{T}_e^*) + \sum_{i \in V} (\sum_{e \neq e':s(e)=s(e')=i} \widetilde{T}_e \widetilde{T}_{e'}^*).$$

Since

$$\sum_{i \in V} (\sum_{e:s(e)=i} \widetilde{T}_e \widetilde{T}_e^*) = \sum_{e \in E} \widetilde{T}_e \widetilde{T}_e^* = 1,$$

we need to verify only that $\widetilde{T}_e \widetilde{T}_{e'}^* = 0$ for any edges $e \neq e'$ with $s(e) = s(e') = i$. Indeed, when we apply the formulas (4.7) and (4.8) we obtain

$$(\widetilde{T}_e \widetilde{T}_{e'}^* \psi)(x) = \chi_{R_e}(x) \chi_{D_{e'}}(\sigma(x)) \rho_m(\sigma(x), \sigma_e)^{-1/2} \rho_m(\sigma(x), \sigma_{e'})^{1/2} \psi(x).$$

We claim that $\chi_{R_e}(x) \chi_{D_{e'}}(\sigma(x)) = 0$ for all x. We note that $e \neq e'$ and $s(e) = s(e')$ implies $r(e) \neq r(e')$. If $x \in R_e$, then $x = (x_0, e, x_1, x_2, \dots)$ and $\sigma(x) = (x_0', x_1, x_2, \dots)$ where $s(x_1) = r(e)$. On the other hand, $D_{e'} = \bigcup_{f:s(f)=r(e')} R_f$, and any point $y = (y_i) \in D_{e'}$ has the property $s(y_1) = r(e') \neq r(e)$. This proves the claim.

We will now prove that the second formula in (4.10) is also true. We have

$$T_i^* T_i = \sum_{e:s(e)=i} \widetilde{T}_e^* \widetilde{T}_e + \sum_{e \neq e':s(e)=s(e')=i} \widetilde{T}_e^* \widetilde{T}_{e'} = \sum_{e:s(e)=i} \widetilde{T}_e^* \widetilde{T}_e$$

because the fact that $\widetilde{T}_e^* \widetilde{T}_e = 0$ can be proved as above. Next, since $\widetilde{T}_e^* \widetilde{T}_e = \sum_f \widetilde{a}_{e,f} \widetilde{T}_f \widetilde{T}_f^*$, we get

$$T_i^* T_i = \sum_{e:s(e)=i} \sum_{f:s(f)=r(e)} \widetilde{T}_f \widetilde{T}_f^*$$

On the other hand,

$$\sum_{j \in V} a_{i,j} T_j T_j^* = \sum_{j \in V} a_{i,j} (\sum_{f:s(f)=j} \widetilde{T}_f)(\sum_{f':s(f')=j} \widetilde{T}_{f'}^*)$$

$$= \sum_{j \in V} a_{i,j} \sum_{f:s(f)=j} \widetilde{T}_f \widetilde{T}_f^*$$

We use now the correspondence $a_{i,j} \leftrightarrow (e, s(e) = i, r(e) = j)$, so that the latter has the form

$$\sum_{j \in V} a_{i,j} T_j T_j^* = \sum_{e:s(e)=i} \sum_{f:s(f)=j} \widetilde{T}_f \widetilde{T}_f^*.$$

This completes the proof of the proposition.   □

The following corollary easily follows from the results proved above and from relation (4.9).

COROLLARY 4.13. *Let $B = B(V, E)$, A, $\widetilde{A}$, and $\{\sigma_e : e \in A\}$ be as in Theorem 4.12. For an admissible map $\alpha : E \to E$, let $\widetilde{\pi}$ and $\widetilde{\pi}_\alpha$ be the representations of $\mathcal{O}_{\widetilde{A}}$ defined by (4.5) and (4.6). Then the unitary equivalence of the representations $\widetilde{\pi}$ and $\widetilde{\pi}_\alpha$ of $\mathcal{O}_{\widetilde{A}}$ implies the unitary equivalence of the corresponding representations of $\mathcal{O}_A$.*

5. Monic representations of $\mathcal{O}_A$

Let $\mathcal{O}_A$ be the Cuntz-Krieger algebra defined by its generators $S_1, \ldots, S_n$ according to relations (2.18). Here $A = (a_{i,j})$ is a 0-1 primitive matrix. We recall (see Subsection 2.5) that $\mathcal{O}_A$ contains the commutative subalgebra $\mathcal{D}_A$ generated by $\{S_I S_I^* : I = i_1 \cdots i_k, k \in \mathbb{N}\}$ which is isomorphic to $C(X_A)$ (or, equivalently, $C(X_B)$ where B is the stationary Bratteli diagram with matrix A).

We define and study a class of representations of $\mathcal{O}_A$ called monic representations. This class was originally considered in [**DJ14a**] for representations of the Cuntz algebra $\mathcal{O}_N$.

DEFINITION 5.1. We say that a representation π of $\mathcal{O}_A$ on a Hilbert space $\mathcal{H}$ is *monic* if there is a cyclic vector $\xi \in \mathcal{H}$ for the abelian subalgebra $\mathcal{D}_A$, i.e.,

$$\overline{\mathrm{span}}\{T_I T_I^* : I \text{ finite word}\} = \mathcal{H}$$

where $T_I = \pi(S_I)$.

EXAMPLE 5.2. (1) Let B be a stationary 0-1 Bratteli diagram with matrix A, and let $\{\sigma_e : e \in E\}$ be the s.f.s. defined by B as in Example 2.17. We *claim* that the representation of $\mathcal{O}_{\widetilde{A}}$ defined in Theorem 4.12 is monic. Indeed, it is not hard to verify that $\widetilde{T}_e \widetilde{T}_e^*$ is the projection on $L^2(X_B, m)$ given by multiplication by the characteristic function χ_{R_e}, $e \in E$. Then, for any finite path $w = (e_1, \ldots, e_k) \in E^*$, the projection $\widetilde{T}_w \widetilde{T}_w^*$ corresponds to the multiplication by the characteristic function $\chi_{[\overline{w}]}$ of the cylinder set $[\overline{w}] = [(e_0', e_1, \ldots, e_k)]$. Since $C(X_B)$ is dense in $L^2(X_B, m)$, we see that $\{\widetilde{T}_e : e \in E\}$ satisfies Definition 5.1.

It follows from Lemma 2.24 that the representation of $\mathcal{O}_A$ defined as in Theorem 4.12 is also monic because the projection $T_i T_i^*$ corresponds to the multiplication by the characteristic function $\chi_{[i]}$ where $[i]$ denotes all paths in X_A beginning at i.

(2) The definition of s.f.s. based on a topological Markov chain X_A (see Example 2.20) gives us another example of a monic representation. Indeed, we again can use the formulas from Theorem4.12 and show that the constant function that equals 1 is cyclic for the representation of $C(X_A)$.

DEFINITION 5.3. Let $\{\sigma_i : i \in \Lambda\}$ be a s.f.s. on a probability measure space (X, m) with $\sigma_i : D_i \to R_i$ and a coding map σ. We say that a collection $(m, (f_i)_{i \in \Lambda})$ is a *monic system* if for any $i \in \Lambda$ one has $m \circ \sigma_i^{-1} << m$ and

$$(5.1) \qquad \rho_m(x, \sigma_i^{-1}) = |f_i|^2$$

for some functions $f_i \in L^2(X, m)$ such that $f_i \neq 0$ for m-a.e. $x \in R_i$. A monic system is called *nonnegative* if $f_i \geq 0$.

REMARK 5.4. Relation (5.1) can be written also as follows

$$\rho_m(\sigma x, \sigma_i)^{-1} = |f_i|^2.$$

To see this, we use the formulas $\rho_m(\sigma_i^{-1}x, \sigma_i)\rho_m(x, \sigma_i^{-1}) = 1$ and $\sigma_i^{-1}x = \sigma x, x \in R_i$. This formula agrees with that used in the definition of T_i and T_i^* (see (2.19) and (2.20)).

LEMMA 5.5. *Suppose that a saturated s.f.s.* $\{\sigma_i : i \in \Lambda\}, \sigma_i : D_i \to R_i$, *defined on* (X, m) *satisfies condition (C-K) from Definition 2.16; and let* σ *be a coding map for* $\{\sigma_i : i \in \Lambda\}$. *Then, for any monic system* $(m, (f_i)_{i \in \Lambda})$, *the operators*

$$(5.2) \qquad\qquad T_i f = f_i \cdot f \circ \sigma, \quad f \in L^2(x, m), \ i \in \Lambda$$

generates a representation of $\mathcal{O}_A$ *on* $L^2(X, m)$ *where* A *is defined in* $(2.9)^1$.

The representation of $\mathcal{O}_A$ defined by $\{T_i : i \in \Lambda\}$ as in Lemma 5.5 will be called *associated to a monic system*.

PROOF. This lemma can be proved in the same manner as [**DJ14a**, Theorem 2.7]. We first notice that $f_i(x) \neq 0$ if and only if $x \in R_i = \sigma_i(D_i)$. Since $\sigma \circ \sigma_i = 1$, we observe that T_i is a partial isometry:

$$||T_i f||^2 = \int_X |f_i|^2 |f \circ \sigma|^2 \, dm = \int_X |f \circ \sigma|^2 \, d(m\sigma_i^{-1}) = \int_X |f|^2 \, dm.$$

Next, if $i \neq j$, then

$$< T_i f, T_j g >= \int_X \overline{f}_i f_j \overline{f \circ \sigma}(g \circ \sigma) \, dm = 0$$

because f_i and f_j are supported by disjoint sets.

To find T_i^*, we define for each $i \in \Lambda$

$$(5.3) \qquad\qquad g_i(x) = \frac{f_i(x)}{|f_i(x)|^2} \quad \text{if } x \in R_i, \quad g_i(x) = 0 \text{ if } x \notin R_i.$$

Then

$$< T_i^* f, g >= \int_X \overline{f} T_i g \, dm = \int_X \overline{f}(g \circ \sigma)g_i |f_i|^2 \, dm = \int_X \overline{(f \circ \sigma_i)}(g \circ \sigma_i) g \, dm,$$

so that

$$T_i^* f = \overline{(g \circ \sigma_i)}(f \sigma_i).$$

It follows from the proved formulas that $T_i T_i^*$ is the projection on $L^2(X, m)$ given by

$$(5.4) \qquad\qquad T_i T_i^* f = f_i \overline{(g_i \circ \sigma_i \circ \sigma)}(f \circ \sigma_i \circ \sigma) = \chi_{R_i} f$$

(we use here (5.3) and the relation $\sigma_i \circ \sigma(x) = x$ for $x \in R_i$). Since $\bigcup_{i \in \Lambda} R_i = X$ and this union is disjoint, we have

$$\sum_{i \in \Lambda} T_i T_i^* = 1.$$

To finish the proof, we check that for any $i \in \Lambda$

$$(5.5) \qquad\qquad T_i^* T_i = \sum_{j \in \Lambda} a_{i,j} T_j T_j^*.$$

---

^{1}Here we changed the notation and used A instead of $\widetilde{A}$.

One can show similarly to (5.4) that for any $f \in L^2(X, m)$

$$T_i^* T_i f = \chi_{D_i} f.$$

It is clear that

$$\sum_{j \in \Lambda} a_{i,j} T_j T_j^* f = \sum_{j \in \Lambda_i} T_j T_j^* f = \sum_{j \in \Lambda_i} \chi_{R_j} f = \chi_{D_i} f$$

and therefore (5.5) is proved.

$\square$

Question: Under what conditions on a s.f.s. $\{\sigma_i : i \in \Lambda\}$ the representation $T_i f = f_i \cdot f \circ \sigma$ is monic?

We now consider a class of monic systems that is naturally related to $\mathcal{O}_A$. Let X_A be the topological Markov chain, and let $\Sigma = \{\sigma_i : i = 1, \ldots, n\}$ be the s.f.s. defined in Example 2.20. A monic system $(m, (f_i))$ on X_A defined by Σ is called *inherent*.

THEOREM 5.6. *Let π be a representation of $\mathcal{O}_A$ defined by a monic system according to (5.2). Then π is monic if and only if it is unitarily equivalent to a representation associated to an inherently monic system.*

PROOF. The fact that π is a representation of $\mathcal{O}_A$ is proved in Lemma 5.5. Suppose that π is generated by $\{T_i : i = 1, \ldots, n\}$ and acts on $L^2(X_A, \mu)$. Assume that π is associated to an inherent monic system $\Sigma = \{\sigma_i : i = 1, \ldots, n\}$. By direct computation we can show that the projection $T_I T_I^*$ is a multiplication by the cylinder set $\chi_{[I]}(x)$ where I is any finite word on the alphabet $(1, \ldots, n)$. It suffices to note that the function $f \equiv 1$ is cyclic for D_A.

To prove the converse, we will follow [**DJ14a**]. If π is monic on $\mathcal{H}$, then there exists a cyclic vector $\xi \in \mathcal{H}$ for $\mathcal{D}_A$. It follows from the isomorphism of $\mathcal{D}_A$ and $C(X_A)$ that there exists a Borel measure μ on X_A such that

$$< \xi, \pi(f)\xi > = \int_{X_A} f \, d\mu, \quad f \in C(X_A).$$

Moreover, $\pi(\mathcal{D}_A)$ is unitarily equivalent to the representation of $C(X_A)$ acting on $L^2(X, \mu)$ by multiplication operators.

Consider the map $W : C(X_A) \to \mathcal{H}$ defined by the relation $W(f) = \pi(f)\xi$. Clearly, W is linear and isometric, so that W can be extended to an isometry from $L^2(X_A, \mu)$ to $\mathcal{H}$. We notice that this isometry is onto because π is monic.

We now define the operators $\widehat{T}_i := W^* T_i W$ ($i = 1, \ldots, n$) acting on $L^2(X_A, \mu)$. We check that $\widehat{T}_i$ satisfy relation (5.2). Set up $f_i = \widehat{T}_i 1$. In what follows we will use the relations

$$T_i^* \pi(f) T_i = \pi(f \circ \sigma_i), \quad T_i \pi(f) = \pi(f \circ \sigma) T_i$$

which can be verified first on characteristic functions of cylinder sets.

We have

$$\int_{X_A} |f_i|^2 f \, d\mu = < \widetilde{T}_i, f\widetilde{T}_i >_{L^2(X, \mu)} = < T_i \xi, \pi(f) T_i \xi >_{\mathcal{H}}$$

$$= < \xi, \pi(f \circ \sigma_i)\xi >_{\mathcal{H}} = \int_{X_A} f \circ \sigma_i \, d\mu.$$

Hence, we deduce that $\mu \circ \sigma_i^{-1} \ll \mu$ and $\dfrac{d\mu \circ \sigma_i^{-1}}{d\mu} = |f_i|^2$.

Next, if $f \in C(X_A)$, then

$$\widetilde{T}_i f = W^* T_i \pi(f)\xi = W^* \pi(f \circ \sigma)T_i\xi = W^* \pi(f \circ \sigma)WW^* T_i W 1 = (f \circ \sigma)f_i.$$

This proves (5.2). The relation $\sum_i T_i T_i^* = 1$ implies that the support of f_i is the set $R_i = \{(x_k) \in X_A : x_0 = i\}$. $\qquad\square$

COROLLARY 5.7. *Let $(m, (f_i))$ and $(m', (f_i'))$ be two inherent monic systems on X_A. Then the representations of $\mathcal{O}_A$ associated to these monic systems are equivalent if and only if the measures m and m' are equivalent, and there exists a function h on X_A such that*

$$\frac{dm'}{dm}(x) = |h(x)|^2, \quad f_i' = h \circ \sigma f_i h^{-1}, \ i = 1, \ldots, n.$$

The *proof* is the same as that of [**DJ14a**, Theorem 2.9].

References

[Ak12]   Hasan Akin, *An upper bound of the directional entropy with respect to the Markov measures*, Internat. J. Bifur. Chaos Appl. Sci. Engrg. **22** (2012), no. 11, 1250263, 6, DOI 10.1142/S021812741250263X. MR3006334

[BKM09]   S. Bezuglyi, J. Kwiatkowski, and K. Medynets, *Aperiodic substitution systems and their Bratteli diagrams*, Ergodic Theory Dynam. Systems **29** (2009), no. 1, 37–72, DOI 10.1017/S0143385708000230. MR2470626 (2009m:37020)

[BKMS10]   S. Bezuglyi, J. Kwiatkowski, K. Medynets, and B. Solomyak, *Invariant measures on stationary Bratteli diagrams*, Ergodic Theory Dynam. Systems **30** (2010), no. 4, 973–1007, DOI 10.1017/S0143385709000443. MR2669408 (2012g:37019)

[BJ14]   S. Bezuglyi, P. Jorgensen, *Markov measures on Bratteli diagrams*, in preparation.

[BrRo81]   Ola Bratteli and Derek W. Robinson, *Operator algebras and quantum-statistical mechanics. II*, Springer-Verlag, New York-Berlin, 1981. Equilibrium states. Models in quantum-statistical mechanics; Texts and Monographs in Physics. MR611508 (82k:82013)

[BrJKR00]   Ola Bratteli, Palle E. T. Jørgensen, Ki Hang Kim, and Fred Roush, *Non-stationarity of isomorphism between AF algebras defined by stationary Bratteli diagrams*, Ergodic Theory Dynam. Systems **20** (2000), no. 6, 1639–1656, DOI 10.1017/S0143385700000912. MR1804950 (2001k:46104)

[BrJKR02a] Ola Bratteli, Palle E. T. Jorgensen, Ki Hang Kim, and Fred Roush, *Computation of isomorphism invariants for stationary dimension groups*, Ergodic Theory Dynam. Systems **22** (2002), no. 1, 99–127, DOI 10.1017/S0143385702000044. MR1889566 (2004c:37019)

[BrJKR02b] Ola Bratteli, Palle E. T. Jorgensen, Ki Hang Kim, and Fred Roush, *Corrigendum to the paper: "Decidability of the isomorphism problem for stationary AF-algebras and the associated ordered simple dimension groups"* [*Ergodic Theory Dynam. Systems* **21** *(2001), no. 6, 1625–1655; MR1869063 (2002h:46088)*], Ergodic Theory Dynam. Systems **22** (2002), no. 2, 633, DOI 10.1017/S0143385702000317. MR1898809

[BrJO2004] Ola Bratteli, Palle E. T. Jorgensen, and Vasyl' Ostrovs'kyĭ, *Representation theory and numerical AF-invariants. The representations and centralizers of certain states on $\mathcal{O}_d$*, Mem. Amer. Math. Soc. **168** (2004), no. 797, xviii+178, DOI 10.1090/memo/0797. MR2030387 (2005i:46069)

[Cu77]   Joachim Cuntz, *Simple C^*-algebras generated by isometries*, Comm. Math. Phys. **57** (1977), no. 2, 173–185. MR0467330 (57 #7189)

[Cu79]   J. Cuntz, *Noncommutative Haar measure and algebraic finiteness conditions for simple C^*-algebras* (English, with French summary), Algèbres d'opérateurs et leurs applications en physique mathématique (Proc. Colloq., Marseille, 1977), Colloq. Internat. CNRS, vol. 274, CNRS, Paris, 1979, pp. 113–133. MR560629 (81f:46070)

[CuKr80]   Joachim Cuntz and Wolfgang Krieger, *A class of C^*-algebras and topological Markov chains*, Invent. Math. **56** (1980), no. 3, 251–268, DOI 10.1007/BF01390048. MR561974 (82f:46073a)

[DHS99]   F. Durand, B. Host, and C. Skau, *Substitutional dynamical systems, Bratteli diagrams and dimension groups*, Ergodic Theory Dynam. Systems **19** (1999), no. 4, 953–993, DOI 10.1017/S0143385799133947. MR1709427 (2000i:46062)

[DJ12]   Dorin Ervin Dutkay and Palle E. T. Jorgensen, *Spectral measures and Cuntz algebras*, Math. Comp. **81** (2012), no. 280, 2275–2301, DOI 10.1090/S0025-5718-2012-02589-0. MR2945156

[DPS14]   Dorin Ervin Dutkay, Gabriel Picioroaga, and Myung-Sin Song, *Orthonormal bases generated by Cuntz algebras*, J. Math. Anal. Appl. **409** (2014), no. 2, 1128–1139, DOI 10.1016/j.jmaa.2013.07.012. MR3103223

[DJ14a]   Dorin Ervin Dutkay and Palle E. T. Jorgensen, *Monic representations of the Cuntz algebra and Markov measures*, J. Funct. Anal. **267** (2014), no. 4, 1011–1034, DOI 10.1016/j.jfa.2014.05.016. MR3217056

[DJ14b]   D.E. Dutkay, P.E.T. Jorgensen, *Representations of Cuntz algebras associated to quasi-invariant Markov measures*, http://arxiv.org/abs/1401.7582.

[Ell89]   George A. Elliott, *Book Review: Partially ordered abelian groups with interpolation*, Bull. Amer. Math. Soc. (N.S.) **21** (1989), no. 1, 200–204, DOI 10.1090/S0273-0979-1989-15822-9. MR1567798

[EllMu93]   George A. Elliott and Daniele Mundici, *A characterisation of lattice-ordered abelian groups*, Math. Z. **213** (1993), no. 2, 179–185, DOI 10.1007/BF03025717. MR1221712 (94e:06010)

[F97]   A. H. Forrest, *K-groups associated with substitution minimal systems*, Israel J. Math. **98** (1997), 101–139, DOI 10.1007/BF02937330. MR1459849 (99c:54056)

[GPS95]   Thierry Giordano, Ian F. Putnam, and Christian F. Skau, *Topological orbit equivalence and C^*-crossed products*, J. Reine Angew. Math. **469** (1995), 51–111. MR1363826 (97g:46085)

[HPS92]   Richard H. Herman, Ian F. Putnam, and Christian F. Skau, *Ordered Bratteli diagrams, dimension groups and topological dynamics*, Internat. J. Math. **3** (1992), no. 6, 827–864, DOI 10.1142/S0129167X92000382. MR1194074 (94f:46096)

[Hi80]   Takeyuki Hida, *Brownian motion*, Applications of Mathematics, vol. 11, Springer-Verlag, New York-Berlin, 1980. Translated from the Japanese by the author and T. P. Speed. MR562914 (81a:60089)

[JP12]   P. E. T. Jorgensen and A. M. Paolucci, *Markov measures and extended zeta functions*, J. Appl. Math. Comput. **38** (2012), no. 1-2, 305–323, DOI 10.1007/s12190-011-0480-5. MR2886683

[Ka48]   Shizuo Kakutani, *On equivalence of infinite product measures*, Ann. of Math. (2) **49** (1948), 214–224. MR0023331 (9,340e)

[MaPa11]   Matilde Marcolli and Anna Maria Paolucci, *Cuntz-Krieger algebras and wavelets on fractals*, Complex Anal. Oper. Theory **5** (2011), no. 1, 41–81, DOI 10.1007/s11785-009-0044-y. MR2773056 (2012e:46116)

[Me06]   Konstantin Medynets, *Cantor aperiodic systems and Bratteli diagrams* (English, with English and French summaries), C. R. Math. Acad. Sci. Paris **342** (2006), no. 1, 43–46, DOI 10.1016/j.crma.2005.10.024. MR2193394 (2006g:37011)

[SG77]   G. E. Shilov and B. L. Gurevich, *Integral, measure and derivative: a unified approach*, Revised English edition, Dover Publications, Inc., New York, 1977. Translated from the Russian and edited by Richard A. Silverman; Dover Books on Advanced Mathematics. MR0466463 (57 #6342)

[Sk97]   Christian Skau, *Orbit structure of topological dynamical systems and its invariants*, Operator algebras and quantum field theory (Rome, 1996), Int. Press, Cambridge, MA, 1997, pp. 533–544. MR1491140 (99a:46104)

DEPARTMENT OF MATHEMATICS, INSTITUTE FOR LOW TEMPERATURE PHYSICS, KHARKIV 61103, UKRAINE
  Current address: Department of Mathematics, University of Iowa, Iowa City, 52242 Iowa
  E-mail address: bezuglyi@gmail.com

DEPARTMENT OF MATHEMATICS, UNIVERSITY OF IOWA, IOWA CITY, IOWA 52442
  E-mail address: palle-jorgensen@uiowa.edu

Contemporary Mathematics
Volume **650**, 2015
http://dx.doi.org/10.1090/conm/650/13041

A survey of amenability theory for direct-limit groups

Matthew Dawson and Gestur Ólafsson

ABSTRACT. We survey results from amenability theory with an emphasis on
applications to harmonic analysis on direct-limit groups.

1. Introduction

The class of all topological groups is far too general to admit many useful general
theorems. For that reason, a few simplifying assumptions are almost always made
in the literature when studying them, so that the groups are "nice" enough. The
assumptions come in two broad classes: separability assumptions, which assure that
the topology is "rich enough," as well as compactness and countability assumptions,
which assure that the group is not "too big." For instance, topological groups are
nearly always assumed to be Hausdorff.

However, the most important assumption commonly made, especially for the
purpose of harmonic analysis, is local compactness. Locally compact groups (and
only locally compact groups) admit Borel measures invariant under the group action. This, in turn, provides the existence of a group C^*-algebra that carries all of
the information from the representation theory of the group.

Most of the literature on representation theory and harmonic analysis is written from the point of view of separable locally compact groups, even in places
where it may not be necessary to make that requirement. Unfortunately, infinite-
dimensional Lie groups are never locally compact. Hence there is no Haar measure
and no hope of a Plancherel formula. The standard tools of harmonic analysis, such
as convolutions and group C^*-algebras, appear to break down completely.

While harmonic analysis may appear at first glance to be at a dead end for
infinite-dimensional groups, a surprising amount of progress has been made on
their representation theory. For instance, all separable unitary representations of
the full unitary group $U(\mathcal{H})$ of a separable Hilbert space $\mathcal{H}$ have been classified
([**21**]). For direct limits of classical Lie groups, there have been several important
results (see, for instance, [**17**]).

More recently, there has also been some progress in finding a good context for
studying harmonic analysis on direct-limit groups. For instance, [**11**] constructed

2010 *Mathematics Subject Classification.* Primary 43-02, 43A07; Secondary 43A90.
Key words and phrases. Invariant means, amenability, infinite-dimensional groups, spherical
functions and representations.
The research of the second author was supported by NSF grant DMS-1101337. The research
of M. Dawson was partially supported by the NSF VIGRE grant at LSU.

a suitable group C^*-algebra for certain direct-limit groups. See also [14]. A very natural construction has also been used by Olshanski, Borodin, Kerov, and Vershik to prove a sort of Plancherel theorem for certain direct-limit groups ([3, 13]). The basic ideas of the construction of the regular representation seem to originate from a paper by Pickrell ([20]).

We briefly summarize the construction here for the purposes of comparison with the ideas discussed in this paper. One begins with an increasing chain $\{G_k\}_{k\in\mathbb{N}}$ of compact groups and considers the direct-limit group $G \equiv \varinjlim G_k = \cup_{k\in\mathbb{N}}G_k$. Denote the inclusion maps by $i_n : G_n \to G_{n+1}$. Next, one attempts to construct projections $p_n : G_{n+1} \to G_n$ for each n such that p_n is G_n-equivariant and $p_n \circ i_n =$ Id. It may not be possible to construct such a collection of projections that are continuous, but they should at least be measurable. These projections determine a projective limit space $\overline{G} = \varprojlim G_n$, in which G is a nowhere-dense subset.

Next, one considers the normalized Haar measure μ_n on G_n for each n. Because each projection p_n is G_n-equivariant and the normalized Haar measure on a compact group is unique, one sees that $p_n^*(\mu_{n+1}) = \mu_n$ for each $n \in \mathbb{N}$. This in turn produces a G_n-equivariant isometric embedding $L^2(G_n) \to L^2(G_{n+1})$. Under the right technical conditions, the projective family of probability measures $\{\mu_n\}_{n\in\mathbb{N}}$ induces a limit measure μ on $\overline{G} = \varprojlim G_k$ by Kolmogorov's theorem in such a way that $L^2(\overline{G}) = \varinjlim L^2(G_n)$, where the latter injective limit is taken in the category of Hilbert spaces.

Finally, one notes that G acts by translations on $L^2(\overline{G})$, producing a unitary representation which may then be decomposed. This representation is in a natural way a generalization for the regular representation defined in terms of the Haar measure for a locally compact group. This program has been carried out for the infinite unitary group $U(\infty) = \cup_{n\in\mathbb{N}}U(n)$ (see [3]) and the infinite symmetric group $S_\infty = \cup_{n\in\mathbb{N}}S_n$ (see [13]). The disadvantage of this approach, however, is that it gives information about functions on $\overline{G}$, which is a very different space from G (in particular, the set G considered as a subset of $\overline{G}$ has measure 0).

An alternate path towards harmonic analysis on direct limits of compact groups is provided by the theory of invariant means and amenability. Amenability theory is actually a very old subject within mathematics. Unfortunately, many of the most interesting results are true only for locally compact groups. Nevertheless, we will see that it is possible to define unitary "regular representations" for direct limits of compact groups which provide a certain decomposition theory for functions defined on the group G itself (see also [2]). Amenability theory has also been used to develop a generalization of the construction of induced representations (see [16, 28]).

This paper aims to collect in one place some of the most relevant results from amenability theory as applied to direct-limit groups. We begin in Section 2 with a review of the basic functional-analytic and topological properties of means. Section 3 discusses some of the basic properties of amenable groups, including fixed-point theorems. In Section 4 we use amenability to explore the question of which Hilbert space representations of a group are unitarizable. In Section 5 we construct a generalization of the regular representation using invariant means and explore some properties of these representations. Section 6 reviews the theory of almost-periodic functions. Section 7 shows the resulting "Plancherel Theorem" for direct limits of locally compact abelian groups. Finally, in Section 8 we discuss the application of invariant means to spherical analysis on direct limits of Gelfand pairs. In particular, we show how several well-known results about such pairs can be motivated by invariant means.

For treatments of the classical theory of invariant means on locally compact groups, we refer the reader to [**18, 19**]. Brief overviews of amenability theory may also be found in [**1, 2, 5**]. For more functional-analytic approaches to studying functions on a (not-necessarily locally compact) topological group, see [**2, 11, 14**].

2. Means on Topological Groups

Consider the space $l^\infty(G)$ of all bounded functions $f : G \to \mathbb{C}$, with norm $\|f\| = \sup_{g \in G} f(g)$ and involution given by $f^*(g) = \overline{f(g)}$ for all $g \in G$. Then $l^\infty(G)$ is a commutative C^*-algebra. In fact, $l^\infty(G)$ is a representation of G under the L given by $L_g f(h) = f(g^{-1}h)$ for $g \in G$ and $f \in l^\infty(G)$. One can also consider the action R given by $R_g f(h) = f(hg)$.

Because we are interested in continuous representations of G, it is natural to consider the space $\mathrm{RUC}_b(G)$ of functions $f \in l^\infty(G)$ such that the map $G \mapsto l^\infty(G)$, $g \mapsto L_g(f)$ is continuous. These functions are precisely the bounded uniformly continuous functions on G for the right-uniformity (see [**12**] for more on uniform spaces). Similarly, one defines $\mathrm{LUC}_b(G)$ to be the space of functions $f \in l^\infty(G)$ such that $G \mapsto l^\infty(G)$, $g \mapsto R_g(f)$ is continuous. Finally, we define the space of bi-uniformly continuous functions to be $\mathrm{UC}_b(G) = \mathrm{RUC}_b(G) \cap \mathrm{LUC}_b(G)$. One has that L provides a continuous representation of G on $\mathrm{RUC}_b(G)$, that R is a continuous representation of G on $\mathrm{LUC}_b(G)$.

Now suppose that $\mathcal{A}$ is a closed C^*-subalgebra of the space $l^\infty(G)$ of all bounded functions on G which contains the constant functions. A **mean** on $\mathcal{A}$ is a continuous linear functional $\mu \in \mathcal{A}^*$ such that

(1) $\mu(\mathbf{1}) = 1$
(2) $\mu(f) \geq 0$ if $f \in \mathcal{A}$ and $f \geq 0$.

In other words, μ is a state for the C^*-algebra $\mathcal{A}$. It immediately follows that if $f \in \mathcal{A}$ with $m \leq f(g) \leq M$ for all $g \in G$, then

$$m \leq \mu(f) \leq M.$$

We write $\mathfrak{M}(\mathcal{A})$ for the space of all all means on $\mathcal{A}$. Note that $\mathfrak{M}(\mathcal{A})$ is contained in the closed unit ball of the dual space $\mathcal{A}^*$ of all continuous linear functionals on $\mathcal{A}$. Furthermore, it is clear that $\mathfrak{M}(\mathcal{A})$ is a weak-$*$ closed, convex subset of $B_1(\mathcal{A}^*)$. From the Banach-Alaoglu Theorem, it follows that $\mathfrak{M}(\mathcal{A})$ is weak-$*$ compact, convex subset of $\mathcal{A}^*$. We warn the reader, however, that unless $\mathcal{A}$ is separable, it is not true in general that $\mathfrak{M}(\mathcal{A})$ is sequentially compact. This subtlety has some important consequences, as we will later see.

For each $g \in G$, we may define a mean $\delta_g \in \mathfrak{M}(\mathcal{A})$ by $\delta_g(f) = f(g)$ for each $f \in \mathcal{A}$. We refer to these means as **point evaluations**. We will soon see that these point evaluations generate, in a certain sense all means on $\mathcal{A}$. We begin with a lemma about means on $l^\infty(G)$.

LEMMA 2.1. *The means $\delta_g \in \mathfrak{M}(l^\infty(G))$ for $g \in G$ are precisely the extremal points of $\mathfrak{M}(l^\infty(G))$.*

PROOF. Suppose that $\mu \in \mathfrak{M}(l^\infty(G))$ is an extremal point. Suppose that $A \subset G$, and define $\mu_A(f) = \mu(1_A f)$. If $\mu_A(\mathbf{1}) \neq 0$, then we see that $\widetilde{\mu}_A = \frac{1}{\mu_A(\mathbf{1})}\mu_A$

is a mean in $\mathfrak{M}(l^\infty(G))$. We define the zero set of μ to be the set

$$Z_\mu = \bigcup_{A \subseteq G \text{ s.t.} \mu_A(\mathbf{1})=0} A.$$

We can then define the support of μ to be the set

$$\operatorname{supp}\mu = G \backslash Z_\mu.$$

Note that $\mu_B(\mathbf{1}) \neq 0$ for all non-empty $B \subseteq \operatorname{supp}\mu$ and that $\mu(f) = \mu(1_{\operatorname{supp}\mu}f)$ for all $f \in \mathcal{A}$.

It is clear that $\operatorname{supp}\mu \neq \emptyset$; in fact, if $\operatorname{supp}\mu = \emptyset$, then $\mu(\mathbf{1}) = 0$, which contradicts the assumption that $\mu(\mathbf{1}) = 1$.

Now suppose that $\operatorname{supp}\mu \neq \emptyset$ contains at least two elements. Then we can write $\operatorname{supp}\mu = A \,\dot\cup\, B$, where $A, B \neq \emptyset$. It follows that

(2.1) $$\mu = \mu_A + \mu_B = \mu_A(\mathbf{1})\widetilde{\mu}_A + \mu_B(\mathbf{1})\widetilde{\mu}_B$$

Then $\widetilde{\mu}_A(1) + \widetilde{\mu}_B(1) = \mu(1_{\operatorname{supp}\mu}1) = 1$. Furthermore, $\mu_A \neq \mu_B$ (for instance, $\mu_A(1_A) = 1$ but $\mu_B(1_A) = 0$). Thus 2.1 contradicts the assumption that μ is an extremal point. It follows that $\operatorname{supp}\mu = \{g\}$ for some $g \in G$, and hence $\mu = \delta_g$. □

THEOREM 2.2. *The convex hull of the point evaluations are weak-$*$ dense in* $\mathfrak{M}(\mathcal{A})$.

PROOF. Because $\mathfrak{M}(l^\infty(G))$ is a compact, convex subset of $l^\infty(G)^*$ under the weak-$*$ topology, the result follows for the case $\mathcal{A} = l^\infty(G)$ by Lemma 2.1 and the Krein-Milman theorem.

For each $f \in l^\infty(G)$, we see that $f = (f + ||f||_\infty\mathbf{1}) - ||f||_\infty\mathbf{1}$, where $f + ||f||_\infty\mathbf{1} \geq 0$ and $||f||_\infty\mathbf{1} \in \mathcal{A}$. Thus, the hypotheses of the M. Riesz Extension Theorem are satisfied, and it follows that every positive functional on $\mathcal{A}$ may be extended to a positive functional on $l^\infty(G)$. In particular, every mean on $\mathcal{A}$ may be extended to a mean on $l^\infty(G)$. Because the convex hull of point evaluations is weak-$*$ dense in $l^\infty(G)$, the theorem follows. □

We warn the reader that despite Theorem 2.2, one can not in general say that an arbitrary mean in $\mathfrak{M}(A)$ is a weak-$*$ limit of convex combinations of point evaluations unless $\mathfrak{M}(\mathcal{A})$ is first-countable.

Means may be thought of as a generalization of the notion of probability measures on G. In fact, it possible to view means on $\mathcal{A}$ as legitimate probability measures on a certain compactification of G.

Denote by $\widehat{A}$ the space of all characters on G under the weak-$*$ topology on $\mathcal{A}^*$. Then $\widehat{A}$ is a compact Hausdorff space. Recall that the Gelfand Transform provides an isomorphism $\widehat{} : \mathcal{A} \to C(\widehat{A})$, given by $\widehat{f}(\lambda) = \lambda(f)$ for each $f \in \mathcal{A}$ and $\lambda \in \widehat{A}$. It follows that there is a linear isomorphism between the states (that is, means) of $\mathcal{A}$ and the states of $C(\widehat{A})$. But the states of $C(\widehat{A})$ are precisely the Radon measures on $\widehat{A}$ by the Riesz Representation Theorem.

Thus, the means on $\mathcal{A}$ correspond bijectively to *measures* on $\widehat{A}$. For a given mean μ on $\mathcal{A}$, we will denote the corresponding measure on $\widehat{A}$ by $\widehat{\mu}$. Due to the fact that

$$\mu(f) = \int_{\widehat{A}} \widehat{f}(x)d\widehat{\mu}(x),$$

we will occasionally use the notation

$$\mu(f) \equiv \int_G f(x) d\mu(x)$$

for $\mu \in \mathfrak{M}(\mathcal{A})$ and $f \in \mathcal{A}$. This notation is slightly misleading because μ is not, in fact, a measure on G. Nevertheless, we note that μ does share some properties with integrals; namely, it is linear and satisfies the inequality $|\mu(f)| \leq \mu(|f|) \leq ||f||_\infty$.

LEMMA 2.3. *The set $G_\mathcal{A}$ is dense in $\widehat{\mathcal{A}}$.*

PROOF. We denote the point evaluation δ_g by $\widehat{g}$ when we wish to consider it as an element of $\widehat{\mathcal{A}}$. Note that, when considered as a mean in $\mathfrak{M}(\mathcal{A})$, the point evaluation δ_g corresponds under the Gelfand transform to the point measure $\delta_{\widehat{g}}$ on $\widehat{\mathcal{A}}$; in fact, if $f \in \mathcal{A}$, then $\delta_g(f) = f(g) = \widehat{f}(\widehat{g})$. It follows from Theorem 2.2 that the convex hull of the point measures $\delta_{\widehat{g}}$ is dense in the space of probability measures on $\widehat{\mathcal{A}}$. In particular, every point measure δ_x on $\widehat{\mathcal{A}}$, where $x \in \widehat{\mathcal{A}}$, is in the closed convex hull of the point measures $\delta_{\widehat{g}}$.

Suppose that $x \in \widehat{\mathcal{A}}$ is not in the closure of $G_\mathcal{A}$. Then Urysohn's Lemma implies the existence of a continuous function $h \in C(\widehat{\mathcal{A}})$ such that $h = 0$ on the restriction to the closure of $\widehat{G}_\mathcal{A}$ but $h(x) = 1$. Hence $\delta_x(h) = 1$ but $\mu(h) = 0$ for every measure μ in the closed convex hull of the point measures $\delta_{\widehat{g}}$, which contradicts the weak-$*$ density noted above. $\qquad\square$

It is clear that $i_\mathcal{A}$ is injective if and only if $\mathcal{A}$ separates points on G. We will later see some examples for which $i_\mathcal{A}$ is far from being injective.

We end this section with three topological results about the compactification $i_\mathcal{A} : G \to \widehat{\mathcal{A}}$.

LEMMA 2.4. *If $\mathcal{A} \subseteq C(G)$, then $i_\mathcal{A}$ is continuous.*

PROOF. A basis for the weak-$*$ topology on $\widehat{\mathcal{A}}$ is given by the neighborhoods

$$B^\epsilon_{f_1,\dots f_k}(\delta_h) = \left\{ x \in \widehat{\mathcal{A}} \,\middle|\, |x(f_i) - \delta_h(f_i))| < \epsilon \text{ for all } 1 \leq i \leq k \right\},$$

where $f_i, \dots f_k \in \mathcal{A}$, $h \in G$, and $\epsilon > 0$, provide a basis for the weak-$*$ topology on $\mathcal{A}$ (here we use that $G_\mathcal{A}$ is dense in $\widehat{\mathcal{A}}$). Pick $g \in i_\mathcal{A}^{-1}(B^\epsilon_{f_1,\dots f_k}(\delta_h))$. Then $|f_i(g) - f_i(h)| < \epsilon$ for $1 \leq i \leq k$. Since $f_1, \dots f_k \in C(G)$, it follows that there is an open neighborhood V of g such that $|f_i(a) - f_i(h)| < \epsilon$ for all $a \in V$. It is clear that $V \subseteq i_\mathcal{A}^{-1}(B^\epsilon_{f_1,\dots f_k}(h))$ and we are done. $\qquad\square$

THEOREM 2.5. *Suppose that $\mathcal{A}$ separates closed subsets of G (for instance, if $\mathcal{A} = l^\infty(G)$ or if G has a normal topology and $C(G) \subseteq \mathcal{A}$). If U and V are disjoint closed subsets of G, then $i_\mathcal{A}(U)$ and $i_\mathcal{A}(V)$ have disjoint closures in $\widehat{\mathcal{A}}$.*

PROOF. Suppose that U and V are disjoint closed subsets of G and that $f : G \to [0,1]$ is an element of $\mathcal{A}$ such that $f|_U = 0$ and $f|_V = 1$. Thus, for any $g \in U$ and $h \in V$, one has that $\delta_g(f) = 0$ and $\delta_h(f) = 1$. It immediately follows that $x(f) = 0$ for each $x \in \overline{i_\mathcal{A}(U)}$ and that $y(f) = 1$ for any $y \in \overline{i_\mathcal{A}(V)}$. The result then follows. $\qquad\square$

THEOREM 2.6. *Suppose that for each closed set $F \subseteq G$ and each point $g \in G \backslash F$ there is a function $f \in \mathcal{A}$ such that $f|_F = 0$ but $f(g) = 1$. Then $i_\mathcal{A}$ is a homeomorphism onto its image.*

PROOF. Pick a closed set $F \subseteq G$, a point $g \notin F$ and a function $f \in \mathcal{A}$ such that $f|_F = 0$ but $f(g) = 1$. Then $\delta_g(f) = 1$ and $x(f) = 0$ for all x in the closure $\overline{i_\mathcal{A}(F)}$. Then $\delta_g \notin \overline{i_\mathcal{A}(F)}$ for all $g \in G \backslash F$. In particular, $i_\mathcal{A}(F)$ is closed in the relative topology of $G_\mathcal{A}$. Thus, $i_\mathcal{A}$ is a closed map onto its image. Since $\mathcal{A}$ separates points on G, we have that $i_\mathcal{A}$ is a continuous injection. The result follows. □

COROLLARY 2.7. *The map* $i_{\mathrm{RUC}_b(G)} : G \to \widehat{\mathrm{RUC}_b(G)}$ *is a homeomorphism onto its image.*

PROOF. Because G is a topological group, it is also a uniform space using the left action of the group on itself. Its topology is thus given by a family of semimetrics. It follows that there are sufficient right-uniformly continuous functions on G to separate closed sets from points. The result then follows from Theorem 2.6. □

3. Amenable Groups

For the rest of this article, we assume that $\mathcal{A} \subseteq \mathrm{RUC}_b(G)$. One can then define a continuous left-action of G on $\mathfrak{M}(\mathcal{A})$ by setting $g \cdot \mu(f) = \mu(L_{g^{-1}}f)$. We say that a mean $\mu \in \mathfrak{M}(\mathcal{A})$ is an **invariant mean on** $\mathcal{A}$ if $g \cdot \mu = \mu$ for all $g \in G$ (that is, if $\mu(L_g f) = \mu(f)$ for all $g \in G$). One says that G is an **amenable group** if there is a nontrivial invariant mean in $\mathfrak{M}(\mathrm{RUC}_b(G))$.

It is not difficult to show that every character in $\widehat{\mathcal{A}}$ is a positive functional on $\mathcal{A}$ and that, in fact, $\widehat{\mathcal{A}}$ is a closed subset of $\mathfrak{M}(\mathcal{A})$ in the weak-$*$ topology. Furthermore, the G-action on $\mathfrak{M}(\mathcal{A})$ restricts to a continuous left-action of G on $\widehat{\mathcal{A}}$, and so it is possible to ask whether there are any G-invariant characters in $\widehat{\mathcal{A}}$. Recalling from Section 2 the correspondence between Radon measures on $\widehat{\mathcal{A}}$ and means on $\mathcal{A}$, we see that there is a one-to-one correspondence between G-invariant means in $\mathfrak{M}(\mathcal{A})$ and G-invariant measures on $\widehat{\mathcal{A}}$.

If there is a nontrivial invariant mean in $\mathfrak{M}(\mathrm{RUC}_b(G))$, then we say that G is **amenable**. The term "amenable" is due to M. M. Day, who discovered a very powerful alternate characterization of amenablility in terms of the affine actions of a group on compact convex sets. (An **affine action** of a group G on a convex subset of K of a vector space V is an action $G \times K \to K$, $(g,v) \mapsto g \cdot v$ such that $g \cdot (tv + (1-t)w) = t(g \cdot v) + (1-t)(g \cdot w)$ for all $v, w \in K$, $g \in G$, and $t \in [0,1]$.)

THEOREM 3.1 (Day's Fixed Point Theorem [**7**]). *Let G be a topological group. The following are equivalent:*

(1) *G is amenable.*
(2) *Every compact Hausdorff space on which G acts continuously admits a G-invariant Radon probability measure.*
(3) *Every continuous affine action of G on a compact convex subset K of a locally convex vector V space has a fixed point.*

PROOF. We closely follow the proof in [**1**, Theorem G.1.7]. We begin with (1) $\implies$ (2). Let K be a compact Hausdorff space with a continuous G-action, and let $\mu \in \mathfrak{M}(\mathrm{RUC}_b(G))$ be a G-invariant mean. Fix a point $v \in K$ and consider the continuous map $p : G \to K$ by $p(g) = g \cdot v$. We define a measure μ_K on K by setting

$$\mu_K(\phi) = \mu(\phi \circ p)$$

for each $\phi \in C(X)$. One shows that $\phi \circ p \in \mathrm{RUC}_b(G)$. It is clear that μ_K defines a continuous positive functional on $C(X)$ with total mass $\|\mu_K\| = \mu_K(\mathbf{1}) = 1$. Thus μ_K defines a G-invariant probability measure on K by the Riesz representation theorem.

To prove (2) $\implies$ (3), let K be a compact convex subset of a locally convex vector space V, and suppose that G acts affinely on K. In particular, by (2) there is a G-invariant probability measure μ_K.

Now let μ be a Radon measure on K. By [**23**], there is $b_\mu \in K$ such that

$$\langle b_\mu, \lambda \rangle = \int_K \langle v, \lambda \rangle d\mu_K(v)$$

for each $\lambda \in V^*$. One refers to b_μ as the **barycenter** of μ.

Note that the space of all Radon measures on K is identical to the space of all means on $C(K)$. Suppose that $\mu = c_1 \delta_{v_1} + \cdots + c_k \delta_{v_k}$, where $c_1, \ldots c_k \geq 0$ with $c_1 + \cdots + c_k = 1$, and where $v_1, \ldots, v_k \in K$. $\sum_{i=1}^k c_i v_i$. If $g \in G$, then $g \cdot \mu = c_1 \delta_{g \cdot v_1} + \cdots c_k \delta_{g \cdot v_k}$. Thus $b_{g \cdot \mu} = g \cdot b_\mu$. Since the convex hull of point measures on K is weak-$*$ dense in the space of all measures on K, it follows that $b_{g \cdot \mu} = g \cdot b_\mu$ for all $g \in G$ and all Radon measures μ on K.

In particular, for the G-invariant measure μ_K on K, we see that $g \cdot b_{\mu_K} = b_{g \cdot \mu_K} = b_{\mu_K}$. Hence b_{μ_K} is a G-fixed point in K.

Finally, to see that (3) $\implies$ (1), we need only note that G acts continuously on the compact convex subset $\mathfrak{M}(\mathrm{RUC}_b(G))$ of the vector space $\mathrm{RUC}_b(G)^*$. Thus, (3) immediately implies the existence of a G-invariant mean in $\mathfrak{M}(\mathrm{RUC}_b(G))$.   □

We immediately arrive at the following corollary:

COROLLARY 3.2. *Suppose that G is an amenable group. Then there is a right-G-invariant mean on $\mathrm{LUC}_b(G)$. Furthermore, there is a mean in $\mathfrak{M}(\mathrm{UC}_b(G))$ which is both right- and left-G-invariant and invariant under the transformation $f \mapsto f^\vee$ given by $f^\vee(x) = f(x^{-1})$.*

The following theorem collects many of the most important lemmas for constructing amenable groups.

THEOREM 3.3. *([**1**, Proposition G.2.2], [**5**, Theorem 449C, Corollary 449F]) Let G be a topological group. Then*

(1) *If G is compact, then G is amenable.*
(2) *If G abelian, then G is amenable.*
(3) *If G is amenable and H is a closed normal subgroup of G, then G/H is amenable.*
(4) *If G is amenable and H is an open subgroup of G, then H is amenable.*
(5) *If G is amenable and H is a dense subgroup of G, then H is amenable.*
(6) *If H is a closed normal subgroup of G such that H and G/H are amenable, then G is amenable.*

PROOF. To prove (1), we note that if G is compact, then $\mathrm{RUC}_b(G) = C(G)$. Thus Haar measure provides an invariant mean on $\mathrm{RUC}_b(G)$.

Statement (2) is the Markov-Kakutani fixed-point theorem. See [**1**, Theorem G.2.1] for a proof.

To prove (3), we suppose that G is amenable and H is a closed normal subgroup. Write $p : G \to G/H$ for the canonical quotient map. We claim that if

$f \in \mathrm{RUC}_b(G/H)$, then $f \circ p \in \mathrm{RUC}_b(G)$. In fact, if $||L_{gH}f - f||_\infty < \epsilon$ for all gH in some neighborhood V of eH, then $||L_g(f \circ p) - f \circ p||_\infty < \epsilon$ for all $g \in p^{-1}(V)$, and the claim follows. Furthermore, $||f \circ p||_\infty = ||f||_\infty$. Now let μ_G be an invariant mean on $\mathrm{RUC}_b(G)$. Then we may define a mean $\mu_{G/H}$ on $\mathrm{RUC}_b(G/H)$ by setting $\mu_{G/H}(f) = \mu_G(f \circ p)$ for all $f \in \mathrm{RUC}_b(G)$. It is clear that $\mu_{G/H}$ is G/H-invariant. Thus G/H is amenable.

See [**5**, Corollary 449F] for the proofs of (4) and (5).

To prove (6), we suppose that H is a closed normal subgroup of G such that H and G/H are amenable. Let G act continuously and affinely on a compact convex subset K of a locally-convex vector space V. We denote by K^H the set of all H-fixed points in K. Because H is amenable, Day's theorem shows that K^H is nonempty. It is not difficult to show that K^H is a closed, convex subset of V. Furthermore, the action of G on K^H factors through to a well-defined, continuous action of G/H on K^H defined by $gH \cdot v = g \cdot v$ for all $v \in K$ and $gH \in G/H$. Finally, K^H must possess a G/H-fixed point x because G/H by Day's Theorem because G/H is amenable. Thus x is a G-fixed in K. Hence G is amenable. $\square$

It is well-known that every closed subgroup of an amenable *locally compact* group is amenable. We caution the reader, however, that this result is not true for groups which are not locally-compact (see [**1**, p. 457]).

There is one more well-known method of constructing amenable groups which is of particular interest to us here: any direct limit of amenable groups is again amenable.

THEOREM 3.4. ([**19**, Proposition 13.6]). *Suppose that I is a linearly-ordered index set and that $\{G_n\}_{n \in I}$ is an increasing chain of amenable subgroups (that is, $G_n \leq G_m$ if $n \leq m$). Then $G_\infty \equiv \varinjlim G_n = \cup_{n \in I} G_n$ is amenable.*

PROOF. For each $n \in I$, choose an invariant mean m_n for G_n. We then define a functional $\mu_n \in \mathrm{RUC}_b(G_\infty)^*$ by

$$\mu_n(f) = m_n(f|_{G_n})$$

for each $f \in \mathrm{RUC}_b(G_\infty)$. Because $f|_{G_n} \in \mathrm{RUC}_b(G_n)$ for all $f \in \mathrm{RUC}_b(G_\infty)$, it is clear that each μ_n is a mean on G_∞. Furthermore, we see that $\mu_n(L_g f) = \mu_n(f)$ whenever $g \in G_n \leq G_\infty$. Thus, any weak-$*$ cluster point of the set $\{\mu_n\}_{n \in I} \subseteq \mathrm{RUC}_b(G_\infty)^*$ would be invariant under $G_\infty = \cup_{n \in I} G_n$. Furthermore, by the Banach Alaoglu theorem, the unit ball in $\mathrm{RUC}_b(G_\infty)^*$ is weak-$*$ compact and thus our sequence must possess a cluster point. $\square$

Because $\mathrm{RUC}_b(G_\infty)$ is not separable when G_∞ is not compact, the unit ball in $\mathrm{RUC}_b(G_\infty)^*$ is not guaranteed to be weak-$*$ *sequentially* compact. Thus there is no reason to expect that $\{\mu_n\}_{n \in \mathbb{N}} \subseteq \mathrm{RUC}_b(G_\infty)^*$ will possess a convergent sequence. In fact, an application of the Axiom of Choice is required to construct an invariant mean on G_∞.

An immediate corollary of Theorem 3.4 is that every group formed as a direct limit of compact groups is amenable.

4. Unitarizability

In this section we look at some applications of invariant means to the theory of unitary representations for topological groups. For a given topological group, it is often very difficult to determine which representations of the group on a Hilbert space are in fact equivalent to unitary representations. For amenable groups, however, there is a very succinct solution to this question:

THEOREM 4.1. ([**19**, Proposition 17.5]). *Suppose that G is an amenable group and that π is a continuous representation of G on a separable Hilbert space $\mathcal{H}$. Then π is equivalent to a unitary representation if and only if it is uniformly bounded (that is, $\sup_{g \in U_\infty} ||\pi(g)|| < \infty$).*

PROOF. Suppose that π is equivalent to a unitary representation. Then there is an invertible bounded intertwining operator $T \in \mathrm{GL}(\mathcal{H})$ unitarizes π. It follows that $T\pi(g)T^{-1}$ is unitary and thus $||\pi(g)|| \le ||T||||T^{-1}||$ for all $g \in U_\infty$, and thus π is uniformly bounded.

To prove the converse, let $M = \sup_{g \in U_\infty} ||\pi(g)||$. Note that one also has that $M = \sup_{g \in U_\infty} ||\pi(g)^{-1}||$. It follows that

$$M^{-1}||u|| \le ||\pi(g)u|| \le M||u||$$

for all $g \in U_\infty$.

Now let μ be a bi-invariant mean on G. We denote the inner product on $\mathcal{H}$ by $\langle \cdot, \cdot \rangle_{\mathcal{H}}$. Note that $g \mapsto \langle \pi(g)u, \pi(g)v \rangle_{\mathcal{H}}$ is a uniformly continuous, bounded function on G (since π is strongly continuous and uniformly bounded). We may thus define a new inner product $\langle \cdot, \cdot \rangle_\mu$ on $\mathcal{H}$ by

$$\langle u, v \rangle_\mu = \int_G \langle \pi(g)u, \pi(g)v \rangle_{\mathcal{H}} d\mu(g)$$

for all $u, v \in \mathcal{H}$, where for clarity we have used the "integral" notation for means that was introduced in Section 2. It is clear that $\langle \cdot, \cdot \rangle_\mu$ provides a positive semi-definite Hermitian form on $\mathcal{H}$.

Note that for $u \in \mathcal{H} \backslash \{0\}$ one has that

$$0 < M^{-2}||u||_{\mathcal{H}}^2 \le ||u||_\mu^2 = \int_G ||\pi(g)u||_{\mathcal{H}}^2 d\mu(g) \le M^2 ||u||_{\mathcal{H}}^2.$$

Thus $\langle \cdot, \cdot \rangle_\mu$ is strictly positive-definite and continuous with respect to $\langle \cdot, \cdot \rangle_{\mathcal{H}}$. □

For a compact group G, every continuous representation is uniformly bounded, and hence the above theorem amounts to the fact that *every* continuous representation of G on a Hilbert space is equivalent to a unitary representation. Unfortunately, the same does not hold true for direct limits of compact groups, as we now briefly demonstrate.

Consider the group $U_\infty = \mathrm{SU}(\infty) = \varinjlim \mathrm{SU}(2n)$. For each $n \in \mathbb{N}$, consider the standard representation π_n of $\mathrm{SU}(2n)$ on $\mathcal{H}_n = \mathbb{C}^{2n}$ (that is, $\pi_n(g)v = g \cdot v$ for all $g \in \mathrm{SU}(2n)$). By taking the direct limit, we may form a unitary representation $\pi = \varinjlim \pi_n$ of $\mathrm{SU}(\infty)$ on the Hilbert space $\mathcal{H} = \ell^2(\mathbb{C}) = \overline{\varinjlim \mathbb{C}^{2n}}$ of square-summable sequences of complex numbers. Note that $\mathrm{SU}(2n)$ acts trivially on the orthogonal complement of $\mathcal{H}_n$. It follows that $\pi|_{\mathrm{SU}(2n)}$ decomposes into a direct sum of the standard representation π_n and infinitely many copies of the trivial irreducible representation. That is,

$$\pi|_{\mathrm{SU}(2n)} = \pi_n \oplus \infty \cdot 1_{\mathrm{SU}(2n)},$$

where $1_{\mathrm{SU}(2n)}$ denotes the trivial irreducible representation of $\mathrm{SU}(2n)$ on $\mathbb{C}$.

Now let $V_1 = \mathcal{H}_1$ and define $V_n = \mathcal{H}_n \ominus \mathcal{H}_{n-1}$ for each $n > 1$. Note that $\dim V_n = 2$ for each $n \in \mathbb{N}$. We now completely discard unitarity and choose some new inner product $\langle,\rangle_{V_n}$ on V_n under which $||\pi(g)|_{V_n}|| \geq n$ for some $g \in \mathrm{SU}(2n)$. For instance, if $\pi(g)v = w$, where $v, w \in V_n$ are linearly independent, then we can choose any inner product $\langle,\rangle_{V_n}$ on V_n such that $||v||_{V_n} = 1$ and $||w||_{V_n} = n$.

Next we define for each $n \in \mathbb{N}$ the finite-dimensional Hilbert space

$$\mathcal{K}_n = \bigoplus_{i=1}^{n} V_i,$$

where each V_i is given the new inner product we just defined. As vector spaces, $\mathcal{K}_n = \mathcal{H}_n$, but they possess different inner products. Now $\{(\pi_n, \mathcal{K}_n)\}_{n\in\mathbb{N}}$ forms a direct system of continuous Hilbert representations. We consider the representation $(\widetilde{\pi}_\infty, \mathcal{K}_\infty) = (\varinjlim \pi_n, \overline{\varinjlim \mathcal{K}_n})$. Note that $\pi|_{\mathrm{SU}(2n)}$ and $\widetilde{\pi}|_{\mathrm{SU}(2n)}$ possess the same irreducible subrepresentations for each $n \in \mathbb{N}$. Finally, it is clear that $\widetilde{\pi}$ is not uniformly bounded (since $\sup_{g\in\mathrm{SU}(2n)} ||\pi(g)|| \geq n$ for each $n \in \mathbb{N}$), and is therefore not unitarizable.

5. Invariant Means and Regular Representations

Unitary representations for locally compact groups are, of course, closely related to harmonic analysis. For instance, if G is a locally compact group, then decomposing the unitary regular representation of G on $L^2(G)$ is one of the foundational problems in harmonic analysis. While groups which are not locally compact do not possess Haar measures, one can develop an L^2-theory using invariant means (these definitions may also be found in, for instance, [2]).

In particular, suppose that G is a topological group and $\mathcal{A}$ is a closed C^*-subalgebra of $\mathrm{RUC}_b(G)$. For each invariant mean $\mu \in \mathfrak{M}(\mathcal{A})$, one can construct a Hilbert space $L^2_\mu(G)$ as follows. Define a pre-Hilbert seminorm on $\mathcal{A}$ by

$$\langle f, g \rangle_\mu = \mu(f\bar{g}),$$

and set $\mathcal{K}_\mu = \{f \in \mathcal{A} | \langle f, f \rangle_\mu = 0\}$. Then $L^2_\mu(G)$ is defined to be the Hilbert-space completion of $\mathcal{A}/\mathcal{K}_\mu$.

LEMMA 5.1. $\mathcal{K}_\mu$ is a closed subspace of $\mathcal{A}$.

PROOF. Suppose that $\{f_k\}_{n\in\mathbb{N}}$ is a sequence in $\mathcal{K}_\mu$ which converges to f in $\mathcal{A}$. Fix $\epsilon > 0$, and choose N such that $||f_k - f||_\infty < \epsilon$ for $n \geq N$. Then

$$\begin{aligned}\langle f, f \rangle_\mu &= \langle f_k + (f_k - f), f_k + (f_k - f) \rangle_\mu \\ &= \langle f_k, f_k \rangle_\mu + \langle f_k - f, f_k - f \rangle_\mu + 2\Re\langle f_k, f_k - f \rangle_\mu \\ &\leq 0 + \epsilon^2 + 2(||f||_\infty + \epsilon)\epsilon,\end{aligned}$$

where we use the fact that $\langle g, h \rangle_\mu = \mu(g\bar{h}) \leq ||g||_\infty ||h||_\infty$ for all $g, h \in \mathcal{A}$. Because $\epsilon > 0$ was arbitrary, it follows that $f \in \mathcal{K}_\mu$. $\square$

Because $\mathcal{A} \subseteq \mathrm{RUC}_b(G)$, one sees that the left-regular representation of G on $\mathcal{A}$ defined by $L_g f(h) = f(g^{-1}h)$ is continuous.

THEOREM 5.2. The regular representation L of G on $\mathcal{A}$ descends to a continuous unitary representation on $L^2_\mu(G)$.

PROOF. Because μ is an invariant mean, it follows that $\mathcal{K}_\mu$ is a closed invariant subspace of $\mathcal{A}$. Hence, L descends to a continuous representation of G_∞ on $UCB(G_\infty)/\mathcal{K}_\mu$.

Since $\langle f, f \rangle_\mu \leq ||f||_\infty^2$ for all $f \in \mathcal{A}$, we see that L is a continuous representation in the pre-Hilbert space topology on $\mathcal{A}$ and descends to a continuous representation in the pre-Hilbert space topology on $\mathcal{A}/\mathcal{K}_\mu$. Furthermore, L acts by isometries in those pre-Hilbert space topologies due to the fact that μ is an invariant mean on G. Thus L extends to a continuous unitary representation on the Hilbert-space completion $L_\mu^2(G)$ of $\mathcal{A}/\mathcal{K}_\mu$. $\qquad\square$

One nice aspect of this approach is that it allows the consideration of an L^2-theory of harmonic analysis that is intrinsic to the group, in the sense that it actually provides a decomposition of functions on the group G and not on a larger G-space, such as with the projective-limit construction mentioned earlier. The disadvantage is that it depends heavily on the choice of invariant mean μ and C^*-algebra $\mathcal{A}$, as we shall see. Furthermore, the fact that the axiom of choice is necessary in many cases to construct invariant means implies that it may not be possible to construct an explicit decomposition of $L_\mu^2(G)$.

It is also possible that $L_\mu^2(G)$ gives the trivial representation of G, in which case very little information may be obtained about the functions on G. In general, as the next theorem demonstrates, one can gain information on the size of $L_\mu^2(G)$ (and therefore gauge how much information may be gleaned about functions on G) by determining the support of the corresponding measure $\widehat{\mu}$ on $\widehat{\mathcal{A}}$.

THEOREM 5.3. *For a G-invariant mean μ on $\mathcal{A}$, on has the equivalance of unitary representations*

$$L_\mu^2(G) \cong_G L^2(\widehat{\mathcal{A}}, \widehat{\mu}),$$

where $L^2(\widehat{\mathcal{A}}, \widehat{\mu})$ denotes the unitary representation of G corresponding to the action of G on $\widehat{\mathcal{A}}$ under the G-invariant measure $\widehat{\mu}$ on $\widehat{\mathcal{A}}$.

PROOF. The map $\mathcal{A} \to C(\widehat{\mathcal{A}})$, $f \mapsto \widehat{f}$ is clearly a G-intertwining operator. The fact that it extends to the required unitary intertwining operator follows from the fact that $\mu(f) = \int_{\widehat{\mathcal{A}}} \widehat{f}(x) d\widehat{\mu}(x)$ for all $f \in \mathcal{A}$. $\qquad\square$

COROLLARY 5.4. *([**2**, Remark 3.11]) One has $\dim L_\mu^2(G) = 1$ if and only if there is a G-invariant character $x \in \widehat{\mathcal{A}}$ such that $\widehat{\mu} = \delta_x$. In that case, one has the decomposition*

$$\mathcal{A} = \mathbb{C}\mathbf{1} \oplus \mathcal{K}_\mu,$$

where $\mathbb{C}\mathbf{1}$ denotes the space of constant functions on G.

A group G is said to be **extremely amenable** if there is a G-invariant character in $\widehat{\mathrm{RUC}_b(G)}$. While it is known that no locally-compact groups are extremely amenable, it has been shown (see [**9**, **10**]) that $\mathrm{SO}(\infty) = \cup_{n \in \mathbb{N}}\mathrm{SO}(n)$, $\mathrm{SU}(\infty) = \cup_{n \in \mathbb{N}}\mathrm{SU}(n)$, and the infinite symmetric group $S_\infty = \cup_{n \in \mathbb{N}}S_n$ are extremely amenable. One can also show that direct products of extremely amenable groups are again extremely amenable (see [**5**, Theorem 449C]).

It is evident that if G is an amenable group, then by Day's theorem, the closure of each G-orbit in $\widehat{\mathcal{A}}$ gives rise to an invariant mean in $\widehat{\mathcal{A}}$. Unfortunately, it is very difficult to study such orbits because of the unavoidable use of the axiom of choice in the construction of $\widehat{\mathcal{A}}$. In the next section we look at a special case in which one may determine all of the invariant means on $\mathcal{A}$ and say something about the decomposition of the representation $L^2_\mu(G)$.

6. Almost Periodic Functions

In general, there is no uniqueness property for invariant means similar to the uniqueness of Haar measures. In fact, for many groups G it has been shown that the set of all invariant means on $\mathrm{RUC}_b(G)$ has cardinality $2^{2^{|G|}}$ (see [**19**]). However, one might hope for the existence of subalgebras $\mathcal{A}$ of $\mathrm{RUC}_b(G)$ which possess a unique invariant mean $\mu \in \mathfrak{M}(\mathcal{A})$. In particular, the algebra of weakly almost periodic functions on a group always satisfies this property. In some cases it is even possible to explicitly determine the value of this mean and write down an explicit decomposition of the unitary regular representation $L^2_\mu(G)$.

DEFINITION 6.1. *A continuous function $f \in C(G)$ is said to be **almost periodic** if the set $\{L_g f\}_{g \in G}$ is relatively compact in the norm topology of $C(G)$. We denote by $\mathrm{AP}(G)$ the space of almost periodic functions on G.*

Von Neumann proved the following result about almost periodic functions:

THEOREM 6.2. *(Von Neumann [**25**]) If $f \in \mathrm{AP}(G)$, then the closed convex hull $\overline{\mathrm{co}}(\{L_g f\}_{g \in G})$ of the set of G-translates contains exactly one constant function. We denote value of this constant function by $M(f)$. Furthermore, M is a G-invariant mean on $\mathrm{AP}(G)$.*

By continuity arguments, one sees that any invariant mean on $\mathrm{AP}(G)$ must take the same value for every function in $\overline{\mathrm{co}}(\{L_g f\}_{g \in G})$. Because this set contains a unique constant function, one sees that for any invariant mean $\mu \in \mathfrak{M}(f)$, one has $\mu(f) = M(f)$. In fact, it is also not difficult to see that every invariant mean λ on $\mathrm{RUC}_b(G)$ must have the property that $\lambda|_{\mathrm{AP}(G)} = M$.

In fact, it is possible to put a topological group structure on $\widehat{\mathrm{AP}(G)}$ so that $i_{\mathrm{AP}(G)} : G \to \widehat{\mathrm{AP}(G)}$ is a continuous homomorphism:

THEOREM 6.3. *([**15**, p. 166, 168]) $\mathrm{AP}(G)$ is a G-invariant closed C^*-subalgebra of $\mathrm{RUC}_b(G)$. Furthermore, the product defined by $\delta_g \cdot \delta_h = \delta_{gh}$ on $i_{\mathrm{AP}}(G)$ extends to a compact topological group structure on $\widehat{\mathrm{AP}(G)}$.*

For the sake of clarity, we introduce the notation $G_c \equiv \widehat{\mathrm{AP}(G)}$. Note that the Gelfand isomorphism $\widehat{\ } : \mathrm{AP}(G) \to C(G_c)$ sets up a correspondence between almost periodic functions on G and continuous functions on G_c. Furthermore, one sees that the invariant mean on $\mathrm{AP}(G)$ corresponds to the Haar measure on G_c. That is, $M(f) = \int_{G_c} \widehat{f}(x) dx$ for all $f \in \mathrm{AP}(G)$.

In the notation of Section 4, we see that $\mathcal{K}_M = 0$ because $\int_{G_c} \widehat{f}(x) dx > 0$ whenever $f \in \mathrm{AP}(G)$ and $f > 0$. In particular, the Gelfand transform extends to a unitary G-intertwining operator $\widehat{\ } : L^2_M(G) \to L^2(G_c)$, where $L^2(G_c)$ is defined using the Haar measure on G_c. Finally, because $G_{\mathrm{AP}(G)}$ is a dense subgroup of $G_c = \widehat{\mathrm{AP}(G)}$, it follows that a subspace of $L^2(G_c)$ is G-invariant if and only if it is G_c-invariant.

The following result now follows immediately from the Peter-Weyl theorem:

THEOREM 6.4. *The representation $L^2(G_c)$ decomposes into a direct sum of finite-dimensional representations of G.*

COROLLARY 6.5. *If G possesses no nontrivial finite-dimensional unitary representations, then $\mathrm{AP}(G)$ contains only constant functions on G.*

For instance, it follows that $SL(n, \mathbb{R})$ and $\mathrm{SU}(\infty) = \cup_{n\in\mathbb{N}}\mathrm{SU}(n)$ have no nontrivial almost-periodic functions, so the decomposition theory for almost periodic functions provides no information for such groups.

The situation is much more interesting if G is abelian. Denote by $\widehat{G}$ the group of all continuous characters of G. Note that any character $\chi \in \widehat{G}$ is almost periodic; in fact, $L_g\chi = \chi(g^{-1})\chi$, so that $\{L_g\chi\}_g \in G$ is a compact subset of a one-dimensional vector space. Thus, every character $\chi \in \widehat{G}$ corresponds to a continuous function $\widehat{\chi}$ on $G_c = \widehat{\mathrm{AP}(G)}$. Because $\widehat{\chi}$ is a character on the dense subgroup $G_{\mathrm{AP}(G)}$, it follows that $\widehat{\chi}$ is in fact a continuous character of G_c. In other words, we have shown that:

THEOREM 6.6. *The Gelfand transform restricts to an continuous surjective group homomorphism $\hat{} : \widehat{G} \to \widehat{G}_c$.*

In fact, one has that $\hat{}$ is also an isomorphism of abstract groups as long as the characters of G separate points on G. This is true for locally compact abelian groups and, as we shall see in the next section, for direct or inverse limits of locally compact abelian groups.

7. Direct Limits of Abelian Groups

The famous Pontryagin Duality Theorem asserts that for any locally compact abelian group G, there is a canonical topological group isomorphism between G and $\widehat{\widehat{G}}$ given by identifying $g \in G$ with the character $\widehat{g}$ given by $\widehat{g}(\chi) = \chi(g)$ for all $\chi \in \widehat{G}$. In fact, it is possible to extend this result to the case of direct limits as we now show.

Suppose that $G = \varinjlim G_n = \cup_{n\in\mathbb{N}}G_n$ is a strict direct limit of locally compact abelian groups. There are natural continuous projections $p_n : \widehat{G}_{n+1} \to \widehat{G}_n$ given by $p_n(\chi) = \chi|_{G_n}$ for each character $\chi \in \widehat{G}_n$. In fact, it is clear that p_n is a homomorphism, and one can show that it is surjective. We may thus construct a projective-limit group $\varprojlim \widehat{G}_n$. One then proves the following theorem:

THEOREM 7.1. *([**24**, p. 45]) The group $G = \varinjlim G_n$ satisfies the Pontryagin duality. In fact, $\widehat{\varinjlim G_n} = \varprojlim \widehat{G}_n$ and $\widehat{\varprojlim \widehat{G}_n} = \varinjlim G_n$.*

In particular, we see that $\widehat{G}$ separates points on the direct-limit group $G = \varinjlim G_n$ and thus that the groups $\widehat{G}$ and $\widehat{G}_c$ are isomorphic as abstract groups. Thus $L^2_M(G)$ decomposes into a direct sum of irreducible representations as follows:

$$L^2_M(G) \cong \bigoplus_{\chi\in\widehat{G}} \mathbb{C}\chi,$$

where $\mathbb{C}\chi$ is the one-dimensional subspace of $L^2_M(G)$ generated by the character $\chi \in \widehat{G}$. Note that $\widehat{G}$ may contain an uncountable number of characters on G, in which case $L^2_M(G)$ is an nonseparable Hilbert space.

We end this section with an example. Consider the Torus group $\mathbb{T} = S^1$. For each n, consider n^{th} Cartesian power $\mathbb{T}^n$. By using the embedding $\mathbb{T}^n \to \mathbb{T}^{n+1}$ given by $z \mapsto (z, 1)$, one can construct the direct-limit group $\mathbb{T}^\infty = \varinjlim \mathbb{T}^n$. We recall that $\widehat{T} = \mathbb{Z}$ and that $\widehat{T^n} = \mathbb{Z}^n$. Hence, Theorem 7.1 implies that $\widehat{T^\infty} = \varprojlim \mathbb{Z}^n$, where the projections $\mathbb{Z}^{n+1} \to \mathbb{Z}^n$ are the canonical projections given by $(z, m) \mapsto z$ for all $m \in \mathbb{Z}$ and $z \in \mathbb{Z}^n$. Thus $\widehat{T^\infty}$ is isomorphic to the the group $\mathbb{Z}^{\mathbb{N}}$ of all sequences of integers. Thus,

$$L^2_M(\mathbb{T}^\infty) \cong \bigoplus_{\sigma \in \mathbb{Z}^{\mathbb{N}}} \mathbb{C}\chi_\sigma,$$

where $\chi_\sigma \in \widehat{T^\infty}$ is the character corresponding to $\sigma \in \mathbb{Z}^n$. In particular, $L^2_M(\mathbb{T}^\infty)$ is far from being separable.

Because $\mathbb{T}^\infty$ is abelian, there exist invariant means on $\mathrm{RUC}_b(\mathbb{T}^\infty)$. For any such invariant mean μ, one can construct the corresponding regular representation $L^2_\mu(\mathbb{T}^\infty)$. One has that $L^2_M(\mathbb{T}^\infty)$ is a subrepresentation of $L^2_\mu(\mathbb{T}^\infty)$ for any mean μ. However, due to the necessity of applying the axiom of choice to construct such an invariant mean, it is not clear whether or not it is possible to say much about the orthogonal complement $L^2_\mu(\mathbb{T}^\infty) \ominus L^2_M(\mathbb{T}^\infty)$ for a mean μ.

8. Spherical Functions and Direct Limits

In this section we see how invariant means may be used to describe the behavior of spherical functions on direct limits of Gelfand pairs. In the classical theory, spherical functions are critically important in studying harmonic analysis on Gelfand pairs.

We remind the reader that a **Gelfand pair** is a pair (G, K) of groups, where G is locally compact and K is compact, such that the convolution algebra on the space $L^1(K \backslash G / K)$ of Haar-integrable bi-K-invariant functions on G is abelian. Riemannian symmetric pairs provide the most important examples of Gelfand spaces. A **spherical function** on G is function $\phi \in C(G)$ such that

$$(8.1) \qquad \int_K \phi(xky)dk = \phi(x)\phi(y)$$

for all $x, y \in G$, where again integration over the compact group K is with the normalized Haar measure. An irreducible unitary representation $(\pi, \mathcal{H})$ of G is said to be a **spherical representation** if $\mathcal{H}^K \neq \{0\}$, where $\mathcal{H}^K$ is the space of all vectors $v \in \mathcal{H}$ such that $\pi(k)v = v$ for all $k \in K$. In fact, for a Gelfand pair (G, K), one can show that $\dim \mathcal{H}^K = 1$ for every irreducible unitary spherical representation $(\pi, \mathcal{H})$ of G.

One can show that the positive-definite spherical functions on G are precisely the matrix coefficients

$$\phi_\pi(g) = \langle \pi(g)v, v \rangle,$$

where π is an irreducible unitary spherical representation of G and v is a unit vector in $\mathcal{H}^K \backslash \{0\}$. This connection between spherical functions and spherical representations allows one to determine the Plancherel decomposition of the quasiregular representation of G on $L^2(G/K)$. Proofs of these classical theorems may be found, for instance, in [**8**].

Suppose one has an increasing family of locally compact groups $\{G_n\}_{n \in \mathbb{N}}$ and an increasing family of compact groups $\{K_n\}_{n \in \mathbb{N}}$ such that $K_n \leq G_n$ for each $n \in \mathbb{N}$ and (G_n, K_n) is a Gelfand pair. Let $G_\infty = \varinjlim G_n$ and $K_\infty = \varinjlim K_n$. If we make the additional assumption that $G_n \cap K_{n+1} = K_n$ for all $n \in \mathbb{N}$, then there is a well-defined G_n-equivariant inclusion $G_n/K_n \to G_{n+1}/K_{n+1}$ given by $gK_n \mapsto gK_{n+1}$ for all $g \in G_n$, and we can write $G_\infty/K_\infty = \varinjlim G_n/K_n$. In this case we say that (G_∞, K_∞) is a **direct-limit spherical pair**. This definition provides a natural infinite-dimensional generalization of the notion of a Gelfand pair.

It is natural to say that an irreducible unitary representation $(\pi, \mathcal{H})$ of G_∞ is a **spherical representation** if the space $\mathcal{H}^{K_\infty}$ of K_∞-fixed vectors in $\mathcal{H}$ is nontrivial. It is possible to show (see [**17**, Theorem 23.6]) that, just as for finite-dimensional Gelfand pairs, $\dim \mathcal{H}^{K_\infty} = 1$ for every irreducible unitary spherical representation $(\pi, \mathcal{H})$.

The proper definition of a spherical function is slightly more subtle, because there is no Haar measure on K_∞ over which to integrate. However, because K_∞ is a direct limit of compact groups, it is amenable, and we may generalize (8.1) by replacing the Haar measure on K with an invariant mean μ on K_∞. That is, we say that a continuous function $\phi \in G_\infty$ is a **spherical function with respect to** μ if

$$(8.2) \qquad \int_{K_\infty} \phi(xky) d\mu(k) = \phi(x)\phi(y)$$

for all $x, y \in G$.

At this point, we remind the reader that, if we define for each $n \in \mathbb{N}$ a mean $\mu_n \in \mathfrak{M}(\mathrm{RUC}_b(K_\infty))$ by $\mu_n(f) = \int_{K_n} (f|_{K_n})(k) dk$, then any weak-$*$ cluster point of $\{\mu_n\}_{n \in \mathbb{N}}$ is an invariant mean on K_∞. While this sequence has many cluster points, none of which may be constructed as functionals on all of $\mathrm{RUC}_b(K_\infty)$ without recourse to the Axiom of Choice, what we can say is that any such K_∞-invariant mean μ that is a weak-$*$ cluster point of $\{\mu_n\}_{n \in \mathbb{N}}$ must have the property that

$$\mu(f) = \lim_{n \to \infty} \int_{K_n} (f|_{K_n})(k) dk$$

for all functions $f \in \mathrm{RUC}_b(K_\infty)$ such that the limit on the right-hand side of the equation exists. We now fix such invariant mean μ in the closure of $\{\mu_n\}_{n \in \mathbb{N}}$. For any spherical function $\varphi \in C(G_\infty)$, it follows that if φ satisfies

$$(8.3) \qquad \lim_{n \to \infty} \int_{K_n} \varphi(xky) d\mu(k) = \varphi(x)\varphi(y),$$

then φ is spherical for μ.

In fact, Olshanski defines a function $f \in C(G_\infty)$ to be spherical if it satisfies (8.3). Note that this condition is stronger than requiring f to be spherical for every invariant mean in the closure of $\{\mu_n\}_{n \in \mathbb{N}}$. However, we will show in Theorem 8.2 that these two conditions are in fact equivalent.

First we need a lemma about projection operators. If G is a topological group, K is a compact subgroup, and $(\pi, \mathcal{H})$ is any unitary representation of G, then the orthogonal projection operator $P : \mathcal{H} \to \mathcal{H}^K$ may be written as

$$P(v) = \int_K \pi(k) v \, dk$$

for all $v \in \mathcal{H}$. In fact, using invariant means it is possible to describe the projection operator $P : \mathcal{H} \to \mathcal{H}^{K_\infty}$ for a unitary representation $(\pi, \mathcal{H})$ of G_∞ in a completely analogous fashion, as the next lemma shows. The proof is extremely similar to the proof in the finite-dimensional context, although some care must be taken due to the fact that means do not satisfy the same properties as integrals.

LEMMA 8.1. *Suppose that* $(\pi, \mathcal{H})$ *is a unitary representation of a group* G *and that* K *is a subgroup of* G *that is amenable. Let* μ *be an invariant mean in* $\mathfrak{M}(\mathrm{RUC}_b(K_\infty))$. *Then there is a bounded operator* $P \in \mathrm{B}(\mathcal{H})$ *such that*

$$\langle w, Pv \rangle = \int_K \langle w, \pi(k)v \rangle d\mu(k)$$

for all $v, w \in \mathcal{H}$. *Furthermore,* P *is a projection from* $\mathcal{H}$ *onto* $\mathcal{H}^K$. *Finally,* P *is the orthogonal projection if* μ *is an inversion-invariant mean.*

PROOF. Recall that the matrix coefficient function $k \mapsto \langle w, \pi(k)v \rangle$ is uniformly continuous for all $v, w \in \mathcal{H}$. Now fix $v \in \mathcal{H}$. Because π is unitary, we have that $|\langle w, \pi(k)v \rangle| \leq ||w|| ||v||$ for all $k \in K$ and $w \in \mathcal{H}$. Hence

$$(8.4) \qquad \left| \int_K \langle w, \pi(k)v \rangle d\mu(k) \right| \leq ||w|| ||v||$$

because μ is a mean. Thus the map $w \mapsto \int_K \langle w, \pi(k)v \rangle d\mu(k)$ defines a bounded linear functional on $\mathcal{H}$ and by the Riesz representation theorem there is a unique vector $Pv \in \mathcal{H}$ such that

$$\langle w, Pv \rangle = \int_K \langle w, \pi(k)v \rangle d\mu(k)$$

for all $w \in \mathcal{H}$. It is clear that $v \mapsto Pv$ is linear. Furthermore, $P \in \mathrm{B}(\mathcal{H})$ by (8.4).

Next we see that $Pv \in \mathcal{H}^K$ for all $v \in \mathcal{H}$. In fact, for all $h \in K$, we have that

$$\begin{aligned}
\langle w, \pi(h)Pv \rangle &= \langle \pi(h^{-1})w, Pv \rangle \\
&= \int_K \langle \pi(h^{-1})w, \pi(k)v \rangle d\mu(k) \\
&= \int_K \langle w, \pi(hk)v \rangle d\mu(k) \\
&= \int_K \langle w, \pi(k)v \rangle d\mu(k) = \langle w, Pv \rangle,
\end{aligned}$$

and thus $\pi(h)Pv = Pv$ for all $h \in K$. Similarly, $Pv = v$ for all $v \in \mathcal{H}^K$. In fact,

$$\begin{aligned}
\langle w, Pv \rangle &= \int_K \langle w, \pi(k)v \rangle d\mu(k) \\
&= \int_K \langle w, v \rangle d\mu(k) = \langle w, v \rangle
\end{aligned}$$

for all $w \in \mathcal{H}$ and $v \in \mathcal{H}^K$. Thus P is a projection onto $\mathcal{H}^K$.

It only remains to be shown that P is self-adjoint if μ is inversion-invariant. For any $v, w \in \mathcal{H}$, we have

$$
\begin{aligned}
\langle w, Pv \rangle &= \int_K \langle w, \pi(k)v \rangle d\mu(k) \\
&= \int_K \langle \pi(k^{-1})w, v \rangle d\mu(k) \\
&= \int_K \langle \pi(k)w, v \rangle d\mu(k) \\
&= \int_K \overline{\langle v, \pi(k)w \rangle} d\mu(k) \\
&= \overline{\int_K \langle v, \pi(k)w \rangle d\mu(k)} = \overline{\langle v, Pw \rangle},
\end{aligned}
$$

where we have used the fact that μ is inversion-invariant and that $\mu(\overline{f}) = \overline{\mu(f)}$ for all $f \in \mathrm{RUC}_b(K)$. $\qquad\square$

We are now ready to show that the condition (8.2) is independent of the choice of mean. Again, the proof is almost entirely analogous to the proof in the finite-dimensional context. We remark that Olshanski showed in [**17**, Theorem 23.6] that conditions (3) and (4) in the following theorem are equivalent using a different method.

THEOREM 8.2. *Suppose that (G_∞, K_∞) is a direct-limit Gelfand pair, and let $\varphi : G \to \mathbb{C}$ be a positive-definite function such that $\varphi(e) = 1$. Then the following are equivalent:*

(1) *φ is spherical for every invariant mean $\mu \in \mathfrak{M}(\mathrm{RUC}_b(K_\infty))$.*
(2) *There exists an invariant mean $\mu \in \mathfrak{M}(\mathrm{RUC}_b(K_\infty))$ with respect to which φ is spherical.*
(3) *There exists an irreducible unitary spherical representation $(\pi, \mathcal{H})$ of G_∞ such that*

$$
\varphi(g) = \langle \pi(g)v, v \rangle,
$$

   where $v \in \mathcal{H}^{K_\infty}$ is a unit vector.
(4) *φ satisfies*

$$
\lim_{n \to \infty} \int_{K_n} \phi(xky)dk = \phi(x)\phi(y).
$$

PROOF. Because φ is a positive-definite function with $\varphi(e) = 1$, we can use the Gelfand-Naimark-Segal construction to construct a unitary representation $(\pi, \mathcal{H})$ of G_∞ and a cyclic unit vector $v \in \mathcal{H}$ such that $\varphi(g) = \langle \pi(g)v, v \rangle$.

That (1) $\Longrightarrow$ (2) is clear. To prove that (2) $\Longrightarrow$ (3), suppose that φ is spherical for an invariant mean μ. Then we claim that φ is right-K_∞-invariant. In fact, we have that

$$
\begin{aligned}
\varphi(xh) &= \int_{K_\infty} \varphi(xhk)d\mu(k) \\
&= \int_{K_\infty} \varphi(xh)d\mu(k) = \varphi(x)
\end{aligned}
$$

for any $h \in K_\infty$, where we use that $\varphi(e) = 1$. The proof that φ is right-K_∞-invariant is identical. It follows that

$$\langle \pi(k)v, \pi(g)v \rangle = \langle \pi(g^{-1}k)v, v \rangle$$
$$= \langle \pi(g^{-1})v, v \rangle = \langle v, \pi(g)v \rangle$$

for all $g \in G$ and $k \in K$. Because v is a cyclic vector in $\mathcal{H}$, we see that $\pi(k)v = v$ for all $k \in K$.

It remains to be shown that $\dim \mathcal{H}^{K_\infty} = 1$ and thus that π is irreducible. But

$$\langle P(\pi(y)v), \pi(x^{-1})v \rangle = \int_K \langle \pi(k)\pi(y)v, \pi(x^{-1})v \rangle d\mu(k)$$
$$= \int_K \phi(xky) d\mu(k)$$
$$= \varphi(x)\varphi(y)$$
$$= \langle \varphi(y)v, \pi(x^{-1})v \rangle$$

for all $x, y \in G$, where $P : \mathcal{H} \to \mathcal{H}^{K_\infty}$ is the projection operator defined in Lemma 8.1. Because v is cyclic, it follows that $P(\pi(y)v) = \varphi(y)v$ for all $y \in G$. Using again the fact that v is cyclic, we see that $\dim(\text{range } P) = 1$. In other words, $\dim \mathcal{H}^{K_\infty} = 1$, and thus $\mathcal{H}$ is irreducible.

Finally we prove (3) $\Longrightarrow$ (1). Suppose that π is an irreducible spherical representation with $v \in \mathcal{H}^{K_\infty}$ and that μ is an invariant mean in $\mathfrak{M}(\text{RUC}_b(K_\infty))$. We need to show that $\langle \varphi(g)v, v \rangle$ is spherical with respect to μ.

As before, we consider the projection $P : \mathcal{H} \to \mathcal{H}^{K_\infty}$ defined in Lemma 8.1. Since $P(\pi(y)v) \in \mathcal{H}^K$ and $\dim \mathcal{H}^K = 1$, it follows that $P(\pi(y)v) = cv$ for some nonzero $c \in \mathbb{C}$. But then

$$c = \langle P(\pi(y)v), v \rangle$$
$$= \int_{K_\infty} \langle \pi(ky)v, v \rangle d\mu(k)$$
$$= \langle \pi(y)v, \pi(k^{-1})v \rangle = \langle \pi(y)v, v \rangle,$$

since v is K_∞-invariant. Hence

$$\int_{K_\infty} \varphi(xky) dk = \int_{K_\infty} \langle \pi(xky)v, v \rangle d\mu(k)$$
$$= \left\langle \int \pi(k)\pi(y)v, \pi(x^{-1})v \, d\mu(k) \right\rangle$$
$$= \langle P(\pi(y)v), \pi(x^{-1})v \rangle$$
$$= \langle \langle \pi(y)v, v \rangle v, \pi(x^{-1})v \rangle$$
$$= \langle \pi(x)v, v \rangle \langle \pi(y)v, v \rangle$$
$$= \varphi(x)\varphi(y).$$

Thus ϕ is spherical for μ.

We have already seen that (4) $\Longrightarrow$ (2) (see the discussion surrounding (8.3)). Finally, we demonstrate that (3) $\Longrightarrow$ (4). Suppose that $\varphi(g) = \langle \pi(g)v, v \rangle$, where π is an irreducible spherical representation with $v \in \mathcal{H}^{K_\infty}$. We know from the preceding paragraph that ϕ is a spherical function with respect to every invariant mean on $\text{RUC}_b(K_\infty)$.

For each $n \in \mathbb{N}$, we consider the orthogonal projection operator $P_n : \mathcal{H} \to \mathcal{H}^{K_n}$, which may be written as

$$P_n(v) = \int_{K_n} \pi(k)v \, dk$$

for all $v \in \mathcal{H}$. Note also that $\mathcal{H}^{K_\infty} = \cap_{n \in \mathbb{N}} \mathcal{H}^{K_n}$. Consider the orthogonal projection $P : \mathcal{H} \to \mathcal{H}^{K_\infty}$. Then $P_n \to P$ in the strong operator topology on $\mathcal{H}$. In other words, we have that

$$P(v) = \lim_{n \to \infty} \int_{K_n} \pi(k)v \, dk$$

for all $v \in \mathcal{H}$. Now let μ be a K_∞-invariant mean on $\mathrm{RUC}_b(K_\infty)$ which is also inversion invariant. Then from Lemma 8.1 we have that the orthogonal projection $P : \mathcal{H} \to \mathcal{H}^{K_\infty}$ satisfies

$$\langle Pv, w \rangle = \int_{K_\infty} \langle \pi(k)v, w \rangle \, d\mu(k).$$

Hence, because ϕ is spherical for μ, we see that

$$
\begin{aligned}
\phi(x)\phi(y) &= \int_{K_\infty} \phi(xky) \, d\mu(k) \\
&= \langle P(\pi(y)v), \pi(x^{-1})v \rangle \\
&= \lim_{n \to \infty} \int_{K_n} \langle \pi(ky)v, \pi(x^{-1})v \rangle \\
&= \lim_{n \to \infty} \int_{K_n} \varphi(xky) \, d\mu(k),
\end{aligned}
$$

and we are done. $\square$

It should be mentioned that for many direct-limit Gelfand pairs, there is a rich collection of spherical functions which have already been classified (see, for instance, [4, 17]). However, some peculiar behaviors arise in this infinite-dimensional context that do not occur in the finite-dimensional theory. Olshanski ([17, Corollary 23.9]) has shown that for the classical direct limits of symmetric spaces (that is, those formed by direct limits of classical matrix groups with embeddings of the form $A \mapsto \begin{pmatrix} A & 0 \\ 0 & 1 \end{pmatrix}$), the product of two spherical functions is again spherical. See also [26, 27] for a different proof in a special case. We recall that the classical direct-limit groups $\mathrm{SO}(\infty)$, $\mathrm{SU}(\infty)$, and their direct products are extremely amenable. In fact, the following corollary of Theorem 8.2 shows how this surprising multiplicative property of spherical functions on G_∞ is related to extreme amenability of K_∞.

COROLLARY 8.3. *If (G_∞, K_∞) is a direct-limit Gelfand pair such that K_∞ is extremely amenable, then the product of two spherical functions is again a spherical function.*

PROOF. Suppose that φ and ψ are spherical functions on G_∞. Because K_∞ is extremely amenable, there is a K_∞-invariant character μ on the C^*-algebra

$\mathrm{RUC}_b(K_\infty)$. Then

$$\int_{K_\infty} (\varphi\psi)(xky)d\mu(k) = \int_{K_\infty} \varphi(xky)d\mu(k) \int_{K_\infty} \psi(xky)d\mu(k)$$
$$= \varphi(x)\varphi(y)\psi(x)\psi(y)$$
$$= (\varphi\psi)(x)(\varphi\psi)(y)$$

Thus $\varphi\psi$ is spherical. $\qquad\square$

We end by remarking that the natural way in which invariant means may be used as a replacement for integration in the context of spherical functions suggests to the authors that there may be other opportunities to apply amenability theory to the study of representations and harmonic analysis on direct-limit groups.

References

[1] B. Bekka, P. de la Harpe, A. Valette, *Kazhdan's property (T)*, New Mathematical Monographs **11**, Cambridge University Press, 2008.

[2] Daniel Beltiţă, *Functional analytic background for a theory of infinite-dimensional reductive Lie groups*, Developments and trends in infinite-dimensional Lie theory, Progr. Math., vol. 288, Birkhäuser Boston, Inc., Boston, MA, 2011, pp. 367–392, DOI 10.1007/978-0-8176-4741-4_11. MR2743769 (2012a:22033)

[3] Alexei Borodin and Grigori Olshanski, *Harmonic analysis on the infinite-dimensional unitary group and determinantal point processes*, Ann. of Math. (2) **161** (2005), no. 3, 1319–1422, DOI 10.4007/annals.2005.161.1319. MR2180403 (2007a:43006)

[4] Matthew Dawson, Gestur Ólafsson, and Joseph A. Wolf, *Direct systems of spherical functions and representations*, J. Lie Theory **23** (2013), no. 3, 711–729. MR3115174

[5] D. H. Fremlin, *Measure theory. Vol. 4*, Torres Fremlin, Colchester, 2006. Topological measure spaces. Part I, II; Corrected second printing of the 2003 original. MR2462372 (2011a:28004)

[6] Mahlon M. Day, *Amenable semigroups*, Illinois J. Math. **1** (1957), 509–544. MR0092128 (19,1067c)

[7] ———, *Fixed point theorems for compact convex sets*, Illinois J. Math. **5** (1961), 585–590. MR0138100 (25 #1547) Correction: Illinois J. Math. **8** (1964), 713.

[8] Gerrit van Dijk, *Introduction to harmonic analysis and generalized Gelfand pairs*, de Gruyter Studies in Mathematics, vol. 36, Walter de Gruyter & Co., Berlin, 2009. MR2640609 (2011e:43012)

[9] Thierry Giordano and Vladimir Pestov, *Some extremely amenable groups related to operator algebras and ergodic theory*, J. Inst. Math. Jussieu **6** (2007), no. 2, 279–315, DOI 10.1017/S1474748006000090. MR2311665 (2008e:22007)

[10] M. Gromov and V. D. Milman, *A topological application of the isoperimetric inequality*, Amer. J. Math. **105** (1983), no. 4, 843–854, DOI 10.2307/2374298. MR708367 (84k:28012)

[11] Hendrik Grundling, *A group algebra for inductive limit groups. Continuity problems of the canonical commutation relations*, Acta Appl. Math. **46** (1997), no. 2, 107–145, DOI 10.1023/A:1017988601883. MR1440014 (98k:22026)

[12] John L. Kelley, *General topology*, D. Van Nostrand Company, Inc., Toronto-New York-London, 1955. MR0070144 (16,1136c)

[13] Sergei Kerov, Grigori Olshanski, and Anatoly Vershik, *Harmonic analysis on the infinite symmetric group*, Invent. Math. **158** (2004), no. 3, 551–642, DOI 10.1007/s00222-004-0381-4. MR2104794 (2006j:43002)

[14] Anthony To-Ming Lau and J. Ludwig, *Fourier-Stieltjes algebra of a topological group*, Adv. Math. **229** (2012), no. 3, 2000–2023, DOI 10.1016/j.aim.2011.12.022. MR2871165 (2012m:43004)

[15] Lynn H. Loomis, *An introduction to abstract harmonic analysis*, D. Van Nostrand Company, Inc., Toronto-New York-London, 1953. MR0054173 (14,883c)

[16] G. Ólafsson, J. A. Wolf, *Separating vector bundle sections by invariant means*, arXiv:1210.5494.

[17] G. I. Ol'shanskiĭ, *Unitary representations of infinite-dimensional pairs* (G, K) *and the formalism of R. Howe*, Representation of Lie groups and related topics, Adv. Stud. Contemp. Math., vol. 7, Gordon and Breach, New York, 1990, pp. 269–463. MR1104279 (92c:22043)

[18] A. T. Paterson, *Amenability*, Amer. Math. Soc. Mathematical Monographs and Surveys, 1988.

[19] Jean-Paul Pier, *Amenable locally compact groups*, Pure and Applied Mathematics (New York), John Wiley & Sons, Inc., New York, 1984. A Wiley-Interscience Publication. MR767264 (86a:43001)

[20] Doug Pickrell, *Measures on infinite-dimensional Grassmann manifolds*, J. Funct. Anal. **70** (1987), no. 2, 323–356, DOI 10.1016/0022-1236(87)90116-9. MR874060 (88d:58017)

[21] Doug Pickrell, *The separable representations of* U(H), Proc. Amer. Math. Soc. **102** (1988), no. 2, 416–420, DOI 10.2307/2045898. MR921009 (89c:22036)

[22] Neil W. Rickert, *Amenable groups and groups with the fixed point property*, Trans. Amer. Math. Soc. **127** (1967), 221–232. MR0222208 (36 #5260)

[23] Walter Rudin, *Functional analysis*, 2nd ed., International Series in Pure and Applied Mathematics, McGraw-Hill, Inc., New York, 1991. MR1157815 (92k:46001)

[24] Yu. S. Samoĭlenko, *Spectral theory of families of selfadjoint operators*, Mathematics and its Applications (Soviet Series), vol. 57, Kluwer Academic Publishers Group, Dordrecht, 1991. Translated from the Russian by E. V. Tisjachnij. MR1135325 (92j:47038)

[25] J. v. Neumann, *Almost periodic functions in a group. I*, Trans. Amer. Math. Soc. **36** (1934), no. 3, 445–492, DOI 10.2307/1989792. MR1501752

[26] Dan Voiculescu, *Sur les représentations factorielles finies de U* (∞) *et autres groupes semblables* (French), C. R. Acad. Sci. Paris Sér. A **279** (1974), 945–946. MR0360924 (50 #13371)

[27] Dan Voiculescu, *Représentations factorielles de type II1 de U*(∞) (French), J. Math. Pures Appl. (9) **55** (1976), no. 1, 1–20. MR0442153 (56 #541)

[28] Joseph A. Wolf, *Principal series representations of infinite dimensional Lie groups, II: construction of induced representations*, Geometric analysis and integral geometry, Contemp. Math., vol. 598, Amer. Math. Soc., Providence, RI, 2013, pp. 257–280, DOI 10.1090/conm/598/11964. MR3156451

DEPARTMENT OF MATHEMATICS, LOUISIANA STATE UNIVERSITY, BATON ROUGE, LIOUSIANA 70803

Current address: Centro de Investigación en Matemáticas, Jalisco s/n, Col. Valenciana, Guajauato, GTO 36240, México

E-mail address: matthew.dawson@cimat.mx

DEPARTMENT OF MATHEMATICS, LOUISIANA STATE UNIVERSITY, BATON ROUGE, LOUISIANA 70803

E-mail address: olafsson@math.lsu.edu

Contemporary Mathematics
Volume **650**, 2015
http://dx.doi.org/10.1090/conm/650/13031

Explicit construction of equivalence bimodules between noncommutative solenoids

Frédéric Latrémolière and Judith A. Packer

ABSTRACT. Let $p \in \mathbb{N}$ be prime, and let θ be irrational. The authors have previously shown that the noncommutative p-solenoid corresponding to the multiplier of the group $\left(\mathbb{Z} \left[\frac{1}{p} \right] \right)^2$ parametrized by $\alpha = (\theta + 1, (\theta + 1)/p, \cdots, (\theta + 1)/p^j, \cdots)$ is strongly Morita equivalent to the noncommutative solenoid on $\left(\mathbb{Z} \left[\frac{1}{p} \right] \right)^2$ coming from the multiplier $\beta = (1 - \frac{\theta+1}{\theta}, 1 - \frac{\theta+1}{p\theta}, \cdots, 1 - \frac{\theta+1}{p^j \theta}, \cdots)$. The method used a construction of Rieffel referred to as the "Heisenberg bimodule" in which the two noncommutative solenoid corresponds to two different twisted group algebras associated to dual lattices in $(\mathbb{Q}_p \times \mathbb{R})^2$. In this paper, we make three additional observations: first, that at each stage, the subalgebra given by the irrational rotation algebra corresponding to $\alpha_{2j} = (\theta + 1)/p^{2j}$ is strongly Morita equivalent to the irrational rotation algebra corresponding to the irrational rotation algebra corresponding to $\beta_{2j} = 1 - \frac{\theta+1}{p^{2j}\theta}$ by a different construction of Rieffel, secondly, that Rieffel's Heisenberg module relating the two non commutative solenoids can be constructed as the closure of a nested sequence of function spaces associated to a multiresolution analysis for a p-adic wavelet, and finally, at each stage, the equivalence bimodule between $A_{\alpha_{2j}}$ and $A_{\beta_{2j}}$ can be identified with the subequivalence bimodules arising from the p-adic MRA. Aside from its instrinsic interest, we believe this construction will guide us in our efforts to show that certain necessary conditions for two noncommutative solenoids to be strongly Morita equivalent are also sufficient.

1. Introduction

In this paper, we continue our study [**LP1**, **LP2**] of the twisted group C^*-algebras for the groups $\left(\mathbb{Z} \left[\frac{1}{p} \right] \right)^2$, where for any prime number p, the group $\mathbb{Z} \left[\frac{1}{p} \right]$ is the additive subgroup of $\mathbb{Q}$ consisting of fractions with denominators given as powers of p:

$$\mathbb{Z} \left[\frac{1}{p} \right] = \left\{ \frac{q}{p^k} : q \in \mathbb{Z}, k \in \mathbb{N} \right\}.$$

The dual of $\mathbb{Z} \left[\frac{1}{p} \right]$ is the p-solenoid Ξ_p, thereby motivating our terminology in calling these C^*-algebras *noncommutative solenoids*. The p-solenoid group Ξ_p is given, up

2010 *Mathematics Subject Classification.* Primary 46L40, 46L80; Secondary 46L08, 19K14.
Key words and phrases. C*-algebras, solenoids, projective modules, p-adic analysis.

to isomorphism, as:

$$\Xi_p = \left\{ (\alpha_j)_{j \in \mathbb{N}} \in \prod_{j \in \mathbb{N}} [0,1) : \forall j \in \mathbb{N} \, \exists m \in \{0, \dots, p-1\} \quad \alpha_{j+1} = p\alpha_j + m \right\},$$

with the group operation given as term-wise addition modulo 1.

Let us fix a prime number p, and let us denote $\left(\mathbb{Z} \left[\frac{1}{p} \right] \right)^2$ by Γ. Let σ be a multiplier on Γ. From our previous work [**LP1**], we know that there exists $\alpha \in \Xi_p$ such that σ is cohomologous to $\Psi_\alpha : \Gamma \times \Gamma \to \mathbb{T}$ defined by:

$$\Psi_\alpha \left(\left(\frac{j_1}{p^{k_1}}, \frac{j_2}{p^{k_2}} \right), \left(\frac{j_3}{p^{k_3}}, \frac{j_4}{p^{k_4}} \right) \right) = e^{2\pi i (\alpha_{(k_1 + k_4)} j_1 j_4)}$$

for all $\left(\frac{j_1}{p^{k_1}}, \frac{j_2}{p^{k_2}} \right), \left(\frac{j_3}{p^{k_3}}, \frac{j_4}{p^{k_4}} \right) \in \Gamma$.

The detailed study of noncommutative solenoids $C^*(\Gamma, \sigma)$ started in an earlier paper [**LP1**], where the authors classified such C^*-algebras up to $*$-isomorphisms, and computed the range of the trace on the K_0-groups of these C^*-algebras. As is not uncommon in these situations, even if there is no unique tracial state on these C^*-algebras, the range of the trace on K_0 is unique, and therefore, the ordering of the range of the K-group is an invariant for strong Morita equivalence.

We conjecture that this necessary condition for strong Morita equivalence is also sufficient; that is, we conjecture that two such noncommutative solenoids are strongly Morita equivalent if and only if the range of the traces of their K_0-groups are order-isomorphic subgroups of $\mathbb{R}$. The proof of the sufficiency is not yet complete; but the main aim of this paper is to demonstrate a method of constructing an equivalence bimodule between two particular such modules using methods due to M. Rieffel [**Rie3**]. Our hope is that the detailed working of this particular example will give us greater insight into the general case. We are also able to note how this particular construction has a relationship to the theory of wavelets and frames for dilations on Hilbert spaces associated to p-adic fields.

The main aim of this paper is to break down the strong Morita equivalence between pairs of noncommutative solenoids $C^*(\Gamma, \Psi_\alpha)$ and $C^*(\Gamma, \Psi_\beta)$, where

$$\alpha = \left(\theta + 1, \frac{\theta + 1}{p}, \cdots, \frac{\theta + 1}{p^j}, \cdots \right)$$

and

$$\beta = \left(1 - \frac{\theta + 1}{\theta}, 1 - \frac{\theta + 1}{p\theta}, \cdots, 1 - \frac{\theta + 1}{p^j \theta}, \cdots \right),$$

for some irrational number $\theta \in (0,1)$, as these pairs of noncommutative solenoids are natural to study in sight of our previous work [**LP2**].

In [**LP2**], we noted that there were two ways to construct projective modules for a noncommutative solenoid $C^*(\Gamma, \Psi_\alpha)$. A first approach exploits the fact that noncommutative solenoids are inductive limits of rotation C*-algebras A_μ, i.e. universal C*-algebra generated by two unitaries U_μ, V_μ with $U_\mu V_\mu = e^{2i\pi\mu} V_\mu U_\mu$. Then, we take a projection P in the initial irrational rotation algebra A_{α_0}, and then consider the projective module $C^*(\Gamma, \Psi_\alpha)P$, noting that $C^*(\Gamma, \Psi_\alpha)$ could be shown to be strongly Morita equivalent to a direct limit of the C^*-algebras $PA_{\alpha_{2j}}P$ in this case. However, there was no certainty as to the structure C^*-algebra $PC^*(\Gamma, \Psi\alpha)P$ arising as a direct limit of the C^*-algebras $PA_{\alpha_{2j}}P$ in this case.

The other method, due in general to Rieffel [**Rie3**], was to consider the self-dual group $M = \mathbb{Q}_p \times \mathbb{R}$, where $\mathbb{Q}_p$ is the field of p-adic numbers (the completion of $\mathbb{Z}\left[\frac{1}{p}\right]$ for the p-adic norm). Let:

$$\eta : ((m_1, m_2), (m_3, m_4)) \in (M \times M) \times (M \times M) \mapsto \langle m_1, m_4 \rangle$$

where $\langle \cdot, \cdot \rangle$ is the dual pairing between M and its dual identified with M. One then embeds the underlying group Γ as a closed subgroup D of $M \times M$ and then constructs a Heisenberg bimodule by making $C_C(M)$ into a $C^*(D, \eta)$ module, where $C^*(D, \eta) \cong C^*(\Gamma, \Psi_\alpha)$ and $C_c(X)$ is the space of compactly supported functions over a locally compact space X. In this situation, we identified the completed module $\overline{C_C(M)}$ as giving an equivalence bimodule between $C^*(D, \eta)$ and another twisted group C^*-algebra $C^*(D^\perp, \overline{\eta})$, which itself turned out to be a noncommutative solenoid [**LP2**].

In this paper, we will show that the first of the two bimodules describe above, $C^*(\Gamma, \Psi_\alpha)P$, is exactly the same as the bimodule $\overline{C_C(M)}$, at least in the situation where the projection P is chosen appropriately and

$$\alpha = \left(\theta + 1, \frac{\theta + 1}{p}, \cdots, \frac{\theta + 1}{p^j}, \cdots \right).$$

Therefore it is possible to identify the direct limit C^*-algebra $PC^*(\Gamma, \Psi_\alpha)P$ as $C^*(D^\perp, \overline{\eta})$, which is isomorphic to $C^*(\Gamma, \Psi_\beta)$ for:

$$\beta = \left(1 - \frac{\theta + 1}{\theta}, 1 - \frac{\theta + 1}{p\theta}, \cdots, 1 - \frac{\theta + 1}{p^j\theta}, \cdots \right).$$

A key tool in showing that the bimodules are the same are what we call *Haar multiresolution structures* for $L^2(\mathbb{Q}_p)$, a concept adapted from the Haar multiresolution analyses of V. Shelkovich and M. Skopina used to construct p-adic wavelets in $L^2(\mathbb{Q}_p)$ ([**ShSk**], [**AES**]). This new multiresolution structures give a clearer path to embedding the directed system of bimodules $A_{\alpha_{2j}} \cdot P$ into $\overline{C_C(\mathbb{Q}_p \times \mathbb{R})}$.

The construction given in this special example also leads us to the definition of the new notion of *projective multiresolution structures*, which are related to B. Purkis's *projective multiresolution analyses* for irrational rotation algebras ([**Pur**]), but are not the same. In our notion of projective multiresolution structures, unlike in the case or projective multiresolution analyses, the underlying C^*-algebras are changing along with the projective modules being studied. Therefore, projective multiresolution structures seem well suited to studying projective modules associated to C^*-algebras constructed via a limiting process, as is the case with our noncommutative solenoids.

2. Review of the construction of noncommutative solenoids

Let p be prime, and let:

$$\mathbb{Z}\left[\frac{1}{p}\right] = \bigcup_{j=0}^{\infty} (p^{-j}\mathbb{Z}) \subset \mathbb{Q}.$$

We recall [**LP1**] that the *noncommutative solenoids* are the twisted group C^*-algebras $C^*\left(\left(\mathbb{Z}\left[\frac{1}{p}\right]\right)^2,\sigma\right)$ for $\sigma \in Z^2\left(\left(\mathbb{Z}\left[\frac{1}{p}\right]\right)^2,\mathbb{T}\right)$, a multiplier on $\Gamma = \left(\mathbb{Z}\left[\frac{1}{p}\right]\right)^2$ with values in the circle group $\mathbb{T}$.

NOTATION 2.1. In this paper, we will fix a prime number p and denote $\left(\mathbb{Z}\left[\frac{1}{p}\right]\right)$ as Γ to ease notations.

The non-trivial multipliers on $\Gamma = \left(\mathbb{Z}\left[\frac{1}{p}\right]\right)^2$ were calculated in [**LP1**]:

THEOREM 2.2. [**LP1**] *Let* $\Gamma = \left(\mathbb{Z}\left[\frac{1}{p}\right]\right)^2$. *If* σ *is a multiplier on* Γ, *then there exists:*

$$\alpha \in \Xi_p = \left\{(\alpha_j)_{j\in\mathbb{N}} \in \prod_{j\in\mathbb{N}}[0,1) : \forall j \in \mathbb{N} \exists m_j \in \{0,\ldots,p-1\} \quad \alpha_{j+1} = p\alpha_j + m_j\right\}$$

such that σ *is cohomologous to the multiplier* $\Psi_\alpha : \Gamma \times \Gamma \to \mathbb{T}$ *defined for all by:*

$$\Psi_\alpha\left(\left(\frac{j_1}{p^{k_1}},\frac{j_2}{p^{k_2}}\right),\left(\frac{j_3}{p^{k_3}},\frac{j_4}{p^{k_4}}\right)\right) = e^{2\pi i(\alpha_{(k_1+k_4)}j_1j_4)}$$

for all $\left(\frac{j_1}{p^{k_1}},\frac{j_2}{p^{k_2}}\right),\left(\frac{j_3}{p^{k_3}},\frac{j_4}{p^{k_4}}\right) \in \Gamma$.

We showed in [**LP1**] that for α, $\beta \in \Xi_p$, the cohomology classes of Ψ_α and Ψ_β are equal in $H^2(\Gamma,\mathbb{T})$ if and only if $\alpha_j = \beta_j$ for all $j \in \mathbb{N}$. As a topological group, $H^2(\Gamma,\mathbb{T}) = \Xi_p$ can be identified with the p-solenoid:

$$\mathcal{S}_p = \left\{(z_n)_{n\in\mathbb{N}} \in \prod_{n\in\mathbb{N}}\mathbb{T} : z_{n+1}^p = z_n\right\},$$

yet our additive version Ξ_p makes it easier to do modular arithmetic calculations concerning the range of the trace on projections that are of use in K-theory.

Let $\Gamma = \left(\mathbb{Z}\left[\frac{1}{p}\right]\right)^2$, and let $\alpha \in \Xi_p$. Recall that the twisted group C^*-algebra $C^*(\Gamma,\Psi_\alpha)$ is the C^*-completion of the involutive Banach algebra $\ell^1(\Gamma,\Psi_\alpha)$, where the convolution of two functions $f_1, f_2 \in \ell^1(\Gamma)$ is given by setting for all $\gamma \in \Gamma$:

$$f_1 * f_2(\gamma) = \sum_{\gamma_1\in\Gamma} f_1(\gamma_1)f_2(\gamma-\gamma_1)\Psi_\alpha(\gamma_1,\gamma-\gamma_1),$$

while the involution is given for all $f \in \ell^1(\Gamma)$ and $\gamma \in \Gamma$ by

$$f^*(\gamma) = \overline{\Psi_\alpha(\gamma,-\gamma)f(-\gamma)}.$$

These C^*-algebras, originally viewed as twisted group algebras for countable discrete torsion-free abelian groups, also have a representation as transformation group C^*-algebras, and in [**LP1**], necessary and sufficient conditions for any two such algebras to be simple were given, as well as a characterization of their $*$-isomorphism classes in terms of elements in Ξ_p.

The group $\mathcal{S}_p \times \mathcal{S}_p$ or, equivalently, $\Xi_p \times \Xi_p$, as the dual group of Γ, has a natural dual action on $C^*(\Gamma,\Psi_\alpha)$. So by work of Hoegh-Krohn, Landstad, and Störmer [**HKLS**], there is always an invariant trace on $C^*(\Gamma,\Psi_\alpha)$ that is unique in the simple case. For α_0 irrational, our noncommutative solenoids are always simple, although we recall from [**LP1**] that there are *aperiodic rational* simple noncommutative solenoids.

Since it will be important in what follows, we review the construction of noncommutative solenoids as direct limit algebras of rotation algebras that was described in detail in [**LP1**].

Recall from [**EH**] that, for $\theta \in [0,1)$, the rotation algebra A_θ is the universal C^*-algebra generated by unitaries U, V satisfying

$$UV = e^{2\pi i \theta} VU.$$

A_θ is simple if and only if θ is irrational. For $\theta \neq 0$ these C^*-algebras are called *noncommutative tori*.

The noncommutative solenoids $C^*(\Gamma, \Psi_\alpha)$ are direct limits of rotation algebras:

THEOREM 2.3. ([**LP1**]) *Let p be prime and $\alpha \in \Xi_p$. Let A_θ denote the rotation C^*-algebra for the rotation of angle $2\pi i \theta$. For all $n \in \mathbb{N}$, let φ_n be the unique *-morphism from $A_{\alpha_{2n}}$ into $A_{\alpha_{2n+2}}$ given by:*

$$\begin{cases} U_{\alpha_{2n}} & \longmapsto & U^p_{\alpha_{2n+2}} \\ V_{\alpha_{2n}} & \longmapsto & V^p_{\alpha_{2n+2}} \end{cases}$$

Then:

$$A_{\alpha_0} \xrightarrow{\varphi_0} A_{\alpha_2} \xrightarrow{\varphi_1} A_{\alpha_4} \xrightarrow{\varphi_2} \cdots$$

converges to $C^*(\Gamma, \Psi_\alpha)$, where $\Gamma = \left(\mathbb{Z}\left[\frac{1}{p}\right] \right)^2$ and Ψ_α is as defined as in Theorem (2.2).

Since the C^*-algebras A_θ are viewed as noncommutative tori, and we have written each $C^*(\Gamma, \Psi_\alpha)$ as a direct limit algebra of noncommutative tori, we feel justified in calling the C^*-algebras $C^*(\Gamma, \Psi_\alpha)$ *noncommutative solenoids*. With this in mind, we change the notation for our C^*-algebras:

NOTATION 2.4. Let p be a prime number. Let $\Gamma = \left(\mathbb{Z}\left[\frac{1}{p}\right] \right)^2$, and for a fixed $\alpha \in \Xi_p$, let Ψ_α be the multiplier on Γ defined in Theorem (2.2).

Henceforth we denote the twisted group C^*-algebra $C^*(\Gamma, \Psi_\alpha)$ by $\mathcal{A}^{\mathcal{S}}_\alpha$ and call the C^*-algebra $\mathcal{A}^{\mathcal{S}}_\alpha$ a *noncommutative solenoid*.

3. Directed systems of equivalence bimodules: a method to form equivalence bimodules between direct limits of C^*-algebras

In this section, we improve a result from [**LP2**]. We remark that B. Abadie and M. Achigar also considered directed sequences X_n of Hilbert A_n bimodules for a directed sequence of C^*-algebras $\{A_n\}$ in Section 2 of [**AA**], but their approach is somewhat different, in part because their aim (constructing C^*-correspondences for a direct limit C^*-algebra) is different.

We first define an appropriate notion of directed system of equivalence bimodules.

DEFINITION 3.1. Let

$$A_0 \xrightarrow{\varphi_0} A_1 \xrightarrow{\varphi_1} A_2 \xrightarrow{\varphi_2} \cdots$$

and

$$B_0 \xrightarrow{\psi_0} B_1 \xrightarrow{\psi_1} B_2 \xrightarrow{\psi_2} \cdots$$

be two directed systems of unital C*-algebras, whose *-morphisms are all unital maps. A sequence $(X_n, i_n)_{n \in \mathbb{N}}$ is a *directed system of equivalence bimodules adapted to the sequences* $(A_n)_{n \in \mathbb{N}}$ *and* $(B_n)_{n \in \mathbb{N}}$ when X_n is an A_n-B_n equivalence bimodule whose A_n and B_n-valued inner products are denoted respectively by $\langle \cdot, \cdot \rangle_{A_n}$ and $\langle \cdot, \cdot \rangle_{B_n}$, for all $n \in \mathbb{N}$, and such that the sequence

$$X_0 \xrightarrow{i_0} X_1 \xrightarrow{i_1} X_2 \xrightarrow{i_2} \cdots$$

is a directed sequence of modules satisfying

$$\langle i_n(f), i_n(g) \rangle_{B_{n+1}} = \psi_n(\langle f, g \rangle_{B_n}), \ f, \ g \in X_n,$$

and

$$i_n(f \cdot b) = i_n(f) \cdot \psi_n(b), \ f \in X_n, \ b \in B_n,$$

with analogous but symmetric equalities holding for the X_n viewed as left-A_n Hilbert modules.

The purpose of Definition (3.1) is to provide all the needed structure to construct equivalence bimodules on the inductive limits of two directed systems of Morita equivalent C*-algebras. To this end, we first define a natural structure of Hilbert C*-module on the inductive limit of a directed system of equivalence bimodules. We will use the following notations:

NOTATION 3.2. The norm of any normed vector space E is denoted by $\| \cdot \|_E$ unless otherwise specified.

NOTATION 3.3. The inductive limit of a given directed sequence:

$$A_0 \xrightarrow{\varphi_0} A_1 \xrightarrow{\varphi_1} A_2 \xrightarrow{\varphi_2} \cdots$$

of C*-algebras is denoted by $\mathcal{A} = \lim_{n \to \infty} (A_n, \varphi_n)$, and is constructed as follows. We first define the algebra of predictable tails:

$$A_\infty = \left\{ (a_j)_{j \in \mathbb{N}} \in \prod_{j \in \mathbb{N}} A_j : \exists N \in \mathbb{N} \ \forall n \geq N \quad \varphi_n \circ \varphi_{n-1} \circ \cdots \circ \varphi_N(a_N) = a_n \right\}.$$

We then define the C*-seminorm:

(3.1) $$\|(a_j)_{j \in \mathbb{N}}\|_\mathcal{A} = \limsup_{n \to \infty} \|a_j\|_{A_j},$$

for all $(a_j)_{j \in \mathbb{N}} \in A_\infty$. The quotient of A_∞ by the ideal $\{(x_n)_{n \in \mathbb{N}} : \|(x_n)_{n \in \mathbb{N}}\|_\mathcal{A} = 0\}$ is denoted by $\mathcal{A}_{\text{pre}}$. Of course, $\| \cdot \|_\mathcal{A}$ induces a C*-norm on $\mathcal{A}_{\text{pre}}$, which we denote again by $\| \cdot \|_\mathcal{A}$. The completion of $\mathcal{A}_{\text{pre}}$ for this norm is the inductive limit C*-algebra $\mathcal{A} = \lim_{n \to \infty} (A_n, \varphi_n)$.

For any $p \leq q \in \mathbb{N}$, we denote $\varphi_q \circ \varphi_{q-1} \circ \cdots \circ \varphi_p$ by $\varphi_{p,q}$, and we note that for any $j \in \mathbb{N}$, there is a canonical *-morphism $\varphi_{\infty,j} : A_j \to \lim_{j \to \infty} A_j$ mapping $a \in A_j$ to the class of $(0, \ldots, 0, a_j, \varphi_j(a_j), \varphi_{j+1,j}(a_j), \ldots)$, where a_j appears at index j. Last, the canonical surjection from A_∞ onto $\mathcal{A}_{\text{pre}}$ is denoted by $\pi_\mathcal{A}$.

THEOREM 3.4. *Let* $(X_n, i_n)_{n \in \mathbb{N}}$ *be a directed system of equivalence bimodules adapted to two directed sequences* $(A_n, \varphi_n)_{n \in \mathbb{N}}$ *and* $(B_n, \psi_n)_{n \in \mathbb{N}}$ *of unital C*-algebras.*

Let $\mathcal{A}$ *and* $\mathcal{B}$ *be the respective inductive limit of* $(A_n, \varphi_n)_{n \in \mathbb{N}}$ *and* $(B_n, \psi_n)_{n \in \mathbb{N}}$. *For any* $n, m \in \mathbb{N}$ *with* $n \leq m$, *we denote* $i_m \circ i_{m-1} \circ \cdots \circ i_n$ *by* $i_{n,m}$ *and the canonical *-morphism from* A_n *to* $\mathcal{A}$ *by* $\varphi_{n,\infty}$.

Let $\mathcal{A}_{\mathrm{pre}} = \bigcup_{n \in \mathbb{N}} \varphi_{n,\infty}(A_n)$ be the dense pre-C^* subalgebra in $\mathcal{A}$ generated by the images of A_n by $\varphi_{n,\infty}$ for all $n \in \mathbb{N}$.

For any two $(x_n)_{n \in \mathbb{N}}, (y_n)_{n \in \mathbb{N}}$ in $\prod_{n \in \mathbb{N}} X_n$, we set:

$$(x_n)_{n \in \mathbb{N}} \cong (y_n)_{n \in \mathbb{N}} \iff \lim_{n \to \infty} \|x_n - y_n\|_{X_n} = 0.$$

Let:

$$X_\infty = \{(x_n)_{n \in \mathbb{N}} : \exists N \in \mathbb{N} \, \forall n \geq N \quad i_{N,n}(x_N) = x_n\},$$

and let:

$$\mathcal{X}_{\mathrm{pre}} = X_\infty / \!\cong .$$

We denote the canonical surjection from X_∞ onto $\mathcal{X}_{\mathrm{pre}}$ by π.

For all $x = \pi\left((x_n)_{n \in \mathbb{N}}\right), y = \pi\left((y_n)_{n \in \mathbb{N}}\right) \in \mathcal{X}_{\mathrm{pre}}$ we set:

$$\langle x, y \rangle_{\mathcal{A}} = \pi_{\mathcal{A}}\left(\left(\langle x_n, y_n \rangle\right)_{n \in \mathbb{N}}\right).$$

Then $\mathcal{X}_{\mathrm{pre}}$ is a $\mathcal{A}_{\mathrm{pre}}$-$\mathcal{B}_{\mathrm{pre}}$ bimodule and $\langle \cdot, \cdot \rangle_{\mathcal{A}}$ is a $\mathcal{A}$-valued preinner product on $\mathcal{X}_{\mathrm{pre}}$.

The completion of $\mathcal{X}_{\mathrm{pre}}$ for the norm associated with the inner product $\langle \cdot, \cdot \rangle_{\mathcal{A}}$ is the directed limit $\lim_{n \to \infty} X_n$, which is an equivalence bimodule between $\mathcal{A}$ and $\mathcal{B}$, and is canonically isomorphic, as a Hilbert bimodule, to the completion of $\mathcal{X}_{\mathrm{pre}}$ for $\langle \cdot, \cdot \rangle_{\mathcal{B}}$.

PROOF. Let $x \in \mathcal{X}_{\mathrm{pre}}$ and let $(x_n)_{n \in \mathbb{N}}, (y_n)_{n \in \mathbb{N}} \in X_\infty$ such that $\pi\left((x_n)_{n \in \mathbb{N}}\right) = \pi\left((y_n)_{n \in \mathbb{N}}\right) = x$. Let $b \in \mathcal{A}_{\mathrm{pre}}$ and let $(a_n)_{n \in \mathbb{N}}, (b_n)_{n \in \mathbb{N}} \in A_\infty$ such that:

$$\pi_{\mathcal{A}}\left((a_n)_{n \in \mathbb{N}}\right) = \pi_{\mathcal{A}}\left((b_n)_{n \in \mathbb{N}}\right) = b,$$

where we use Notation (3.3): in particular, $\pi_{\mathcal{A}}$ is the canonical surjection from A_∞ onto $\mathcal{A}_{\mathrm{pre}}$.

We begin with the observation that, by definition of X_∞ and A_∞, there exists $N \in \mathbb{N}$ such that, for all $n \geq N$, we have at once $i_{N,n}(x_N) = x_n$ and $\varphi_{N,n}(a_N) = a_n$. Now, by Definition (3.1), we have for all $n \geq N$:

$$i_{N,n}(a_N x_N) = \varphi_{N,n}(a_N) i_{N,n}(x_N) = a_n x_n.$$

Thus $(a_n x_n)_{n \in \mathbb{N}} \in X_\infty$. The same of course holds for $(b_n y_n)_{n \in \mathbb{N}}$.

Moreover:

$$\|a_n x_n - b_n y_n\|_{X_n} \leq \|a_n\|_{A_n} \|x_n - y_n\|_{X_n} + \|a_n - b_n\|_{A_n} \|y_n\|_{X_n}$$

for all $n \in \mathbb{N}$, from which it follows immediately that:

$$\lim_{n \to \infty} \|a_n x_n - b_n y_n\|_{X_n} = 0.$$

Hence, $\pi((a_n x_n)_{n \in \mathbb{N}}) = \pi((b_n y_n)_{n \in \mathbb{N}}) \in \mathcal{X}_{\mathrm{pre}}$. We thus define without ambiguity:

$$b \cdot x = \pi\left((a_n x_n)_{n \in \mathbb{N}}\right) \in \mathcal{X}_{\mathrm{pre}}.$$

It is now routine to check that $\mathcal{X}_{\mathrm{pre}}$ thus becomes a $\mathcal{A}_{\mathrm{pre}}$-left module.

Now, let x and y in $\mathcal{X}_{\mathrm{pre}}$, and choose $(x_n)_{n \in \mathbb{N}}, (y_n)_{n \in \mathbb{N}} \in X_\infty$ such that $\pi((x_n)_{n \in \mathbb{N}}) = x$ and $\pi((y_n)_{n \in \mathbb{N}}) = y$. By definition of X_∞, there exists $N \in \mathbb{N}$ such that for all $n \geq N$, we have both $x_n = i_{N,n}(x_N)$ and $y_n = i_{N,n}(y_N)$. Therefore, by Definition (3.1), we have:

$$\varphi_{N,n}\left(\langle x_N, y_N \rangle_{A_N}\right) = \langle i_{N,n}(x_N), i_{N,n}(y_N) \rangle_{A_n} = \langle x_n, y_n \rangle_{A_n}$$

and thus $(\langle x_n, y_n \rangle)_{n \in \mathbb{N}} \in A_\infty$. Moreover, if $(x'_n)_{n \in \mathbb{N}}, (y'_n)_{n \in \mathbb{N}} \, in \, X_\infty$ are chosen so that $\pi((x'_n)_{n \in \mathbb{N}} = x$ and $\pi((y'_n)_{n \in \mathbb{N}}) = y$, then:

$$
\begin{aligned}
\| \langle x_n, y_n \rangle_{A_n} - \langle x'_n, y'_n \rangle_{A_n} \|_{A_n} &\leq \| \langle x_n, y_n - y'_n \rangle_{A_n} \|_{A_n} + \| \langle x_n - x'_n, y'_n \rangle_{A_n} \|_{A_n} \\
&\leq \|x_n\|_{X_n} \|y_n - y'_n\|_{X_n} + \|y'_n\|_{X_n} \|x_n - x'_n\|_{X_n} \\
&\xrightarrow{n \to \infty} 0,
\end{aligned}
$$

and thus, once again, we may define without ambiguity:

$$
\langle x, y \rangle_{\mathcal{A}} = \pi \left((\langle x_n, y_n \rangle)_{n \in \mathbb{N}} \right).
$$

It is a routine matter to check that $\langle \cdot, \cdot \rangle_{\mathcal{A}}$ is a pre-inner product on $\mathcal{X}_{\mathrm{pre}}$, as defined in [**Rie4**].

A similar construction endows $\mathcal{X}_{\mathrm{pre}}$ with a $\mathcal{B}$-right module structure and with an associated pre-inner product $\langle \cdot, \cdot \rangle_{\mathcal{B}}$.

Now, let $(x_n)_{n \in \mathbb{N}}, (y_n)_{n \in \mathbb{N}}, (z_n)_{n \in \mathbb{N}} \in X_\infty$. Since, for each $n \in \mathbb{N}$, the bimdodule X_n is an equivalence bimodule between A_n and B_n, we get:

(3.2)

$$
\begin{aligned}
\langle \pi((x_n)_{n \in \mathbb{N}}), \pi((y_n)_{n \in \mathbb{N}}) \rangle_{\mathcal{A}} \, \pi((z_n)_{n \in \mathbb{N}}) &= \pi \left(\left(\langle x_n, y_n \rangle_{A_n} z_n \right)_{n \in \mathbb{N}} \right) \\
&= \pi \left(\left(x_n \langle y_n, z_n \rangle_{B_n} \right)_{n \in \mathbb{N}} \right) \\
&= \pi((x_n)_{n \in \mathbb{N}}) \, \pi_{\mathcal{B}} \left(\langle (y_n), (z_n) \rangle_{\mathcal{B}} \right) \\
&= \pi((x_n)_{n \in \mathbb{N}}) \langle \pi((y_n)_{n \in \mathbb{N}}), \pi((z_n)_{n \in \mathbb{N}}) \rangle_{\mathcal{B}},
\end{aligned}
$$

where, once again, $\pi_{\mathcal{B}}$ is the canonical surjection $B_\infty \twoheadrightarrow \mathcal{B}_{\mathrm{pre}}$.

It then follows easily that the completion $\mathcal{X}$ of $\mathcal{X}_{\mathrm{pre}}$ for the norm associated with $\langle \cdot, \cdot \rangle_{\mathcal{A}}$ is a $\mathcal{A}$ left Hilbert module, and from Equation (3.2), that this completion equals the completion for $\langle \cdot, \cdot \rangle_{\mathcal{B}}$ and is in fact an $\mathcal{A}$-$\mathcal{B}$ bimodule. Keeping the notation for the inner products induced on $\mathcal{X}$ by the two preinner products defined above, we have:

$$
\langle x, y \rangle_{\mathcal{A}} z = x \langle y, z \rangle_{\mathcal{B}}, \; \forall \, x, y, z \in \mathcal{X}.
$$

We also note that the range of $\langle \cdot, \cdot \rangle_{\mathcal{X}}$:

$$
\text{closure of the linear span of } \{ \langle x, y \rangle_{\mathcal{A}} : x, y \in \mathcal{X} \}
$$

is the closure of the linear span of $\{ \langle x, y \rangle_{\mathcal{A}} : y \in \mathcal{X}_{\mathrm{pre}} \}$ by construction. Yet, the latter is dense in $\mathcal{A}$.

Indeed, let $b \in \mathcal{A}_{\mathrm{pre}}$. There exists $(b_n)_{n \in \mathbb{N}} \in A_\infty$ such that $\pi_{\mathcal{A}}((b_n)_{n \in \mathbb{N}}) = b$. Thus there exists $N \in \mathbb{N}$ such that for all $n \geq N$ we have $\varphi_{N,n}(b_N) = b_n$. Without loss of generality, we may assume $b_n = 0$ for $n < N$.

Let $\varepsilon > 0$. Now, since X_N is a full $\mathcal{A}$-left Hilbert module, there exists $x_N^1, y_N^1, \ldots, x_N^m, y_N^m \in X_N$ and $\lambda_1, \ldots, \lambda_m \in \mathbb{C}$ for some $m \in \mathbb{N}$ such that:

$$
\left\| b_N - \sum_{j=1}^m \lambda_j \left\langle x_N^j, y_N^j \right\rangle_{A_N} \right\| \leq \varepsilon.
$$

Fix $j \in \{1, \ldots, m\}$. We set $x_n^j = 0$ for all $n < N$, and we then define $x_n^j = i_{N,n}(x_N^j)$. Thus by construction, $(x_n^j)_{n \in \mathbb{N}} \in X_\infty$. We construct $y^j = (Y_n^j)_{n \in \mathbb{N}} \in X_\infty$ similarly.

Then, for all $n \geq N$, we have, using Definition (3.1),:

$$\left\| b_n - \sum_{j=1}^{m} \lambda_j \left\langle x_n^j, y_n^j \right\rangle_{A_N} \right\| = \left\| \varphi_{N,n}(b_N) - \sum_{j=1}^{m} \lambda_j \left\langle i_{N,n}(x_N^j), i_{N,n}(y_N^j) \right\rangle_{A_N} \right\|$$

$$= \left\| \varphi_{N,n} \left(b_N - \sum_{j=1}^{m} \lambda_j \left\langle x_N^j, y_N^j \right\rangle_{A_N} \right) \right\|_{\mathcal{A}}$$

$$\leq \varepsilon.$$

Thus $(\mathcal{X}, \langle \cdot, \cdot \rangle_{\mathcal{A}})$ is a full left Hilbert $\mathcal{A}$-module. The same reasoning applies to the right Hilbert $\mathcal{B}$-module structure on $\mathcal{X}$.

Following the same approach as used in Equation (3.2), we can thus conclude that all the properties in [**Rie4**, Definition 6.10] are met by the bimodule $\mathcal{X}$ over $\mathcal{A}$ and $\mathcal{B}$, with the inner products $\langle \cdot, \cdot \rangle_{\mathcal{A}}$ and $\langle \cdot, \cdot \rangle_{\mathcal{B}}$. □

4. The explicit construction of equivalence bimodules at each stage

Fix an irrational number θ between 0 and 1, and let

$$\alpha = \left(\theta + 1, \frac{\theta + 1}{p}, \cdots, \frac{\theta + 1}{p^j} = \alpha_j, \cdots \right).$$

We recall that the noncommutative solenoid $C^*(\Gamma, \Psi_\alpha)$ can be viewed as the direct limit of the irrational rotation algebras $\left(A_{\alpha_{2j}} \right)_{j \in \mathbb{N}}$. We know from results in [**LP2**] that $C^*(\Gamma, \Psi_\alpha)$ is strongly Morita equivalent to $C^*(\Gamma, \Psi_\beta)$ where

$$\beta = \left(1 - \frac{\theta + 1}{\theta}, 1 - \frac{\theta + 1}{p\theta}, \cdots, 1 - \frac{\theta + 1}{p^j \theta} = \beta_j, \cdots \right).$$

Since $C^*(\Gamma, \Psi_\beta)$ can be expressed as a direct limit of the C^*-algebras $\left(A_{\beta_{2j}} \right)_{j \in \mathbb{N}}$, we want to analyze the Morita equivalence at each stage more carefully.

The following lemma includes formulas that will prove very useful formulas to us.

LEMMA 4.1. *For all $j \in \mathbb{N}$, let $\alpha_j = \frac{\theta+1}{p^j}$ and $\beta_j = 1 - \frac{\theta+1}{p^j \theta}$. Then the irrational rotation algebra $A_{\alpha_{2j}}$ is strongly Morita equivalent to $A_{\beta_{2j}}$.*

PROOF. We first note that

$$\forall k \in \mathbb{Z} \quad e^{2\pi i k \left(1 - \frac{\theta+1}{p^j \theta} \right)} = e^{-2\pi i k \left(\frac{\theta+1}{p^j \theta} \right)}$$

so that without loss of generality we can assume that $\beta_j = -\left(\frac{\theta+1}{p^j \theta} \right)$ for every $j \in \mathbb{N}$. Recall from the work of M. Rieffel in [**Rie1**], explicated further in [**Rie2**], that A_α is strongly Morita equivalent to A_β if and only if there exists a matrix $\begin{pmatrix} a & b \\ c & d \end{pmatrix} \in GL(2, \mathbb{Z})$ such that

$$\alpha = \frac{a\beta + b}{c\beta + d} \text{ modulo } 1.$$

In this case, if we take $a = d = 1$, $b = 0$, and $c = p^{2^j}$, we have:

$$\frac{\beta_{2j} + 0}{p^{2j}\beta_{2j} + 1} = \frac{-\beta_{2j}}{-p^{2j}\beta_{2j} - 1} = (-1) \cdot [-\frac{\theta+1}{p^{2j}\theta}] \cdot \frac{1}{\frac{\theta+1}{\theta} - 1}$$

$$= \frac{(\theta+1)}{p^{2j}\theta} \cdot \frac{1}{\frac{\theta+1}{\theta} - 1}$$

$$= \frac{\theta+1}{p^{2j}\theta} \cdot [\frac{1}{\frac{\theta+1}{\theta} - 1}] \cdot \frac{\theta}{\theta}$$

$$= \frac{\theta+1}{p^{2j}} \cdot \frac{1}{\theta+1-\theta} = \frac{\theta+1}{p^{2j}} = \alpha_{2j}.$$

This concludes our proof. $\qquad\square$

We now discuss the construction of each equivalence bimodule between $A_{\alpha_{2j}}$ and $A_{\beta_{2j}}$ as defined by M. Rieffel in [**Rie2**]. The following is a direct result of Theorem 1.1 of [**Rie2**]:

PROPOSITION 4.2. Let p be a prime number, $\theta \in [0,1)$ and for all $j \in \mathbb{N}$, let $\alpha_j = \frac{\theta+1}{p^j}$ and $\beta_j = 1 - \frac{\theta+1}{p^j\theta}$.

For $j \geq 0$, let $\mathbf{F}_{\mathbf{p^{2j}}} = \mathbb{Z}/p^{2j}\mathbb{Z}$. Let $G = \mathbb{R} \times \mathbf{F}_{\mathbf{p^{2j}}}$, and consider the closed subgroups

$$H = \{(n, [n]) : n \in \mathbb{Z}\}$$

and

$$K = \{(-n\theta, [n]) : n \in \mathbb{Z}\}$$

of G, where $[\cdot]$ is the canonical surjection $\mathbb{Z} \twoheadrightarrow \mathbf{F}_{p^{2j}}$. Then $A_{\alpha_{2j}}$ is $*$-isomorphic to $C(G/H) \rtimes K$, and $A_{\beta_{2j}}$ is $*$-isomorphic to $C(G/K) \rtimes H$, where the actions are given by translation. Moreover, $C_C(G)$ can be equipped with a left $A_{\alpha_{2j}}$-module action and a left $A_{\alpha_{2j}}$-valued inner product, and a right $A_{\beta_{2j}}$-action and right $A_{\beta_{2j}}$-valued inner product in such a way that $C_C(G)$, suitably completed, becomes a $A_{\alpha_{2j}} - A_{\beta_{2j}}$ equivalence bimodule.

PROOF. This follows from Theorem 1.1 of Rieffel's paper [**Rie2**], with (using the notation there) $a = 1$, $b = 0, q = p^{2j}, p = 1$, $\alpha = \beta_{2j} = -(\frac{\theta+1}{p^{2j}\theta})$, and $\gamma = \frac{1}{p^{2j}\beta_{2j}+1} = -\theta$. We write the left-$A_{\alpha_{2j}}$ action and inner products as they will be useful in the sequel. We remark that our formula for the inner product is modified from Rieffel's because we use the inverse identification of G/H with $\mathbb{T}$ from the one used in [**Rie2**].

For $g \in G$, the class of g in G/H is denoted by $\tilde{g}$. For this computation, we also identify K with $\mathbb{Z}$ via the map $n \mapsto (n, [n])$. For F_1 and F_2 in $C_C(G) = C_C(\mathbb{R} \times \mathbf{F}_{\mathbf{p^{2j}}})$, and for all $(t, [m]) \in G$ and $n \in \mathbb{Z}$, we have:

$$\langle F_1, F_2 \rangle_{A_{\alpha_{2j}}} (\widetilde{(t, [m])}, n)$$

$$= \sum_{\ell \in \mathbb{Z}} F_1(p^{2j}t - m - \ell, [-m-\ell])\overline{F_2(p^{2j}t - m - \ell + n\theta, [-m-\ell-n]))}$$

$$= \sum_{\ell \in \mathbb{Z}} F_1(p^{2j}t - \ell, [-\ell])\overline{F_2(p^{2j}t - \ell + n\theta, [-\ell-n]))}.$$

Moreover, if $f \in C_C(G/H \times K) \subset A_{\alpha_{2j}}$ and $F \in C_C(G) = C_C(\mathbb{R} \times \mathbf{F}_{\mathbf{p}^{2j}})$ we obtain:

$$(f \cdot F)(t, [m]) = \sum_{n \in \mathbb{Z}} f(\widetilde{(t, [m])}, n) F(t + n\theta, [m - n]).$$

We also note for future reference that for fixed $j \in \mathbb{N}$, the generators $U_{\alpha_{2j}}$ and $V_{\alpha_{2j}}$ in $C_C(G/H \times K) \subset A_{\alpha_{2j}}$ satisfying

$$U_{\alpha_{2j}} V_{\alpha_{2j}} = e^{2\pi i \alpha_{2j}} V_{\alpha_{2j}} U_{\alpha_{2j}}$$

are given by

$$U_{\alpha_{2j}}(\widetilde{(r, [k])}, n) = \begin{cases} 0, & \text{if } n \neq 1, \\ 1, & \text{if } n = 1, \end{cases}$$

and

$$V_{\alpha_{2j}}(\widetilde{(t, [m])}, n) = \begin{cases} 0, & \text{if } n \neq 0, \\ e^{2\pi i (t-m)/p^{2j}}, & \text{if } n = 0. \end{cases}$$

One computes that the action of $U_{\alpha_{2j}}$ on $F \in C_C(G)$ is given by:

(4.1) $$(U_{\alpha_{2j}} \cdot F)(t, [m]) = F(t + \theta, [m - 1]),$$

and the action of $V_{\alpha_{2j}}$ on $F \in C_C(G)$ is given by:

(4.2) $$(V_{\alpha_{2j}} \cdot F)(t, [m]) = e^{2\pi i (t-m)/p^{2j}} F(t, [m]).$$

This will be useful in the sequel. $\qquad\square$

We give a corollary to the proposition that will help us in our identification of equivalence bimodules:

COROLLARY 4.3. *Let G, H, and K be as in Proposition 4.2 Let $\phi \in C_C(\mathbb{R})$, fix $[m]$, $m' \in \mathbf{F}_{\mathbf{p}^{2j}}$, and define $\phi \otimes \delta_{m'} \in C_C(G)$ by*

$$\phi \otimes \delta_{m'}(r, [m]) = \begin{cases} 0, & \text{if } [m] \neq [m'], \\ \phi(r), & \text{if } [m] = [m']. \end{cases}$$

Then for $[m_1]$, $[m_2] \in \mathbf{F}_{\mathbf{p}^{2j}}$ and ϕ_1, $\phi_2 \in C_C(\mathbb{R})$,

$$\langle \phi_1 \otimes \delta_{m_1}, \phi_2 \otimes \delta_{m_2} \rangle_{A_{\alpha_{2j}}} (\widetilde{(t, [m])}, n) =$$

$$\begin{cases} 0, & \text{if } [n] \neq [m_1 - m_2], \\ \sum_{\ell \in \mathbb{Z}} \phi_1(p^{2j}t + m_1 - \ell p^{2j}) \overline{\phi_2(p^{2j}t + m_1 - \ell p^{2j} + n\theta)}, & \text{if } [n] = [m_1 - m_2]. \end{cases}$$

PROOF. By our formulas above, we have that

$$\langle \phi_1 \otimes \delta_{m_1}, \phi_2 \otimes \delta_{m_2} \rangle_{A_{\alpha_{2j}}} (\widetilde{(t, [m])}, n)$$

$$= \sum_{\ell \in \mathbb{Z}} \phi_1 \otimes \delta_{m_1}(p^{2j}t - \ell), [-\ell]) \overline{\phi_2 \otimes \delta_{m_2}(p^{2j}t - \ell + n\theta, [-\ell - n])}$$

$$= \sum_{\ell \in \mathbb{Z}} \phi_1(p^{2j}t - \ell) \delta_{m_1}([-\ell]) \overline{\phi_2(p^{2j}t - \ell + n\theta)} \delta_{m_2}([-\ell - n]).$$

We note $[-\ell] = [m_1] \bmod p^{2j}$ only if $\ell = -m_1 \bmod p^{2j}$ so if and only if $\ell = -m_1 + zp^{2j}$ for some $z \in \mathbb{Z}$. Likewise, $[-\ell - n] = [m_1 + zp^{2j}) - n] = [m_2] \bmod p^{2j}$ if and only if $m_1 - n = m_2 \bmod p^{2j}$, so to have any chance of a non-zero outcome we must have $n = m_1 - m_2 + xp^{2j}$ for some $x \in \mathbb{Z}$. It follows that for $n = m_1 - m_2 + xp^{2j}$ where $x \in \mathbb{Z}$ we have:

$$\langle \phi_1 \otimes \delta_{m_1}, \phi_2 \otimes \delta_{m_2} \rangle_{A_{\alpha_{2j}}} (\widetilde{(t, [m])}, n)$$

$$= \sum_{z \in \mathbb{Z}} \phi_1(p^{2j}t + m_1 - zp^{2j}))\overline{\phi_2(p^{2j}t + m_1 - zp^{2j} + n\theta)}\delta_{m_2}([m_1 - n])$$

$$= \sum_{z \in \mathbb{Z}} \phi_1(p^{2j}t + m_1 - zp^{2j})\overline{\phi_2(p^{2j}t + m_1 - zp^{2j} + n\theta)}\delta_{m_2}([m_1 - n]),$$

and this last sum is equal to

$$= \sum_{z \in \mathbb{Z}} \phi_1(p^{2j}t + m_1 - zp^{2j})\overline{\phi_2(p^{2j}t + m_1 - zp^{2j} + n\theta)}$$

if $[n] = [(m_1 - m_2)] \bmod p^{2j}$ and is equal to 0 if $[n] \neq [m_1 - m_2] \bmod p^{2j}$.

That is, we have:

$$\langle \phi_1 \otimes \delta_{m_1}, \phi_2 \otimes \delta_{m_2} \rangle_{A_{\alpha_{2j}}}(\widetilde{(t, [m])}, n) = 0$$

if $n \neq [m_1 - m_2] \bmod p^{2j}$ and

$$\langle \phi_1 \otimes \delta_{m_1}, \phi_2 \otimes \delta_{m_2} \rangle_{A_{\alpha_{2j}}}(\widetilde{(t, [m])}, n) =$$
$$\sum_{z \in \mathbb{Z}} \phi_1(p^{2j}t + m_1 - zp^{2j})\overline{\phi_2(p^{2j}t + m_1 - zp^{2j} + n\theta)},$$

if $n = [m_1 - m_2] \bmod p^{2j}$.   □

From the above Corollary, we obtain the following Theorem, which we will use to identify our bimodules in the sequel:

THEOREM 4.4. *Let* $\phi_1, \phi_2 \in C_C(\mathbb{R})$ *have sufficient regularity; for example, suppose they are* C^∞ *with compact support. Fix* m_1, $m_2 \in \mathbb{Z}$ *and* $j \in \mathbb{N} \cup \{0\}$. *Then:*

- *if* $n \neq [m_1 - m_2] \bmod p^{2j}$ *then*

$$\langle \phi_1 \otimes \delta_{m_1}, \phi_2 \otimes \delta_{m_2} \rangle_{A_{\alpha_{2j}}}(\widetilde{(t, [m])}, n) = 0$$

- *if* $n = [m_1 - m_2] \bmod p^{2j}$ *then:*

$$\langle \phi_1 \otimes \delta_{m_1}, \phi_2 \otimes \delta_{m_2} \rangle_{A_{\alpha_{2j}}}(\widetilde{(t, [m])}, n) =$$
$$\sum_{k_2 \in \mathbb{Z}} \left(\frac{1}{p^{2j}} \int_{-\infty}^{\infty} \phi_1(u)\overline{\phi_2(u + n\theta)}e^{-2\pi i \frac{k_2(u - m_1)}{p^{2j}}} \, du \right) e^{2\pi i k_2 t}.$$

PROOF. If $n \neq [m_1 - m_2] \bmod p^{2j}$, the result is clear, so we concentrate on the case where $n = [m_1 - m_2] \bmod p^{2j}$.

When ϕ_1 and ϕ_2 have sufficient regularity and rapid decay as described in the statement of the theorem, and if $n = m_1 - m_2 + xp^{2j}$ for some $x \in \mathbb{Z}$, then a quick use of the Poisson summation formula shows that:

$$\sum_{z \in \mathbb{Z}} \phi_1(p^{2j}t + m_1 - zp^{2j})\overline{\phi_2(p^{2j}t + m_1 - zp^{2j} + n\theta)}$$

$$= \sum_{z \in \mathbb{Z}} \phi_1(p^{2j}(t - z + \frac{m_1}{p^{2j}}))\overline{\phi_2(p^{2j}(t - z + \frac{m_1 + n\theta}{p^{2j}}))}$$

$$= \sum_{k_2 \in \mathbb{Z}} \left(\int_{-\infty}^{\infty} \phi_1(p^{2j}(y + \frac{m_1}{p^{2j}}))\overline{\phi_2(p^{2j}(y + \frac{m_1 + n\theta}{p^{2j}}))}e^{-2\pi i k_2 y} \, dy \right) e^{2\pi i k_2 t}$$

Now let $u = p^{2j}y + m_1$. Then $\frac{1}{p^{2j}} = dy$ and $\frac{u-m_1}{p^{2j}} = y$. We therefore obtain the above expression equal to:

$$\sum_{k_2 \in \mathbb{Z}} \left(\int_{-\infty}^{\infty} \phi_1(p^{2j}(y + \frac{m_1}{p^{2j}})) \overline{\phi_2(p^{2j}(y + \frac{m_1 + n\theta}{p^{2j}}))} e^{-2\pi i k_2 y} \, dy \right) e^{2\pi i k_2 t}$$

$$= \sum_{k_2 \in \mathbb{Z}} \left(\frac{1}{p^{2j}} \int_{-\infty}^{\infty} \phi_1(u) \overline{\phi_2(u + n\theta)} e^{-2\pi i \frac{k_2(u-m_1)}{p^{2j}}} \, du \right) e^{2\pi i k_2 t},$$

as desired. $\qquad\qquad\qquad\qquad\qquad\qquad\qquad\qquad\qquad\qquad\qquad\qquad\square$

5. Forming projective modules over noncommutative solenoids using p-adic fields

In [**LP2**] it was shown that equivalence bimodules between noncommutative solenoids could be constructed by using the Heisenberg bimodule formulation of M. Rieffel [**Rie3**]. In our case, $\mathbb{Z}\left[\frac{1}{p}\right]$ was embedded as a co-compact 'lattice' in the larger (self-dual) group $M = [\mathbb{Q}_p \times \mathbb{R}]$, with the quotient group M/Γ being exactly the solenoid $\mathcal{S}_p$ in our setting. We review this construction in what follows.

First we discuss the structure of the p-adic field $\mathbb{Q}_p$, which is a locally compact abelian group under addition. Recall that for p prime, the field of p-adic numbers $\mathbb{Q}_p$ is the completion of the rationals $\mathbb{Q}$ for the distance induced by the p-adic absolute value:

$$\forall x \in \mathbb{Q} \setminus \{0\} \quad |x|_p = p^{-n} \text{ if } x = p^n \left(\frac{a}{b}\right) \text{ with } a \text{ and } b \text{ relatively prime with } p,$$

and $|0|_p = 0$. It can be shown that any element of $\mathbb{Q}_p$ can be written as:

$$\sum_{i=k}^{+\infty} a_i p^i, \ a_i \in \{0, 1, \cdots, p-1\},$$

for some $k \in \mathbb{Z}$, where the series is convergent for the p-absolute value $|\cdot|_p$.

The group $\mathbb{Z}_p$ of p-adic integers sits inside $\mathbb{Q}_p$ as a closed compact subgroup, consisting of those p-adic numbers of the form $\sum_{k=0}^{\infty} a_i p^i$ with $a_i \in \{0, \ldots, p-1\}$ for all $i \in \mathbb{N}$.

The quotient of $\mathbb{Q}_p$ by $\mathbb{Z}_p$ is the Prüfer p-group, consisting of all p^{nth}-roots of unity.

The group $\mathbb{Q}_p$ is self-dual: for any character χ of $\mathbb{Q}_p$, there exists a unique $x \in \mathbb{Q}_p$ such that:

$$\chi = \chi_x : q \in \mathbb{Q}_p \longmapsto e^{2i\pi\{x \cdot q\}_p} \in \mathbb{T}$$

where $\{x \cdot q\}_p$ is the fractional part of the product $x \cdot q$ in $\mathbb{Q}_p$, i.e. it is the sum of the terms involving the negative powers of p in the p-adic expansion of $x \cdot q$.

Similarly, for every character χ of $\mathbb{R}$, there exists some unique $r \in \mathbb{R}$ such that:

$$\chi = \chi_r : x \in \mathbb{R} \longmapsto e^{2i\pi r x} \in \mathbb{T}$$

Therefore, for any character χ of M, there exists some unique pair $(x, r) \in \mathbb{Q}_p \times \mathbb{R} = M$, such that:

$$\chi = \chi_{(x,r)} : (q, t) \in M \longmapsto \chi_x(q) \chi_r(t).$$

It is possible to check that the map $(x, r) \in M \mapsto \chi_{(x,r)}$ is a group isomorphism between M and $\hat{M}$, so that $M = \mathbb{Q}_p \times \mathbb{R}$ is indeed self-dual.

As before, let $\Gamma = \left(\mathbb{Z} \left[\frac{1}{p} \right] \right)^2$, and now let $M = [\mathbb{Q}_p \times \mathbb{R}]$. Let $\iota : \Gamma \to M \times \hat{M} \cong M \times M$ be any embedding of Γ into $M \times M$ as a cocompact subgroup. Let the image $\iota(\Gamma)$ be denoted by D. Then D is a discrete co-compact subgroup of $M \times \hat{M}$. Rieffel defined the **Heisenberg multiplier** $\eta : (M \times \hat{M}) \times (M \times \hat{M}) \to \mathbb{T}$ by:

$$\eta((m,s),(n,t)) = \langle m,t \rangle, \quad (m,s),(n,t) \in M \times \hat{M}.$$

Following Rieffel, the **symmetrized version** of η is denoted by the letter ρ, and is the multiplier defined by:

$$\rho((m,s),(n,t)) = \eta((m,s),(n,t))\overline{\eta((n,t),(m,s))}, \quad (m,s),(n,t) \in M \times \hat{M}.$$

We recall the following result of Rieffel ([**Rie3**]), specialized to our noncommutative solenoids to provide the main examples in [**LP2**]:

THEOREM 5.1. *(Rieffel, [**Rie2**], Theorem 2.12, L.-P., [**LP2**], Theorem 5.6) Let M, D, η, and ρ be as above. Then $C_C(M)$ can be given the structure of a left-$C_C(D,\eta)$ module. Moreover, suitably completed with respect to the norm determined by the inner product, $\overline{C_C(M)}$ can be made into a $C^*(D,\eta) - C^*(D^\perp, \overline{\eta})$ Morita equivalence bimodule, where*

$$D^\perp = \{(n,t) \in M \times \hat{M} : \rho((m,s),(n,t)) = 1 \ \forall (m,s) \in D\}.$$

In order to construct explicit bimodules for our examples, we give a detailed formula for η in our case.

DEFINITION 5.2. The Heisenberg multiplier $\eta : [\mathbb{Q}_p \times \mathbb{R}]^2 \times [\mathbb{Q}_p \times \mathbb{R}]^2 \to \mathbb{T}$ is defined by

$$\eta[((q_1,r_1),(q_2,r_2)),((q_3,r_3),(q_4,r_4))] = e^{2\pi i r_1 r_4} e^{2\pi i \{q_1 q_4\}_p},$$

where $\{q_1 q_4\}_p$ is the fractional part of the product $q_1 \cdot q_4$, i.e. the sum of the terms involving the negative powers of p in the p-adic expansion of $q_1 q_4$.

For $\theta \in \mathbb{R}$, $\theta \neq 0$, we define $\iota_\theta : \left(\mathbb{Z} \left[\frac{1}{p} \right] \right)^2 \to [\mathbb{Q}_p \times \mathbb{R}]^2$ by

$$\iota_\theta(r_1, r_2) = [(\iota(r_1), \theta \cdot r_1), (\iota(r_2), r_2)],$$

where $\iota : \mathbb{Z} \left[\frac{1}{p} \right] \to \mathbb{Q}_p$ is the natural embedding, i.e. if $r = \frac{k}{p^j} \geq 0$, so that we can write $r = \sum_{j=M}^N \frac{a_i}{p_i}$ for with integers M, N such that $-\infty < M \leq N < \infty$, and $a_i \in \{0, 1, \cdots, p-1\}$, then

$$\iota\left(\sum_{j=M}^N a_j p^j \right) = \sum_{j=M}^N a_j p^j,$$

and

$$\iota\left(-\sum_{j=M}^N a_j p^j \right) = -[\iota\left(\sum_{j=M}^N a_j p^j \right)] = (p-a_M)p^M + \sum_{j=M+1}^N (p-1-a_j)p^j + \sum_{j=N+1}^\infty (p-1)p^j.$$

For example, $\iota(-1) = -\iota(1) = \sum_{j=0}^\infty (p-1)p^j$. When there is no danger of confusion, for example if $r = \frac{a}{p^k}$ where $a \in \{0, 1, \cdots, p-1\}$, we sometimes use $\frac{a}{p^k}$

instead of $\iota(\frac{a}{p^k})$.

Then

$$\eta(\iota_\theta(r_1,r_2)),\iota_\theta(r_3,r_4)) = e^{2\pi i\{\iota(r_1)\iota_1(r_4)\}_p}e^{2\pi i\theta r_1 r_4}$$
$$= e^{2\pi i r_1 r_4}e^{2\pi i\theta r_1 r_4} = e^{2\pi i(\theta+1)r_1 r_4}.$$

(Here we used the fact that for $r_i, r_j \in \mathbb{Z}\left[\frac{1}{p}\right]$, $\{\iota(r_i)\iota(r_j)\}_p \equiv r_i r_j$ modulo $\mathbb{Z}$.)

One checks that setting $D_\theta = \iota_\theta\left(\mathbb{Z}\left[\frac{1}{p}\right]\right)^2$, the C^*-algebra $C^*(D_\theta,\eta)$ is exactly the same as the noncommutative solenoid $C^*(\Gamma,\alpha)$, for

$$\alpha = \left(\theta+1, \frac{\theta+1}{p}, \cdots, \frac{\theta+1}{p^n}, \cdots\right),$$

i.e. $\alpha_n = \frac{\theta+1}{p^n}$ for all $n \in \mathbb{N}$.

For this particular embedding of $\left(\mathbb{Z}\left[\frac{1}{p}\right]\right)^2$ as the discrete subgroup D inside $M \times \hat{M}$, we calculate that

$$D_\theta^\perp = \left\{\left(\iota(r_1), -\frac{r_1}{\theta}\right), (\iota(r_2), -r_2) : r_1, r_2 \in \mathbb{Z}\left[\frac{1}{p}\right]\right\}.$$

Moreover,

$$\overline{\eta}([(\iota(r_1), -\frac{r_1}{\theta}), (\iota(r_2), -r_2)], [(\iota(r_3), -\frac{r_3}{\theta}), (\iota(r_4), -r_4)]) = e^{-2\pi i(\frac{1}{\theta}+1)r_1 r_4}.$$

It is evident that $C^*(D_\theta^\perp,\eta)$ is also a non-commutative solenoid $C^*(\Gamma,\beta)$ where $\beta_n = 1 - \frac{\theta+1}{p^n\theta}$, and an application of Theorem 5.6 of [**LP2**] shows that $C^*(\Gamma,\alpha)$ and $C^*(\Gamma,\beta)$ are strongly Morita equivalent in this case.

Note that for

$$\alpha = (\alpha_j)_{j\in\mathbb{N}} = \left(\theta+1, \frac{\theta+1}{p}, \cdots, \frac{\theta+1}{p^j}, \cdots\right),$$

and

$$\beta = (\beta_j)_{j\in\mathbb{N}} = \left(1 - \frac{\theta+1}{p^j\theta}\right)_{j\in\mathbb{N}},$$

we have

$$\theta \cdot \tau(K_0(C^*(\Gamma,\Psi_\alpha))) = \tau(K_0(C^*(\Gamma,\Psi_\beta))).$$

We now discuss this example in further detail and relate it to the strong Morita equivalence bimodules constructed in the previous sections.

PROPOSITION 5.3. Consider $C^*(D_\theta,\eta)$ as defined above, where

$$D_\theta = \{((\iota(r_1), \theta\cdot r_1), (\iota(r_2), r_2)) : r_1, r_2 \in \mathbb{Z}\left[\frac{1}{p}\right]\}$$
$$= \left\{\left(\left(\iota(\frac{j_1}{p^{k_1}}), \theta\cdot\frac{j_1}{p^{k_1}}\right), \left(\iota(\frac{j_2}{p^{k_2}}), \frac{j_2}{p^{k_2}}\right)\right) : j_1, j_2 \in \mathbb{Z}, k_1, k_2 \in \mathbb{N}\cup\{0\}\right\}$$
$$\subset M \times \hat{M}.$$

For α as above, define

$$U_{\alpha,j} = \delta_{((\iota(\frac{1}{p^j}), \frac{\theta}{p^j}), (0,0))} \in C^*(D_\theta,\eta)$$

and

$$V_{\alpha,j} = \delta_{((0,0),(\iota(\frac{1}{p^j}),\frac{1}{p^j}))} \in C^*(D_\theta, \eta).$$

Then for all $j \geq 0$, $U_{\alpha,j} = (U_{\alpha,j+1})^p$, $V_{\alpha,j} = (V_{\alpha,j+1})^p$, and

$$U_{\alpha,j} V_{\alpha,j} = e^{2\pi i \frac{\theta+1}{p^{2j}}} V_{\alpha,j} U_{\alpha,j}.$$

Therefore the algebra elements $U_{\alpha,j}$ and $V_{\alpha,j}$ correspond to the algebra elements $U_{\alpha_{2j}}$ and $V_{\alpha_{2j}}$ described in Section 3, when the noncommutative solenoid was shown to be a direct limit of rotation algebras.

PROOF. We calculate

$$\eta(((\iota(\frac{1}{p^j}), \frac{\theta}{p^j}),(0,0)),((0,0),(\iota(\frac{1}{p^j}),\frac{1}{p^j}))) = e^{2\pi i \frac{\theta+1}{p^{2j}}},$$

and

$$\eta(((0,0),(\iota(\frac{1}{p^j}),\frac{1}{p^j})),((\iota(\frac{1}{p^j}),\frac{\theta}{p^j}),(0,0))) = e^{2\pi i \cdot 0} = 1.$$

The identity

$$U_{\alpha,j} V_{\alpha,j} = e^{2\pi i \frac{\theta+1}{p^{2j}}} V_{\alpha,j} U_{\alpha,j}$$

then follows from standard twisted group algebra calculations, as do the other identities.   □

6. The Haar multiresolution analysis for $L^2(\mathbb{Q}_p)$ of Shelkovich and Skopina

In the previous section, it was shown that if we wish to analyze the Heisenberg equivalence bimodule of M. Rieffel between the noncommutative solenoids the noncommutative solenoids $C^*(\Gamma, \Psi_\alpha)$ and $C^*(\Gamma, \Psi_\beta)$, we need to study $\overline{C_C(\mathbb{Q}_p \times \mathbb{R})}$, where the closure is taken in the norms induced by the inner products on either side. It thus makes sense to consider the L^2 closure of $C_C(\mathbb{Q}_p)$, and consider a multiresolution structure for it generated by continuous, compactly supported functions on $\mathbb{Q}_p$, which we will then tensor by $C_C(\mathbb{R})$ to construct our nested sequence of equivalence bimodules in $\overline{C_C(\mathbb{Q}_p \times \mathbb{R})}$.

We first recall the definition due to Shelkovich and Skopina [**ShSk**] and studied further by Albeverio, Evdokimov and Skopina [**AES**] of the Haar multiresolution analysis for dilation and translation in $L^2(\mathbb{Q}_p)$.

DEFINITION 6.1. A collection $\{\mathcal{V}_j\}_{j=-\infty}^{\infty}$ of closed subspaces of $L^2(\mathbb{Q}_p)$ is called a *multiresolution analysis* (MRA) for dilation by p if:

(1) $\mathcal{V}_j \subset \mathcal{V}_{j+1}$ for all $j \in \mathbb{Z}$;
(2) $\cup_{j \in \mathbb{Z}} \mathcal{V}_j$ is dense in $L^2(\mathbb{Q}_p)$;
(3) $\cap_{j \in \mathbb{Z}} \mathcal{V}_j = \{0\}$;
(4) $f \in \mathcal{V}_j$ if and only if $f(p^{-1}(\cdot)) \in \mathcal{V}_{j+1}$;
(5) There exists a "scaling function" $\phi \in \mathcal{V}_0$ such that if we set:

$$\mathcal{I}_p = \{a \in \mathbb{Q}_p : \{a\}_p = 0\},$$

then

$$\mathcal{V}_0 = \overline{\text{span}}\{\phi(q - a) : a \in \mathcal{I}_p\},$$

where $\mathbb{Z}_p$ is the compact open ring of integers in the p-adic field $\mathbb{Q}_p$,. Note that the set $\mathcal{I}_p$ gives a natural family of coset representatives for $\mathbb{Q}_p/\mathbb{Z}_p$, but is not a group.

(Note that lacking the appropriate lattice in $\mathbb{Q}_p$, it is necessary to use the coset representatives $\mathcal{I}_p$ to form the analog of shift-invariant subspaces.)

Using the p-adic Haar wavelet basis of S. Kozyrev (2002), Shelkovich and Skopina in 2009 constructed the following closed subspaces $\{\mathcal{V}_j : j \in \mathbb{Z}\}$ of $L^2(\mathbb{Q}_p)$, which are called a p-adic Haar MRA:

$$\mathcal{V}_j = \overline{\operatorname{span}}\{p^{j/2}\chi_{[\mathbb{Z}_p]}(p^{-j}\cdot -n) : n \in \mathbb{Q}_p/\mathbb{Z}_p\},\ j \in \mathbb{Z}.$$

The scaling function ϕ in this case was shown to be $\phi = \chi_{\mathbb{Z}_p}$. Note that unlike the scaling functions in $L^2(\mathbb{R})$, the scaling functions for MRA's in $L^2(\mathbb{Q}_p)$ are $\mathbb{Z}$-periodic, in general ([ShSk], [AES]).

The key refinement equation for the scaling function in the Haar multiresolution analysis for $L^2(\mathbb{Q}_p)$ is:

$$\chi_{[\mathbb{Z}_p]}(q) = \sum_{n=0}^{p-1}\chi_{[\mathbb{Z}_p]}\left(p^{-1}q - \frac{n}{p}\right),$$

and in fact one can show

$$\chi_{[\mathbb{Z}_p]}(q) = \sum_{n=0}^{p^j-1}\chi_{[\mathbb{Z}_p]}\left(p^{-j}q - \frac{n}{p^j}\right),\ \forall j \geq 0.$$

These identities are key in some of our calculations that follow.

We now slightly modify the definition of MRA for $L^2(\mathbb{Q}_p)$, to obtain a definition that will be more useful in the construction of projective modules over noncommutative solenoids.

DEFINITION 6.2. A collection $\{\widetilde{\mathcal{V}}_j\}_{j=0}^{\infty}$ of closed subspaces of $L^2(\mathbb{Q}_p)$ is called a *multiresolution structure* (MRS) for dilation by p if:

(1) $\widetilde{\mathcal{V}}_j \subset \widetilde{\mathcal{V}}_{j+1}$ for all $j \geq 0$;
(2) $\cup_{j\in\mathbb{Z}}\widetilde{\mathcal{V}}_j$ is dense in $L^2(\mathbb{Q}_p)$;
(3) If $f \in \widetilde{\mathcal{V}}_j$, then $f(p^{-1}(\cdot)) \in \widetilde{\mathcal{V}}_{j+1}$;
(4) There exists a "scaling function" $\phi \in \widetilde{\mathcal{V}}_0$ such that

$$\widetilde{\mathcal{V}}_j = \overline{\operatorname{span}}\left\{\phi\left(p^{-j}q - \iota(a)\right) : \iota(a) \in \iota(\tfrac{1}{p^{2j}}\mathbb{Z}) \subset \iota(\mathbb{Z}[\tfrac{1}{p}]) \subset \mathbb{Q}_p\right\}.$$

Condition (4) in particular says that each $\widetilde{\mathcal{V}}_j$ is invariant under translation by $\iota(\frac{1}{p^{2j}}\mathbb{Z})$. Thus if $\mathcal{V}_0$ is finite dimensional, i.e. if the translates of ϕ by $\iota(\mathbb{Z})$ repeat after a finite number of steps, then each $\widetilde{\mathbb{V}}_j$ will be finite dimensional, as well.

It is evident that by taking the scaling function involved, every multiresolution structure for $L^2(\mathbb{Q}_p)$ gives rise a sequence of subspaces that satisfy all the conditions of multiresolution analysis for $L^2(\mathbb{Q}_p)$ save for the conditions of having the subspaces with negative indices which intersect to the zero subspace. In the case of the Haar multiresolution analysis, we can in fact show that its nonnegative subspaces can be constructed from a multiresolution structure. We do this in the next example.

EXAMPLE 6.3. We choose as our scaling function the Haar scaling function $\phi(q) = \chi_{[\mathbb{Z}_p]}(q) \in L^2(\mathbb{Q})$. Let $\widetilde{\mathcal{V}}_0$ be the one dimensional subspace generated by ϕ. Note ϕ is $\mathbb{Z}$-periodic so that $\widetilde{\mathcal{V}}_0 = \overline{\operatorname{span}}\{\phi(q - \iota(a)) : a \in \mathbb{Z}\}$, and condition

(4) of the definition is satisfied for $j = 0$. Note the refinement equation $\chi_{[\mathbb{Z}_p]}(q) = \sum_{n=0}^{p-1} \chi_{[\mathbb{Z}_p]}(p^{-1}q - \frac{n}{p})$ shows that $\phi \in \widetilde{\mathcal{V}}_1$ so that $\widetilde{\mathcal{V}}_0 \subset \widetilde{\mathcal{V}}_1$. Using mathematical induction, one shows that $\phi(p^{-j}q - \iota(a)) \in \mathcal{V}_{j+1}$ whenever $a \in \frac{1}{p^j}\mathbb{Z}$. It follows that $\widetilde{\mathcal{V}}_j \in \widetilde{\mathcal{V}}_{j+1}$ for all $j \geq 0$, and that dilation by $\frac{1}{p}$ carries $\widetilde{\mathcal{V}}_j$ into, but not onto, $\widetilde{\mathcal{V}}_{j+1}$.

It only remains to verify condition (3), that $\cup_{j \in \mathbb{Z}} \widetilde{\mathcal{V}}_j$ is dense in $L^2(\mathbb{Q}_p)$. Let $f \in L^2(\mathbb{Q}_p)$ and fix $\epsilon > 0$. Since $\cup_{j=0}^{\infty} \overline{\text{span}}\{p^{j/2}\chi_{[\mathbb{Z}_p]}(p^{-j} \cdot -n) : n \in \mathbb{Q}_p/\mathbb{Z}_p\}$ is dense in $L^2(\mathbb{Q}_p)$, we know that $\cup_{j=0}^{\infty} \text{span}\{p^{j/2}\chi_{[\mathbb{Z}_p]}(p^{-j} \cdot -n) : n \in \mathbb{Q}_p/\mathbb{Z}_p\}$ is dense in $L^2(\mathbb{Q}_p)$, so that there exists $J, , M, N \in \mathbb{N}$, $a_1, a_2, \cdots, a_M \in \mathbb{C}$, and $n_1, n_2, \cdots n_M \in \mathbb{Q}_p/\mathbb{Z}_p$ such that

$$\|f - \sum_{i=1}^{M} a_i \chi_{[\mathbb{Z}_p]}(p^{-J} \cdot -n_i)\| < \frac{\epsilon}{2}.$$

By choosing a common denominator, we can find $N \in \mathbb{N}$ and $k_1, k_2, \cdots, k_M \in \mathbb{Z}$ with $|k_i| < p^N$ such that

$$n_i = \iota(\frac{k_i}{p^N}), \ 1 \leq i \leq M,$$

so that

$$\|f - \sum_{i=1}^{M} a_i \chi_{[\mathbb{Z}_p]}(p^{-J} \cdot -\iota(\frac{k_i}{p^N}))\| < \epsilon.$$

We consider a function of the form $\chi_{[\mathbb{Z}_p]}(p^{-J} \cdot -\iota(\frac{k_i}{p^N})}$. We know $p^{-J}q - \iota(\frac{k_i}{p^N}) \in \mathbb{Z}_p$ if and only if $p^{-J}q \in \mathbb{Z}_p + \iota(\frac{k_i}{p^N})$ if and only if $q \in p^J([\mathbb{Z}_p] + \iota(\frac{k_i}{p^N}))$. Depending on the parity of k_i modulo p^N, as k_i runs from 0 to $p^N - 1$ the subsets $p^J([\mathbb{Z}_p] + \iota(\frac{k_i}{p^N}))$ are different disjoint sets whose union is $p^{J-N}\mathbb{Z}_p$. We first assume that $N \geq J$. In this case,

$$p^J[\mathbb{Z}_p] = \bigsqcup_{k=0}^{p^{N-J}-1} p^N[\mathbb{Z}_p + \frac{k}{p^{N-J}})].$$

Therefore,

$$\chi_{p^J[\mathbb{Z}_p]}(q) = \sum_{\ell=0}^{p^{N-J}-1} \chi_{p^N\mathbb{Z}_p + p^J \cdot \ell}(q)$$

and

$$\chi_{p^J[\mathbb{Z}_p] + \iota(\frac{k_i}{p^{N-J}})}(q) = \sum_{\ell=0}^{p^{N-J}-1} \chi_{p^N\mathbb{Z}_p + p^J \cdot \ell + \iota(\frac{k_i}{p^{N-J}})}(q).$$

But this final equation implies that

$$\chi_{p^J[\mathbb{Z}_p] + \iota(\frac{k_i}{p^{N-J}})}) \in \widetilde{\mathcal{V}}_N = \overline{\text{span}}\{\chi_{[\mathbb{Z}_p]}(p^{-N}q - \iota(\frac{k}{p^{2N}})) : 0 \leq k \leq p^{2N} - 1\}$$

$$= \text{span}\{\chi_{[\mathbb{Z}_p]}(p^{-N}q - \iota(\frac{k}{p^{2N}})) : 0 \leq k \leq p^{2N} - 1\}$$

$$= \text{span}\{\chi_{[p^N\mathbb{Z}_p + \iota(\frac{k}{p^N})]}(q) : 0 \leq k \leq p^{2N} - 1\}.$$

Therefore $\chi_{[\mathbb{Z}_p]}(p^{-J} \cdot -\iota(\frac{k_i}{p^N})) \in \widetilde{\mathcal{V}}_N$.

If $N < J$ we can write $p^{-J}q - \iota(\frac{k_i}{p^N}) = p^{-J}a - \iota(\frac{k'}{p^J})$ for some integer $k' < p^J$ and in this case, $p^{-J}q - \iota(\frac{k'}{p^J}) \in \mathbb{Z}_p$ if and only if $p^{-J}q \in \mathbb{Z}_p + \iota(\frac{k'}{p^J})$ if and only if $q \in p^J[\mathbb{Z}_p] + k'$ for $k' \in \{0, 1, \cdots, p^J - 1\}$. But then $\chi_{[\mathbb{Z}_p]}(p^{-J} \cdot -\iota(\frac{k'}{p^J})) \in \widetilde{\mathcal{V}}_J$.

Choosing $K = \max\{J, N\}$, it is clear that $\chi_{[\mathbb{Z}_p]}(p^{-J} \cdot -\iota(\frac{k_i}{p^N})) \in \widetilde{\mathcal{V}}_K$, $1 \le i \le M$, so that $\sum_{i=1}^M a_i \chi_{[\mathbb{Z}_p]}(p^{-J} \cdot -\iota(\frac{k_i}{p^N})) \in \widetilde{\mathcal{V}}_K$.

Therefore, given $f \in L^2(\mathbb{Q}_p)$, and $\epsilon > 0$, we have found $K \ge 0$ and $\phi \in \widetilde{\mathcal{V}}_K$ with
$$\|f - \phi\| < \epsilon.$$
Thus $\cup_{j \in \mathbb{Z}} \widetilde{\mathcal{V}}_j$ is dense in $L^2(\mathbb{Q}_p)$, and we have an example of a multiresolution structure, as desired.

REMARK 6.4. Theorem 10 of [**AES**] gives the somewhat surprising result that the only multiresolution analysis for $L^2(\mathbb{Q}_p)$ generated by a an orthogonal test scaling function is the Haar multiresolution analysis defined above. (A scaling function ϕ is said to be *orthogonal* if $\{\phi(\cdot - a) : a \in I_p\}$ is an orthonormal basis for V_0, and the space $\mathcal{D}$ of locally constant compactly supported functions on $\mathcal{Q}_p$ are called the space of *test functions* on $\mathbb{Q}_p$.) It follows that the Haar MRS of Example 6.3 can be used to construct the unique Haar MRA in $L^2(\mathbb{Q}_p)$ coming from orthogonal test scaling functions. The result in [**AES**] also suggests to us that multiresolution structures might be of use, since these will distinguish between two different orthogonal test scaling functions, whereas the MRA does not. We intend to study the relationship between multiresolution structures and wavelets in $L^2(\mathbb{Q}_p)$ further in a future paper. For the purposes of this paper, we restrict ourselves to the multiresolution structure corresponding to the Haar scaling function.

7. The Haar MRS for $L^2(\mathbb{Q}_p)$ and projective multiresolution structures

We now want to use the Haar multiresolution structure for $L^2(\mathbb{Q}_p)$ of the previous section derived from the p-adic MRA of Shelkovich and Skopina to construct a *projective multiresolution structure* for the given projective module over $C^*(\Gamma, \Psi_\alpha)$.

To do this, we modify a definition of B. Purkis (Ph.D. thesis 2014), [**Pur**], who generalized the notion of *projective multiresolution analysis* of M. Rieffel and the second author to the non-commutative setting.

DEFINITION 7.1. Let $\{C_j\}_{j=0}^\infty$ be a nested sequence of unital C^*-algebras with the direct limit C^*-algebra $\mathcal{C}$ preserving the unit, and let $\mathcal{X}$ be a finitely generated (left) projective $\mathcal{C}$-module. A *projective multiresolution structure* (PMRS) for the pair $(\mathcal{C}, \mathcal{X})$ is a family $\{V_j\}_{j \ge 0}$ of closed subspaces of $\mathcal{X}$ such that

(1) For all $j \ge 0$, V_j is a finitely generated projective C_j-submodule of $\mathcal{X}$; i.e. V_j is invariant under C_j and $\langle V_j, V_j \rangle = C_j \subseteq \mathcal{C}$;
(2) $V_j \subset V_{j+1}$ for all $j \ge 0$
(3) $\bigcup_{j=0}^\infty V_j$ is dense in $\mathcal{X}$.

We note that in the directed systems of equivalence bimodules defined in Section 3, the collection of A_n-modules $\{X_n\}$ is a projective multiresolution structure for the pair $(\mathcal{A}, \mathcal{X})$.

REMARK 7.2. Note also that the difference between projective multiresolution structures and projective multiresolution analyses is the following. For projective multiresolution analyses, one has a fixed C^*-algebra $\mathcal{C}$ and a fixed Hilbert $\mathcal{C}$-module $\mathcal{X}$ (not necessarily finitely generated), along with a sequence $\{V_j\}_{j \geq 0}$ of nested, finitely generated projective Hilbert $\mathcal{C}$-modules such that $\bigcup_{j=0}^{\infty} V_j$ is dense in $\mathcal{X}$. For projective multiresolution structures, each V_j is a finitely generated projective C_j-module, but not necessarily a $\mathcal{C}$-module (indeed, in most examples we will study, each V_j cannot be a $\mathcal{C}$-module, simply because $\mathcal{C}$ is "too big" for V_j to be a $\mathcal{C}$ module).

EXAMPLE 7.3. Fix an integer $p \geq 2$, and consider the directed sequence of C^*-algebras below:

$$C_0 \xrightarrow{\varphi_0} C_1 \xrightarrow{\varphi_1} C_2 \xrightarrow{\varphi_2} \cdots$$

where $C_j = C(\mathbb{T})$ and $\varphi_2(\iota_z) = (\iota_z)^p \in C_{j+1} = C(\mathbb{T})$, where ι_z represents the identity function in $C(\mathbb{T})$, given by $\iota_z(z) = z$. The direct limit of the $\{C_j\}$ is $\mathcal{C} = C(\mathcal{S}_p)$, the commutative C^*-algebra of all continuous complex-valued functions on the p-solenoid $\mathcal{S}_p$. This follows from the fact that as a topological space, $\mathcal{S}_p$ can be constructed as an inverse limit of circles $\{\mathbb{T}\}$.

Now each C_j is a singly generated free left module over itself where the inner product is defined by:

$$\langle f, g \rangle_{C_j} = f \cdot \bar{g}, \ f, \ g \in C_j.$$

Similarly, $C(\mathcal{S}_p)$ is a singly generated free left module over itself. Setting $\mathcal{X} = C(\mathcal{S}_p)$ and $V_j = C_j$ for $j \geq 0$, we obtain a projective multiresolution structure for the pair $(C(\mathcal{S}_p), C(\mathcal{S}_p))$. We note that $V_j = C(\mathbb{T})$ can never be a $C(\mathcal{S}_p)$-module.

EXAMPLE 7.4. Example 7.3 is a special case of the following more general setting, where we take $P = 1$. Suppose we are given a directed limit of C^*-algebras

$$A_0 \xrightarrow{\varphi_0} A_1 \xrightarrow{\varphi_1} A_2 \xrightarrow{\varphi_2} \cdots$$

where each A_j is unital and each φ_j is a unital $*$-monomorphism. Therefore we can consider the $\{A_j\}_{j=0}^{\infty}$ as a nested sequence of subalgebras of the direct limit unital C^*-algebra $\mathcal{A}$. Let P be any full projection in A_0 and let V_j be the left Hilbert A_j module given by $V_j = A_j P$, with inner product defined by $\langle v_j P, w_j P \rangle_{A_j} = v_j P P^* w_j^* = v_j P w_j^*$ for $v_j, \ w_j \in A_j$. Then taking $\mathcal{X} = \mathcal{A} P$, we obtain that $\{A_j P\}_{j=0}^{\infty}$ is a projective multiresolution structure for the pair $(\mathcal{A}, \mathcal{X} = \mathcal{A}P)$.

We now aim to build up a projective multiresolution structure for the pair $(C^*(\Gamma, \alpha), \overline{C_C(\mathbb{Q}_p \times \mathbb{R})})$ defined in the previous section. Our strategy will be as follows: for each j, using the canonical embedding of $A_{\alpha_{2j}}$ in $C^*(\Gamma, \alpha)$ described in Proposition 5.3, we will construct a subspace $V_j \subset \overline{C_C(\mathbb{Q}_p \times \mathbb{R})}$ that is invariant under the actions of $U_{\alpha,j}$ and $V_{\alpha,j}$, hence is an $A_{\alpha_{2j}}$-module, where $A_{\alpha_{2j}}$ viewed as a subalgebra of $C^*(\Gamma, \alpha) = C^*(D_\theta, \eta)$. It will also be the case that that $\langle V_j, V_j \rangle_{C^*(\Gamma, \alpha)}$ is dense in the image of $A_{\alpha_{2j}}$ viewed as a subalgebra of $C^*(\Gamma, \alpha) = C^*(D_\theta, \eta)$.

We first construct the subspaces $V_j \subset \overline{C_C(\mathbb{Q}_p \times \mathbb{R})}$ and calculate the $A_{\alpha_{2j}}$-action and $A_{\alpha_{2j}}$-valued inner product on these subspaces. We recall first that the ring of p-adic integers $\mathbb{Z}_p$ sits inside the p-adic rationals as a compact open subgroup. Similarly, for every $j \geq 0$, the ring $p^j \mathbb{Z}_p$ sits inside $\mathbb{Q}_p$.

DEFINITION 7.5. For $j \geq 0$, let $\mathbf{F}_{\mathbf{p^{2j}}} = \mathbb{Z}/p^{2j}\mathbb{Z}$, and consider the following embedding

$$\rho^{(j)} : C_C(\mathbf{F}_{\mathbf{p^{2j}}} \times \mathbb{R}) \to C_C(\mathbb{Q}_p \times \mathbb{R})$$

defined on generators by:

$$\rho^{(j)}(\chi_{\{m\}} \otimes f)(q,t) = [\sqrt{p}]^j \chi_{[\iota(\frac{m}{p^j})+p^j\mathbb{Z}_p]}(q) f(t), \ 0 \le m \le p^{2j}-1.$$

Denote for $j \ge 0$,

$$V_j = \overline{\rho^{(j)}(C_C(\mathbf{F_{p^{2j}}} \times \mathbb{R}))} \subset \overline{C_C(\mathbb{Q}_p \times \mathbb{R})}.$$

We now state and prove a major lemma of this paper.

LEMMA 7.6. *For $j \ge 0$, let V_j be as defined in Definition 7.5. Let D^j be the subgroup of D defined by*

$$D^j = \left\{ \left(\left(\iota(\frac{k_1}{p^j}), \theta \cdot \frac{k_1}{p^j} \right), \left(\iota(\frac{k_2}{p^j}), \frac{k_2}{p^j} \right) \right) : k_1, k_2 \in \mathbb{Z} \right\}.$$

Consider the C^*-subalgebra $C^*(D^j, \eta^{(j)})$ of $C^*(D_\theta, \eta)$ associated to D^j; note that this subalgebra is generated by $U_{\alpha,j}$ and $V_{\alpha,j}$. Then

(1) *V_j is a $A_{(\theta+1)/p^{2j}} = C^*(D^j, \eta^{(j)})$-module, i.e. V_j is invariant under the action of $C^*(D^j, \eta^{(j)})$;*
(2) *$\langle V_j, V_j \rangle_{C^*(D_\theta, \eta)} \subseteq C^*(D^j, \eta^{(j)})$ (i.e. all inner products on the left-hand side vanish off of D^j.)*

PROOF. We note that given $m_1, m_2 \in \{0, 1, \cdots, p^{2j}-1\}$ and $f_1, f_2 \in C_C(\mathbb{R})$, and $k_1, k_2 \in \mathbb{Z}$ relatively prime to p, and $\ell \in \mathbb{N}$ with $\ell > j$, we have

$$\langle (\rho^{(j)}(\chi_{m_1} \otimes f_1), \rho^{(j)}(\chi_{m_2} \otimes f_2)\rangle_{C^*(D_\theta,\eta)}((\iota(\frac{k_1}{p^\ell}), \frac{\theta \cdot k_1}{p^\ell}), (\iota(\frac{k_2}{p^\ell}), \frac{k_2}{p^\ell})))$$

$$= \langle [\sqrt{p}]^j \chi_{[\iota(\frac{m_1}{p^j})+p^j\mathbb{Z}_p]}(q) f_1(t),$$

$$[\sqrt{p}]^j \chi_{[\iota(\frac{m_2}{p^j})+p^j\mathbb{Z}_p]}(q) f_2(t)\rangle_{C^*(D_\theta,\eta)}(((\iota(\frac{k_1}{p^\ell}), \frac{\theta \cdot k_1}{p^\ell}), (\iota(\frac{k_2}{p^\ell}), \frac{k_2}{p^\ell})))$$

$$= \int_{\mathbb{Q}_p} \int_{\mathbb{R}} ([\sqrt{p}]^j)^2 \chi_{[\iota(\frac{m_1}{p^j})+p^j\mathbb{Z}_p]}(q) f_1(t) e^{-2\pi i \{q \cdot \frac{k_2}{p^\ell}\}_p} e^{-2\pi i t \cdot \frac{k_2}{p^\ell}}$$

$$\chi_{[\iota(\frac{m_2}{p^j})+p^j\mathbb{Z}_p]}(q + \frac{k_1}{p^\ell}) \overline{f_2(t + \frac{k_1 \cdot \theta}{p^\ell})} dq dt$$

$$= p^j \int_{\mathbb{Q}_p} e^{-2\pi i \{q \cdot \frac{k_2}{p^\ell}\}_p} \chi_{[\iota(\frac{m_1}{p^j})+p^j\mathbb{Z}_p]}(q) \chi_{[\iota(\frac{m_2}{p^j}-\frac{k_1}{p^\ell})+p^j\mathbb{Z}_p]}(q) dq$$

$$\int_{\mathbb{R}} e^{-2\pi i t \cdot \frac{k_2}{p^\ell}} f_1(t) \overline{f_2(t + \frac{k_1 \cdot \theta}{p^\ell})} dt.$$

We examine the term

$$\int_{\mathbb{Q}_p} e^{-2\pi i \{q \cdot \frac{k_2}{p^\ell}\}_p} \chi_{[\iota(\frac{m_1}{p^j})+p^j\mathbb{Z}_p]}(q) \chi_{[\iota(\frac{m_2}{p^j}-\frac{k_1}{p^\ell})+p^j\mathbb{Z}_p]}(q) dq.$$

We first remark that the subsets $\{\iota(\frac{m}{p^j}) + p^j \mathbb{Z}_p : 0 \le m \le p^{2j}-1\}$ are pairwise disjoint and their union is equal to $\frac{1}{p^j}\mathbb{Z}_p$. Secondly, $\iota(\frac{k}{p^\ell}) + p^j \mathbb{Z}_p = \iota(\frac{k'}{p^\ell}) + p^j \mathbb{Z}_p$ if and only if $\frac{k-k'}{p^\ell} = 0$ modulo p^j. We also note that

$$\int_{\mathbb{Q}_p} e^{-2\pi i \{q \cdot \frac{k_2}{p^\ell}\}_p} \chi_{[\iota(\frac{m_1}{p^j})+p^j\mathbb{Z}_p]}(q)\chi_{[\iota(\frac{m_2}{p^j}-\frac{k_1}{p^\ell})+p^j\mathbb{Z}_p]}(q)\,dq$$

$$= \int_{\mathbb{Q}_p} e^{-2\pi i \{q \cdot \iota(\frac{k_2}{p^\ell})\}_p} \chi_{[p^j\mathbb{Z}_p]}(q-\iota(\frac{m_1}{p^j}))\chi_{[\iota(\frac{m_2}{p^j}-\frac{k_1}{p^\ell})+p^j\mathbb{Z}_p]}(q)\,dq$$

$$= \int_{\mathbb{Q}_p} e^{-2\pi i \{(q+\iota(\frac{m_1}{p^j})) \cdot \iota(\frac{k_2}{p^\ell})\}_p} \chi_{[p^j\mathbb{Z}_p]}(q)\chi_{[\iota(\frac{m_2}{p^j}-\frac{k_1}{p^\ell})+p^j\mathbb{Z}_p]}(q+\iota(\frac{m_1}{p^j}))\,dq$$

$$= e^{-2\pi i \{\iota(\frac{m_1}{p^j}) \cdot \iota(\frac{k_2}{p^\ell})\}_p} \int_{\mathbb{Q}_p} e^{-2\pi i \{q \cdot \iota(\frac{k_2}{p^\ell})\}_p} \chi_{[p^j\mathbb{Z}_p]}(q)\chi_{[\iota(\frac{m_2-m_1}{p^j}-\frac{k_1}{p^\ell})+p^j\mathbb{Z}_p]}(q)\,dq$$

$$= e^{-2\pi i \{\iota(\frac{m_1}{p^j} \cdot \frac{k_2}{p^\ell})\}_p} \int_{\mathbb{Q}_p} e^{-2\pi i \{q \cdot \iota(\frac{k_2}{p^\ell})\}_p} \chi_{[p^j\mathbb{Z}_p]}(q)\chi_{[\iota(\frac{m_2-m_1}{p^j}-\frac{k_1}{p^\ell})+p^j\mathbb{Z}_p]}(q)\,dq$$

$$= e^{-2\pi i \{\iota(\frac{m_1}{p^j} \cdot \frac{k_2}{p^\ell})\}_p} \int_{\mathbb{Q}_p} e^{-2\pi i \{q \cdot \iota(\frac{k_2}{p^\ell})\}_p} \chi_{[p^j\mathbb{Z}_p]}(q)\chi_{[\iota(\frac{m'}{p^j}-\frac{k_1}{p^\ell})+p^j\mathbb{Z}_p]}(q)$$

(where $m' \in \{0, 1, \cdots p^{2j}-1\}$ is equal to $m_2 - m_1$ modulo p^{2j})

$$= e^{-2\pi i \{\iota(\frac{m_1}{p^j} \cdot \frac{k_2}{p^\ell})\}_p} \int_{\mathbb{Q}_p} e^{-2\pi i \{q \cdot \iota(\frac{k_2}{p^\ell})\}_p} \chi_{[p^j\mathbb{Z}_p]}(q)\chi_{[\iota(\frac{m' \cdot p^{\ell-j}-k_1}{p^\ell})+p^j\mathbb{Z}_p]}(q)\,dq.$$

We make the observation related to the observation above that the subsets

$$\left\{ \iota(\frac{m}{p^\ell}) + p^j\mathbb{Z}_p : 0 \le m \le p^{j+\ell} - 1 \right\}$$

are pairwise disjoint and their union is equal to $\frac{1}{p^\ell}\mathbb{Z}_p$. Therefore in order that our product not be zero we need $m' \cdot p^{\ell-j} - k_1 = 0$ modulo $p^{j+\ell}$; that is we need $m'p^{\ell-j} = k_1 + j \cdot p^{j+\ell}$ for some $j \in \mathbb{Z}$. This means

$$k_1 = p^{\ell-j}(m' - p^{2j}).$$

But this means k_1 is divisible by $p^{\ell-j}$, a positive power of p, which we assumed not to be the case. Therefore

$$\chi_{[p^j\mathbb{Z}_p]}(q)\chi_{[\iota(\frac{m' \cdot p^{\ell-j}-k_1}{p^\ell})+p^j\mathbb{Z}_p]}(q) = 0$$

so that our inner product must be zero off of the subgroup D^j, and $\langle V_j, V_j \rangle$ takes on values only in $C^*(D^j, \eta^{(j)})$.

For future reference we provide a formula for the inner product in the case where $\ell \le j$. As before, we let $m_1, m_2 \in \{0, 1, \cdots, p^{2j}-1\}$ and $f_1, f_2 \in C_C(\mathbb{R})$, and now take k_1, $k_2 \in \mathbb{Z}$ not necessarily relatively prime to p. Then

$$\langle [\sqrt{p}]^j \chi_{[\iota(\frac{m_1}{p^j})+p^j\mathbb{Z}_p]}(q)f_1(t),$$

$$[\sqrt{p}]^j \chi_{[\iota(\frac{m_2}{p^j})+p^j\mathbb{Z}_p]}(q)f_2(t)\rangle_{C^*(D_\theta,\eta)}(((\iota(\frac{k_1}{p^j}),\frac{\theta \cdot k_1}{p^j}),(\iota(\frac{k_2}{p^j}),\frac{k_2}{p^j})))$$

$$= \int_{\mathbb{Q}_p}\int_{\mathbb{R}}([\sqrt{p}]^j)^2\chi_{[\iota(\frac{m_1}{p^j})+p^j\mathbb{Z}_p]}(q)f_1(t)e^{-2\pi i\{q\cdot\iota(\frac{k_2}{p^j})\}_p}e^{-2\pi it\cdot\iota(\frac{k_2}{p^j})}$$

$$\chi_{[\iota(\frac{m_2}{p^j})+p^j\mathbb{Z}_p]}(q+\iota(\frac{k_1}{p^j}))\overline{f_2(t+\frac{k_1\cdot\theta}{p^j})}dqdt$$

$$= p^j\int_{\mathbb{Q}_p}e^{-2\pi i\{q\cdot\iota(\frac{k_2}{p^j})\}_p}\chi_{[\iota(\frac{m_1}{p^j})+p^j\mathbb{Z}_p]}(q)\chi_{[\iota(\frac{m_2}{p^j}-\frac{k_1}{p^j})+p^j\mathbb{Z}_p]}(q)dq$$

$$\int_{\mathbb{R}}e^{-2\pi it\cdot\frac{k_2}{p^j}}f_1(t)\overline{f_2(t+\frac{k_1\cdot\theta}{p^j})}dt.$$

As before we consider the term

$$\int_{\mathbb{Q}_p}e^{-2\pi i\{q\cdot\iota(\frac{k_2}{p^j})\}_p}\chi_{[\iota(\frac{m_1}{p^j})+p^j\mathbb{Z}_p]}(q)\chi_{[\iota(\frac{m_2-k_1}{p^j})+p^j\mathbb{Z}_p]}(q)dq.$$

If $m_1 \neq m_2 - k_1$ modulo p^{2j}, that is, if $k_1 \neq m_2 - m_1$ modulo p^{2j}, then the product of the characteristic functions is equal to 0, since the intersection of the sets involved will be empty. If $k_1 = m_2 - m_1$ modulo p^{2j}, then the integral becomes

$$\int_{\mathbb{Q}_p}e^{-2\pi i\{q\cdot\iota(\frac{k_2}{p^j})\}_p}\chi_{[\iota(\frac{m_1}{p^j})+p^j\mathbb{Z}_p]}(q)dq$$

$$= \int_{\mathbb{Q}_p}e^{-2\pi i\{q\cdot\iota(\frac{k_2}{p^j})\}_p}\chi_{[p^j\mathbb{Z}_p]}(q-\iota(\frac{m_1}{p^j}))dq$$

$$= \int_{\mathbb{Q}_p}e^{-2\pi i\{(q'+\iota(\frac{m_1}{p^j}))\cdot\iota(\frac{k_2}{p^j})\}_p}\chi_{[p^j\mathbb{Z}_p]}(q')dq'$$

$$= e^{-2\pi i\{(\iota(\frac{m_1k_2}{p^{2j}}))\}_p}\int_{p^j\mathbb{Z}_p}e^{-2\pi i\{q'\cdot\iota(\frac{k_2}{p^j})\}_p}1dq'$$

$$= e^{-2\pi i\{(\iota(\frac{m_1k_2}{p^{2j}}))\}_p}\int_{p^j\mathbb{Z}_p}1\cdot 1dq'$$

(since for $q' \in p^j\mathbb{Z}_p$, we know that $\{q'\cdot\iota(\frac{k_2}{p^j})\}_p = 0$,)

$$= \frac{1}{p^j}\cdot e^{-2\pi i\{(\iota(\frac{m_1k_2}{p^{2j}}))\}_p}$$

(since the measure of $p^j\mathbb{Z}_p$ is equal to $\frac{1}{p^j}$.)

Therefore, if $k_1 = m_2 - m_1$ modulo p^{2j}, we have

$$\langle [\sqrt{p}]^j \chi_{[\iota(\frac{m_1}{p^j})+p^j\mathbb{Z}_p]}(q)f_1(t), [\sqrt{p}]^j \chi_{[\iota(\frac{m_2}{p^j})+p^j\mathbb{Z}_p]}(q)f_2(t)\rangle_{C^*(D_\theta,\eta)}(((\iota(\frac{k_1}{p^j}),\frac{\theta\cdot k_1}{p^j}),(\iota(\frac{k_2}{p^j}),\frac{k_2}{p^j})))$$

$$p^j\frac{1}{p^j}\cdot e^{-2\pi i\{(\iota(\frac{m_1k_2}{p^{2j}}))\}_p}\int_{\mathbb{R}}e^{-2\pi it\cdot\frac{k_2}{p^j}}f_1(t)\overline{f_2(t+\frac{k_1\cdot\theta}{p^j})}dt$$

$$= e^{-2\pi i\{(\iota(\frac{m_1k_2}{p^{2j}}))\}_p}\int_{\mathbb{R}}e^{-2\pi it\cdot\frac{k_2}{p^j}}f_1(t)\overline{f_2(t+\frac{k_1\cdot\theta}{p^j})}dt.$$

This gives the formula

$$\langle [\sqrt{p}]^j \chi_{[\iota(\frac{m_1}{p^j})+p^j\mathbb{Z}_p]}(q)f_1(t),$$

$$[\sqrt{p}]^j \chi_{[\iota(\frac{m_2}{p^j})+p^j\mathbb{Z}_p]}(q)f_2(t)\rangle_{C^*(D_\theta,\eta)}(((\iota(\frac{k_1}{p^j}),\frac{\theta\cdot k_1}{p^j}),(\iota(\frac{k_2}{p^j}),\frac{k_2}{p^j})))$$

$$= \begin{cases} 0 \text{ if } k_1 \neq m_2 - m_1 \bmod p^{2j}, \\ e^{-2\pi i\{(\iota(\frac{m_1 k_2}{p^{2j}}))\}_p} \int_{\mathbb{R}} e^{-2\pi it\cdot\frac{k_2}{p^j}} f_1(t)\overline{f_2(t+\frac{k_1\cdot\theta}{p^j})}dt \\ \qquad \text{if } k_1 = m_2 - m_1 \bmod p^{2j}. \end{cases}$$

From this it follows that:

$$\langle p^{-2j} \chi_{[\iota(\frac{m_1}{p^j})+p^j\mathbb{Z}_p]}(q)f_1(t),$$

$$p^{-2j} \chi_{[\iota(\frac{m_2}{p^j})+p^j\mathbb{Z}_p]}(q)f_2(t)\rangle_{C^*(D_\theta,\eta)}(((\iota(\frac{k_1}{p^j}),\frac{\theta\cdot k_1}{p^j}),(\iota(\frac{k_2}{p^j}),\frac{k_2}{p^j})))$$

$$= \begin{cases} 0 \text{ if } k_1 \neq m_2 - m_1 \bmod p^{2j}, \\ \frac{1}{p^{3j}}e^{-2\pi i\{(\iota(\frac{m_1 k_2}{p^{2j}}))\}_p} \int_{\mathbb{R}} e^{-2\pi it\cdot\frac{k_2}{p^j}} f_1(t)\overline{f_2(t+\frac{k_1\cdot\theta}{p^j})}dt \\ \text{if } k_1 = m_2 - m_1 \bmod p^{2j}. \end{cases}$$

We now show that V_j is invariant under the algebra elements $U_{\alpha,j}$ and $V_{\alpha,j}$ so thus is a $C^*(D^j,\eta^{(j)})$-module. Consider the element $\sqrt{p}\chi_{[\iota(\frac{m}{p^j})+p^j\mathbb{Z}_p]}(q)f(t) \in V_j$ where $m \in \{0,1,\cdots,p^{2j}-1\}$ and $f \in C_C(\mathbb{R})$. Then by definition,

$$U_{\alpha,j}([\sqrt{p}]^j \chi_{[\iota(\frac{m}{p^j})+p^j\mathbb{Z}_p]} \otimes f)(q,t) = \chi_{[\iota(\frac{m}{p^j})+p^j\mathbb{Z}_p]}(q+\iota(\frac{1}{p^j}))f(t+\frac{\theta}{p^j})$$

$$= [\sqrt{p}]^j \chi_{[\iota(\frac{m-1}{p^j})+p^j\mathbb{Z}_p]}(q)f(t+\frac{\theta}{p^j}) = [\sqrt{p}]^j \chi_{[\iota(\frac{m'}{p^j})+p^j\mathbb{Z}_p]}(q)f(t+\frac{\theta}{p^j}),$$

where $m' \in \{0,1,\cdots,p^{2j}-1\}$and $m' = m-1$ modulo p^{2j}. Therefore $U_{\alpha,j}(V_j) \subseteq V_j$. Also, by definition,

$$V_{\alpha,j}([\sqrt{p}]^j \chi_{[\iota(\frac{m}{p^j})+p^j\mathbb{Z}_p]} \otimes f)(q,t) = <(\frac{1}{p^j},\frac{1}{p^j}),(q,t)>[\sqrt{p}]^j \chi_{\iota(\frac{m}{p^j})+p^j\mathbb{Z}_p}(q)f(t)$$

$$= e^{2\pi i\frac{t}{p^j}} e^{2\pi i\{q\cdot\iota(\frac{1}{p^j})\}_p}[\sqrt{p}]^j \chi_{[\iota(\frac{m}{p^j})+p^j\mathbb{Z}_p]}(q)f(t)$$

$$= \begin{cases} 0, & \text{if } q \notin \iota(\frac{m}{p^j})+p^j\mathbb{Z}_p, \\ [\sqrt{p}]^j e^{2\pi i\frac{m}{p^{2j}}} e^{2\pi i\frac{t}{p^j}} f(t), & \text{if } q \in \frac{m}{p^j}+p^j\mathbb{Z}_p. \end{cases}$$

$$= [\sqrt{p}]^j \chi_{[\iota(\frac{m}{p^j})+p^j\mathbb{Z}_p]}(q) \cdot e^{2\pi i\frac{p^j t+m}{p^{2j}}} f(t).$$

Therefore $V_{\alpha,j}(V_j) \subseteq V_j$ also, so that V_j is a $C^*(D^j,\eta^{(j)})$-module, as desired. $\square$

We now work on showing that for every $j \geq 0$, the left $A_{\alpha_{2j}}$-module described in Proposition 4.2 is isomorphic as a left $A_{\alpha_{2j}}$-rigged module to the left $C^*(D^j,\eta^{(j)}) \cong A_{\alpha_{2j}}$-module V_j described above in Lemma 7.6.

Fix $j \in \mathbb{N} \cup \{0\}$. Let $G = G = \mathbb{R} \times \mathbf{F}_{\mathbf{p^{2j}}}$ and $\overline{C_C(G)}$ be as defined in Proposition 4.2, and let V_j be as defined above. We define a map $\Psi_j : \overline{C_C(G)} \to V_j$ on a spanning set of $\overline{C_C(G)}$ and V_j by:

$$\Psi_j(f \otimes \delta_m)(q,t) = p^{-j}f(p^jt)\chi_{[\iota(\frac{-m}{p^j})+p^j\mathbb{Z}_p]}(q),$$

where $f \in C_C(\mathbb{R})$ and $m \in \mathbf{F}_{p^{2j}} = \{0, 1, \cdots, p^{2j} - 1\}$.

For f_1, $f_2 \in C_C(\mathbb{R})$ and m_1, $m_2 \in \mathbf{F}_{p^{2j}}$ we obtain:

$$\langle \Psi_j(f_1 \otimes \delta_{m_1}), \Psi_j(f_2 \otimes \delta_{m_1}) \rangle_{C^*(D_\theta, \eta)}(((\iota(\frac{k_1}{p^j}), \frac{\theta \cdot k_1}{p^j}), (\iota(\frac{k_2}{p^j}), \frac{k_2}{p^j})))$$

$$= \langle p^{-j}\chi_{[\iota(\frac{-m_1}{p^j}) + p^j \mathbb{Z}_p]}(q) f_1(p^j t),$$

$$p^{-j}\chi_{[\iota(\frac{-m_2}{p^j}) + p^j \mathbb{Z}_p]}(q) f_2(p^j t) \rangle_{C^*(D_\theta, \eta)}(((\iota(\frac{k_1}{p^j}), \frac{\theta \cdot k_1}{p^j}), (\iota(\frac{k_2}{p^j}), \frac{k_2}{p^j})))$$

$$= \begin{cases} 0 \text{ if } k_1 \neq -m_2 - (-m_1) \bmod p^{2j}, \\ \frac{1}{p^j} e^{-2\pi i \{(\iota(\frac{-m_1 k_2}{p^{2j}}))\}_p} \int_{\mathbb{R}} e^{-2\pi i t \cdot \frac{k_2}{p^j}} f_1(p^j t)\overline{f_2(p^j t + k_1 \cdot \theta)} dt \\ \qquad \text{if } k_1 = -m_2 - (-m_1) \bmod p^{2j}, \end{cases}$$

$$= \begin{cases} 0 \text{ if } k_1 \neq m_1 - m_2 \bmod p^{2j}, \\ \frac{1}{p^j} e^{2\pi i \{(\iota(\frac{m_1 k_2}{p^{2j}}))\}_p} \int_{\mathbb{R}} e^{-2\pi i t \cdot \frac{k_2}{p^j}} f_1(p^j t)\overline{f_2(p^j t + k_1 \cdot \theta)} dt \\ \qquad \text{if } k_1 = m_1 - m_2 \bmod p^{2j}. \end{cases}$$

We also note that for $f \in C_C(\mathbb{R})$ and $m' \in \mathbf{F}_{p^{2j}}$ we have:

$$\Psi_j(U_{\alpha_{2j}} \cdot (f \otimes \delta_{m'})(r, [m]))(q, t)$$
$$= \Psi_j(f_\theta \delta_{m'}(m - 1))(q, t)$$
$$= \Psi_j(f_\theta \otimes \delta_{[m'+1]})(q, t) = p^{-j}\chi_{[\iota(\frac{-(m'+1)}{p^j}) + p^j \mathbb{Z}_p]}(q) f_(\theta(p^j(t))$$
$$= p^{-j}\chi_{[\iota(\frac{-m'-1}{p^j}) + p^j \mathbb{Z}_p]}(q) f(p^j(t) + \theta)$$

whereas,

$$U_{\alpha,j}(\Psi_j(f \otimes \delta_{m'})(r, [m]))(q, t) = U_{\alpha,j}(p^{-j} f(p^j r)\chi_{[\iota(\frac{-m'}{p^j}) + p^j \mathbb{Z}_p]}(q))(q, t)$$

$$= p^{-j} f(p^j(t + \frac{\theta}{p^j})) \chi_{[\iota(\frac{-m'-1}{p^j}) + p^j \mathbb{Z}_p]}(q)$$

$$= p^{-j}\chi_{[\iota(\frac{-m'-1}{p^j}) + p^j \mathbb{Z}_p]}(q) f(p^j t + \theta).$$

Therefore

$$\Psi_j(U_{\alpha_{2j}} \cdot (f \otimes \delta_{m'})(r, [m]))(q, t) = U_{\alpha,j}(\Psi_j(f \otimes \delta_{m'})(r, [m]))(q, t).$$

Similarly, for $f \in C_C(\mathbb{R})$ and $m' \in \mathbf{F}_{p^{2j}}$ we have:

$$\Psi_j(V_{\alpha_{2j}} \cdot (f \otimes \delta_{m'})(r, [m]))(q, t) = \Psi_j(e^{2\pi i (r - m')/p^{2j}} f(r) \otimes \delta_{m'}(m))(q, t)$$

$$= p^{-j} e^{2\pi i (p^j t - m')/p^{2j}} f(p^j(t))\chi_{[\iota(\frac{-m'}{p^j}) + p^j \mathbb{Z}_p]}(q)$$

$$= p^{-j} e^{2\pi i \frac{p^j t - m'}{p^{2j}}} f(p^j(t))\chi_{[\iota(\frac{-m'}{p^j}) + p^j \mathbb{Z}_p]}(q)$$

$$= p^{-j} \chi_{[\iota(\frac{-m'}{p^j}) + p^j \mathbb{Z}_p]}(q) e^{2\pi i \frac{p^j t - m'}{p^{2j}}} f(p^j(t)).$$

On the other hand,

$$V_{\alpha,j}(\Psi_j(f \otimes \delta_{m'})(r,[m]))(q,t) = V_{\alpha,j}(p^{-j}f(p^j r)\chi_{[\iota(\frac{-m'}{p^j})+p^j \mathbb{Z}_p]}(q))(q,t)$$

$$= e^{2\pi i \frac{p^j t - m'}{p^{2j}}} \cdot p^{-j}f(p^j t)\chi_{[\iota(\frac{-m'}{p^j})+p^j \mathbb{Z}_p]}(q)$$

$$= p^{-j}\chi_{[\iota(\frac{-m'}{p^j})+p^j \mathbb{Z}_p]}(q)e^{2\pi i \frac{p^j t - m'}{p^{2j}}}f(p^j(t)).$$

Therefore:

$$\Psi_j(V_{\alpha_{2j}} \cdot (f \otimes \delta_{m'})(r,[m]))(q,t) = V_{\alpha,j}(\Psi_j(f \otimes \delta_{m'})(r,[m]))(q,t).$$

To identify the inner products is slightly trickier, so we consider the following identification of $C^*(D^j, \eta^{(j)})$ with $A_{\alpha_{2j}} = C(\mathbb{T}) \rtimes_{\alpha_{2j}} \mathbb{Z}$: Given $\Lambda \in C_C(D^j, \eta^{(j)})$ which can be viewed as

$$\Lambda(((\iota(\frac{k_1}{p^j}), \frac{\theta \cdot k_1}{p^j}), (\iota(\frac{k_2}{p^j}), \frac{k_2}{p^j}))), \ k_1, \ k_2 \in \mathbb{Z}$$

we define $\Phi : C_C(D^j, \eta^{(j)}) \to C_C(\mathbb{T} \times \mathbb{Z})$ by

$$\Phi(\Lambda)(r,n) = \sum_{k_2 \in \mathbb{Z}} \Lambda(((\iota(\frac{n}{p^j}), \frac{\theta \cdot n}{p^j}), (\iota(\frac{k_2}{p^j}), \frac{k_2}{p^j})))e^{2\pi i k_2 r}$$

$$= \sum_{k_2 \in \mathbb{Z}} \Lambda(((\iota(\frac{n}{p^j}), \frac{\theta \cdot n}{p^j}), (\iota(\frac{k_2}{p^j}), \frac{k_2}{p^j})))e^{2\pi i k_2 r}.$$

We now check that the module morphism given by each Φ_j preserves the inner products:

For f_1, $f_2 \in C_C(\mathbb{R})$ and m_1, $m_2 \in\in \mathbf{F}_{p^{2j}}$ we obtain:

$$\langle \Psi_j(f_1 \otimes \delta_{m_1}),$$

$$\Psi_j(f_2 \otimes \delta_{m_2})\rangle_{C^*(D_\theta,\eta)} (((\iota(\tfrac{k_1}{p^j}), \tfrac{\theta \cdot k_1}{p^j}), (\iota(\tfrac{k_2}{p^j}), \tfrac{k_2}{p^j})))$$

$$= \langle p^{-j} \chi_{[\iota(\frac{-m_1}{p^j}) + p^j \mathbb{Z}_p]}(q) f_1(p^j t),$$

$$p^{-j} \chi_{[\iota(\frac{-m_2}{p^j}) + p^j \mathbb{Z}_p]}(q) f_2(p^j t)\rangle_{C^*(D_\theta,\eta)} (((\iota(\tfrac{k_1}{p^j}), \tfrac{\theta \cdot k_1}{p^j}), (\iota(\tfrac{k_2}{p^j}), \tfrac{k_2}{p^j})))$$

$$= \begin{cases} 0 \text{ if } k_1 \neq -m_2 - (-m_1) \bmod p^{2j}, \\ \frac{1}{p^j} e^{-2\pi i \{(\iota(\frac{-m_1 k_2}{p^{2j}}))\}_p} \int_{\mathbb{R}} e^{-2\pi i t \cdot \frac{k_2}{p^j}} f_1(p^j t) \overline{f_2(p^j t + k_1 \cdot \theta)} dt \\ \text{if } k_1 = -m_2 - (-m_1) \bmod p^{2j}, \end{cases}$$

$$= \begin{cases} 0 \text{ if } k_1 \neq m_1 - m_2 \bmod p^{2j}, \\ \frac{1}{p^j} e^{2\pi i \{(\iota(\frac{m_1 k_2}{p^{2j}}))\}_p} \int_{\mathbb{R}} e^{-2\pi i t \cdot \frac{k_2}{p^j}} f_1(p^j t) \overline{f_2(p^j t + k_1 \cdot \theta)} dt, \\ \text{if } k_1 = m_1 - m_2 \bmod p^{2j} \end{cases}$$

Now since $m_1 k_2 \in \mathbb{Z}$ we can write this as:

$$= \begin{cases} 0 \text{ if } k_1 \neq m_1 - m_2 \bmod p^{2j}, \\ \frac{1}{p^j} \int_{\mathbb{R}} e^{-2\pi i \frac{k_2(p^{2j} t - m_1)}{p^{2j}}} f_1(p^j t) \overline{f_2(p^j t + k_1 \cdot \theta)} dt \\ \text{if } k_1 = m_1 - m_2 \bmod p^{2j} \end{cases}$$

$$= \begin{cases} 0 \text{ if } k_1 \neq m_1 - m_2 \bmod p^{2j}, \\ \frac{p^j}{p^{2j}} \int_{\mathbb{R}} e^{-2\pi i \frac{k_2(p^{2j} t - m_1)}{p^{2j}}} f_1(p^j t) \overline{f_2(p^j t + k_1 \cdot \theta)} dt, \\ \text{if } k_1 = m_1 - m_2 \bmod p^{2j} \end{cases}$$

$$= \begin{cases} 0 \text{if } k_1 \neq m_1 - m_2 \bmod p^{2j}, \\ \frac{1}{p^{2j}} \int_{\mathbb{R}} e^{-2\pi i \frac{k_2(u - m_1)}{p^{2j}}} f_1(u) \overline{f_2(u + k_1 \cdot \theta)} du \\ \text{if } k_1 = m_1 - m_2 \bmod p^{2j} \end{cases}$$

$$= \begin{cases} 0 \text{ if } k_1 \neq m_1 - m_2 \bmod p^{2j}, \\ \frac{1}{p^{2j}} \int_{\mathbb{R}} e^{-2\pi i \frac{k_2(u - m_1)}{p^{2j}}} f_1(u) \overline{f_2(u + k_1 \cdot \theta)} du, \\ \text{if } k_1 = m_1 - m_2 \bmod p^{2j} \end{cases}$$

From this we obtain, for $n = m_1 - m_2$ modulo p^{2j} :

$$\Phi(\langle \Psi_j(f_1 \otimes \delta_{m_1}), \Psi_j(f_2 \otimes \delta_{m_2})\rangle_{C^*(D_\theta,\eta)})(\widetilde{(r, [m'])}, n)$$

$$= \sum_{k_2 \in \mathbb{Z}} \langle \Psi_j(f_1 \otimes \delta_{m_1}), \Psi_j(f_2 \otimes \delta_{m_2})\rangle_{C^*(D_\theta,\eta)} (((\iota(\tfrac{n}{p^j}), \tfrac{\theta \cdot n}{p^j}), (\iota(\tfrac{k_2}{p^j}), \tfrac{k_2}{p^j}))) e^{2\pi i k_2 r}$$

$$= \begin{cases} 0 \text{ if } n \neq m_1 - m_2 \bmod p^{2j}, \\ \sum_{k_2 \in \mathbb{Z}} [\frac{1}{p^{2j}} \int_{\mathbb{R}} f_1(u) \overline{f_2(u + n \cdot \theta)} e^{-2\pi i \frac{k_2(u - m_1)}{p^{2j}}} du] e^{2\pi i k_2 r}, \\ \text{if } n = m_1 - m_2 \bmod p^{2j}. \end{cases}$$

But this, together with the results of Theorem 4.4, proves that for ϕ_1 and ϕ_2 with compact support and sufficiently regular, we have

$$\Phi(\langle \Psi_j(\phi_1 \otimes \delta_{m_1}), \Psi_j(\phi_2 \otimes \delta_{m_2}) \rangle_{C^*(D_\theta, \eta)})((\widetilde{r, [m']}), n)$$
$$= \langle \phi_1 \otimes \delta_{m_1}, \phi_2 \otimes \delta_{m_2} \rangle_{A_{\alpha_{2j}}}((\widetilde{r, [m]}), n),$$

so that the map Ψ_j provides a isomorphism of projective $A_{\alpha_{2j}}$-modules, as desired.

We now are prepared to prove the main theorem of this paper:

THEOREM 7.7. *Fix an irrational* $\theta \in (0,1)$. *Let* $\Xi = \overline{C_C(\mathbb{Q}_p \times \mathbb{R})}$ *be the equivalence bimodule between the noncommutative solenoids* $C^*(\Gamma, \alpha)$ *and* $C^*(\Gamma, \beta)$ *for* $\alpha = (\alpha_0 = \theta, \alpha_1 = \frac{\theta+1}{p}, \alpha_2 = \frac{\theta+1}{p^2}, \cdots, \alpha_j = \frac{\theta+1}{p^j}, \cdots,)$ *and* $\beta = (\beta_0 = 1 - \frac{\theta+1}{\theta}, \beta_1 = 1 - \frac{\theta+1}{p\theta}, \cdots, \beta_j = 1 - \frac{\theta+1}{p^j\theta}, \cdots)$ *constructed in* [**LP2**]. *Let* $\{V_j\}$ *be the finitely generated projective* $A_{\alpha_{2j}}$-*submodules of* Ξ *constructed in Lemma 7.6. Then the collection* $\{V_j\}$ *forms a projective multiresolution structure for the pair* $(C^*(\Gamma, \alpha), \Xi)$. *Moreover, this projective multiresolution structure can be identified with the projective multiresolution structure defined in Example 7.4, where* $A_j = A_{\alpha_{2j}}$, P *is the projection in* A_{α_0} *of trace* θ, $V_j = A_{\alpha_{2j}} \cdot P$ *for all* $j \geq 0$, *and*

$$\mathcal{X} = \lim_{j \to \infty} A_{\alpha_{2j}} \cdot P = C^*(\Gamma, \alpha) \cdot P.$$

PROOF. We refer to Definition 7.1 and note that Lemma 7.6 has established that $V_j = \overline{\rho^{(j)}(C_C(\mathbf{F}_{p^{2j}} \times \mathbb{R}))}$ is a $A_{\alpha_{2j}} = A_{(\theta+1)/p^{2j}} = C^*(D^j, \eta^{(j)})$-module and that $\langle V_j, V_j \rangle_{C^*(\Gamma, \alpha)} \subseteq A_{\alpha_{2j}}$. The discussion immediately preceding the statement of this Theorem established that V_j is a projective $A_{\alpha_{2j}}$-module that can in fact be identified with the projective $A_{\alpha_{2j}}$-module described in Proposition 4.2. Therefore, $\{V_j\}$ forms a projective multiresolution structure for the pair $(C^*(\Gamma, \alpha), \Xi)$.

We now note that the parts of the proof of Lemma 7.6 having to deal with inner products can be easily adapted to show that the right valued inner products $\langle V_j, V_j \rangle_{C^*(D^\perp, \overline{\eta})}$ take on values in precisely the right subalgebra $B_j = C^*(D^{\perp,(j)}, \overline{\eta})$, where $D^\perp$ was calculated in [**LP2**] to be

$$D^\perp = \{((\iota(r_1), -\frac{r_1}{\theta}), (\iota(r_2), -r_2)) : r_1, r_2 \in \mathbb{Z}[\frac{1}{p}]\},$$

and

$$D^{\perp,(j)} = \{((\iota(\frac{k_1}{p^j}), -\frac{k_1}{p^j\theta}), (\iota(\frac{k_2}{p^j}), -\frac{k_2}{p^j})) : k_1, k_2 \in \mathbb{Z}\}.$$

But $C^*(D^{\perp,(j)}, \overline{\eta})$ was calculated in [**LP2**] to be exactly $A_{\beta_{2j}}$, with

$$C^*(D^{\perp,(j)}, \overline{\eta}) \cong \lim_{j \to \infty} C^*(D^{\perp,(j)}, \overline{\eta}) = \lim_{j \to \infty} A_{\beta_{2j}}.$$

By Proposition 2.2. of [**Rie1**], for some $N \in \mathbb{N}$ there exists $\xi_1, \cdots, \xi_N \in V_0$ such that

$$\sum_{i=1}^{N} \langle \xi_i, \xi_i \rangle_{B_0} = \sum_{i=1}^{N} \langle \xi_i, \xi_i \rangle_{A_{\beta_0}} = 1_{A_{\beta_0}}.$$

By Theorem 1.1 of [**Rie2**], V_0 is a (left) projective A_{α_0}-module of trace $|-\theta| = \theta$, and indeed the proof of our own Proposition 4.2 shows that we can take $N = 1$ and find $\xi_1 \in V_0$ with $\langle \xi_1, \xi_1 \rangle_{A_{\beta_0}} = 1_{A_{\beta_0}}$. By Proposition 2.2. of [**Rie1**], we see that

$$\langle \xi_1, \xi_1 \rangle_{A_{\alpha_0}} = P$$

where P is a projection in A_{α_0} with trace θ. Moreover the same proposition shows us that as an $A_{\alpha_0} - A_{\beta_0}$-bimodule, V_0 is isomomorphic to $A_{\alpha_0} \cdot P$ and A_{β_0} is isomorphic to $P \cdot A_{\alpha_0} \cdot P$.

However, since $\xi_1 \in V_0 \subseteq V_j \subseteq \Xi$, we see that for every $j \geq 0$ the same argument works, and the equivalence bimodule

$$A_j = C^*(D^{(j)}, \eta) \; - \; V_j \; - \; B_j = C^*(D^{\perp,(j)}, \overline{\eta})$$

is isomorphic to the bimodule

$$A_j = A_{\alpha_j} \; - \; A_{\alpha_j} \cdot P \; - \; P \cdot A_{\alpha_j} \cdot P \cong B_j \cong A_{\beta_j}$$

and finally, the equivalence bimodule

$$C^*(D, \eta) \; - \; \Xi \; - \; C^*(D^{\perp}, \overline{\eta})$$

is isomorphic to the bimodule

$$C^*(\Gamma, \alpha) \; - \; C^*(\Gamma, \alpha) \cdot P \; - \; P \cdot C^*(\Gamma, \alpha) \cdot P,$$

as we desired to show.

$\square$

REMARK 7.8. The main point here is not that

$$C^*(D, \eta) \; - \; \Xi \; - \; C^*(D^{\perp}, \overline{\eta})$$

is isomorphic to the bimodule

$$C^*(\Gamma, \alpha) \; - \; C^*(\Gamma, \alpha) \cdot P \; - \; P \cdot C^*(\Gamma, \alpha) \cdot P,$$

since that can be observed fairly quickly from Proposition 2.2 of [**Rie1**] and our identification of $C^*(D, \eta)$ with $C^*(\Gamma, \alpha)$. The more interesting point is that this equivalence can be written as a direct limit of strong Morita equivalence bimodules at each stage. Projective multiresolution structures, therefore, do appear to be the correct objects for studying equivalence bimodules between direct limit algebras.

References

[AA]   Beatriz Abadie and Mauricio Achigar, *Cuntz-Pimsner C^*-algebras and crossed products by Hilbert C^*-bimodules*, Rocky Mountain J. Math. **39** (2009), no. 4, 1051–1081, DOI 10.1216/RMJ-2009-39-4-1051. MR2524704 (2010f:46083)

[AES]   S. Albeverio, S. Evdokimov, and M. Skopina, *p-adic multiresolution analysis and wavelet frames*, J. Fourier Anal. Appl. **16** (2010), no. 5, 693–714, DOI 10.1007/s00041-009-9118-5. MR2673705 (2012c:42075)

[EH]   Edward G. Effros and Frank Hahn, *Locally compact transformation groups and C^*-algebras*, Bull. Amer. Math. Soc. **73** (1967), 222–226. MR0233213 (38 #1536)

[HKLS]   R. Høegh-Krohn, M. B. Landstad, and E. Størmer, *Compact ergodic groups of automorphisms*, Ann. of Math. (2) **114** (1981), no. 1, 75–86, DOI 10.2307/1971377. MR625345 (82i:46097)

[LP1]   F. Latrémolière and J. Packer, *Noncommutative solenoids*, Accepted for publication, New York J. Math., ArXiv: 1110.6227.

[LP2]   Frédéric Latrémolière and Judith A. Packer, *Noncommutative solenoids and their projective modules*, Commutative and noncommutative harmonic analysis and applications, Contemp. Math., vol. 603, Amer. Math. Soc., Providence, RI, 2013, pp. 35–53, DOI 10.1090/conm/603/12039. MR3204025

[Lu1]   Franz Luef, *Projective modules over noncommutative tori are multi-window Gabor frames for modulation spaces*, J. Funct. Anal. **257** (2009), no. 6, 1921–1946, DOI 10.1016/j.jfa.2009.06.001. MR2540994 (2010g:46116)

[Lu2]   Franz Luef, *Projections in noncommutative tori and Gabor frames*, Proc. Amer. Math.
         Soc. **139** (2011), no. 2, 571–582, DOI 10.1090/S0002-9939-2010-10489-6. MR2736339
         (2012b:46153)
[Pur]   Benjamin Purkis, *Projective multiresolution analyses over irrational rotation alge-
         bras*, Commutative and noncommutative harmonic analysis and applications, Con-
         temp. Math., vol. 603, Amer. Math. Soc., Providence, RI, 2013, pp. 73–85, DOI
         10.1090/conm/603/12048. MR3204027
[Rie1]   Marc A. Rieffel, *C*-algebras associated with irrational rotations*, Pacific J. Math. **93**
         (1981), no. 2, 415–429. MR623572 (83b:46087)
[Rie2]   Marc A. Rieffel, *The cancellation theorem for projective modules over irrational ro-
         tation C*-algebras*, Proc. London Math. Soc. (3) **47** (1983), no. 2, 285–302, DOI
         10.1112/plms/s3-47.2.285. MR703981 (85g:46085)
[Rie3]   Marc A. Rieffel, *Projective modules over higher-dimensional noncommutative tori*,
         Canad. J. Math. **40** (1988), no. 2, 257–338, DOI 10.4153/CJM-1988-012-9. MR941652
         (89m:46110)
[Rie4]   Marc A. Rieffel, *Induced representations of C*-algebras*, Advances in Math. **13** (1974),
         176–257. MR0353003 (50 #5489)
[ShSk]   Vladimir Shelkovich and Maria Skopina, *p-adic Haar multiresolution analysis and
         pseudo-differential operators*, J. Fourier Anal. Appl. **15** (2009), no. 3, 366–393, DOI
         10.1007/s00041-008-9050-0. MR2511868 (2010f:42077)

DEPARTMENT OF MATHEMATICS, UNIVERSITY OF DENVER, DENVER, COLORADO 80208
E-mail address: frederic@math.du.edu

DEPARTMENT OF MATHEMATICS, CAMPUS BOX 395, UNIVERSITY OF COLORADO, BOULDER,
COLORADO 80309-0395
E-mail address: packer@euclid.colorado.edu

Harmonic Analysis on Manifolds

Contemporary Mathematics
Volume **650**, 2015
http://dx.doi.org/10.1090/conm/650/13043

An application of hypergeometric shift operators to the χ-spherical Fourier transform

Vivian M. Ho and Gestur Ólafsson

ABSTRACT. We study the action of hypergeometric shift operators on the Heckman-Opdam hypergeometric functions associated with the BC_n type root system and some negative multiplicities. Those hypergeometric functions are connected to the χ-spherical functions on Hermitian symmetric spaces U/K where χ is a nontrivial character of K. We apply shift operators to the hypergeometric functions to move negative multiplicities to positive ones. This allows us to use many well-known results of the hypergeometric functions associated with positive multiplicities. In particular, we use this technique to achieve exponential estimates for the χ-spherical functions. The motive comes from the Paley-Wiener type theorem on line bundles over Hermitian symmetric spaces.

1. Introduction

The theory of spherical functions on semisimple Riemannian symmetric spaces is an interesting part of harmonic analysis dating back to the work of Gel'fand, Godement (for the abstract setting), Harish-Chandra (in the concrete setting for a Riemannian symmetric space), and Helgason. Later this theory was generalized as the theory of hypergeometric functions, resp. of Jacobi polynomials, of several variables associated with a root system, by a series of joint work of Heckman and Opdam [**3**, **4**, **10**, **11**]. In the case of spherical functions one can investigate them using both differential and integral operators, while there are only the differential equations at hand for the case of hypergeometric functions. The Heckman-Opdam hypergeometric functions are joint eigenfunctions of a commuting algebra of differential operators associated to a root system and a multiplicity parameter (which is a Weyl group invariant function on the root system). The multiplicities can be arbitrary complex numbers. These hypergeometric functions are holomorphic, Weyl group invariant, and normalized by the value one at the identity. When the root multiplicities do correspond to those of a Riemannian symmetric space, these hypergeometric functions are nothing but the restrictions to a Cartan subspace of spherical functions on the associated Riemannian symmetric space. For an overall study of this subject we refer to the books [**5**, Part I] and [**7**].

2010 *Mathematics Subject Classification.* Primary 43A90, 33C67; Secondary 43A85, 53C35.

Key words and phrases. Hypergeometric shift operator, hypergeometric function, symmetric space, χ-spherical Fourier transform, Paley-Wiener theorem.

The research of the second author was supported by NSF grant DMS-1101337.

One of the new tools which was born with the general theory of hypergeometric functions was the theory of shift operators. Those are generalizations of the classical identity for hypergeometric functions

$$(1.1) \qquad \frac{d}{dz}F(\alpha,\beta,\gamma;z) = \frac{\alpha\beta}{\gamma}F(\alpha+1,\beta+1,\gamma+1;z).$$

Here, $\frac{d}{dz}$ is the simplest example of a shift operator, for the rank one root system BC_1. A shift operator for the root system of type BC_2 was first found by Koornwinder in [9], used to study the Jacobi polynomials. Subsequently some particular higher rank cases were established by several authors. A thorough study of higher rank shift operators in full generality was done by Opdam in [10–12]. We also recommend [5, Part I] as a good resource.

In this article we mainly explore the idea to apply some suitable shift operators to hypergeometric functions associated with the root system of type BC_n $(n \geq 1)$ and certain negative multiplicities. The motivation for this work originates from our article [8] on Paley-Wiener type theorems for line bundles over compact Hermitian symmetric spaces U/K and the needed estimates for the spherical functions. Let χ be a nontrivial character of K. We characterized the χ-bicovariant functions f on U (geometrically equivalent to smooth sections of those line bundles) with small support in terms of holomorphic extendability and exponential growth of their χ-spherical Fourier transforms with the exponent linked to the size of the support of f. This characterization relies on the fact that the χ-spherical functions on U extend holomorphically to their complexifications, and their restrictions on the noncompact dual G are in turn the χ-spherical functions on G. It is well known, see [5], that the χ-spherical functions on G are related to hypergeometric functions with shifted multiplicities, some of which can be negative.

Denote for a moment a hypergeometric function by F. In [13] the author gave a uniform exponential estimate on the growth behavior of F, which is of crucial importance for the Paley-Wiener Theorem, but which requires all multiplicities to be positive, see Proposition 3.5. There are two possible ways now to attach the problem. Either generalize the Opdam estimates in [13] or use the shift operators to reduce the problem to positive multiplicities. The first way was chosen in [8]. Here we discuss the second idea.

Let us give a brief outline of this article. In Section 2 we settle on some necessary definitions and notations. A succinct review of the Heckman-Opdam hypergeometric functions is given in Section 3. To follow, in Section 4 we study some properties of the hypergeometric shift operators of Opdam that will be used in the subsequent sections. Next, Section 5 is devoted to an application of shift operators to the hypergeometric functions associated with the root system of type BC_n and certain nonpositive multiplicities, where root multiplicities do correspond to Hermitian symmetric spaces. Using this technique we will achieve desired estimates for the χ-spherical functions. Finally, in Section 6 we treat the rank one case as an example to strengthen the skills. A remark is given for a nice generalization of the rank one case.

2. Notation and Preliminaries

The material in this section is standard. We refer to [**5**, Part I] for basic notations and definitions. We use the notations from the introduction mostly without
reference.

Let $\mathfrak{a}$ be an Euclidean space of dimension n and $\mathfrak{a}^*$ its dual space. Denote by
$\langle \cdot , \cdot \rangle$ the inner products on $\mathfrak{a}$ and $\mathfrak{a}^*$. Let $\Sigma \subset \mathfrak{a}^*$ be a possibly nonreduced root
system with $\text{rank}(\Sigma) = \dim \mathfrak{a} = n$. In particular Σ spans $\mathfrak{a}^*$. Denote by

$$(2.1) \qquad \Sigma_* = \{\alpha \in \Sigma \mid 2\alpha \notin \Sigma\} \quad \text{and} \quad \Sigma_i = \{\alpha \in \Sigma \mid \tfrac{1}{2}\alpha \notin \Sigma\}.$$

Both Σ_* and Σ_i are reduced root systems. Fix a system Σ^+ of positive roots in Σ.
Set $\Sigma_*^+ = \Sigma_* \cap \Sigma^+$. Denote by $\mathfrak{a}_{\mathbb{C}}$ the complexification of $\mathfrak{a}$:

$$\mathfrak{a}_{\mathbb{C}} = \mathfrak{a} \oplus \mathfrak{b} = \mathfrak{a} \otimes_{\mathbb{R}} \mathbb{C}, \quad \mathfrak{b} = i\mathfrak{a}$$

and by $A_{\mathbb{C}}$ the complex torus with Lie algebra $\mathfrak{a}_{\mathbb{C}}$. We have the polar decomposition
$A_{\mathbb{C}} = AB$ with $A = \exp \mathfrak{a}$ the split form and $B = \exp \mathfrak{b}$ the compact form.

The Weyl group $W = W(\Sigma)$ is generated by the reflections r_α for $\alpha \in \Sigma$. A
multiplicity function $m : \Sigma \to \mathbb{C}$ is a W-invariant function. Write $m_\alpha = m(\alpha)$ [1].
Set $m_\alpha = 0$ if $\alpha \notin \Sigma$. A multiplicity function is said to be positive if $m_\alpha \geq 0$ for
all α. The set of multiplicity functions is denoted by $\mathcal{M}$ and the subset of positive
multiplicity functions is denoted by $\mathcal{M}^+$. Let

$$\rho = \rho(m) = \frac{1}{2} \sum_{\alpha \in \Sigma^+} m_\alpha \alpha.$$

For $\lambda \in \mathfrak{a}_{\mathbb{C}}^*$ and $\alpha \in \mathfrak{a}^*$ with $\alpha \neq 0$, set

$$\lambda_\alpha := \frac{\langle \lambda, \alpha \rangle}{\langle \alpha, \alpha \rangle}.$$

Let P be the weight lattice of Σ, that is,

$$P = \{\lambda \in \mathfrak{a}^* \mid \lambda_\alpha \in \mathbb{Z}, \ \forall \alpha \in \Sigma\}.$$

Write $\exp : \mathfrak{a}_{\mathbb{C}} \to A_{\mathbb{C}}$ for the exponential map and $\log : A_{\mathbb{C}} \to \mathfrak{a}_{\mathbb{C}}$ the multi-valued
inverse. For $\lambda \in \mathfrak{a}_{\mathbb{C}}^*$, we define the function e^λ on $A_{\mathbb{C}}$ by

$$(2.2) \qquad e^\lambda(a) = a^\lambda := e^{\lambda(\log a)}.$$

If $\lambda \in P$, we see that e^λ is well defined and single-valued on $A_{\mathbb{C}}$. So $a \mapsto a^\lambda$ $(\lambda \in P)$
is a character of $A_{\mathbb{C}}$. Since $\exp : \mathfrak{a} \to A$ is a bijection, (2.2) is well defined and
single-valued on A for all $\lambda \in \mathfrak{a}_{\mathbb{C}}^*$. Denote by $\mathbb{C}[P]$ (or $\mathbb{C}[A_{\mathbb{C}}]$) the $\mathbb{C}$-linear span of
e^λ with $\lambda \in P$ satisfying $e^\lambda \cdot e^\mu = e^{\lambda+\mu}$, $(e^\lambda)^{-1} = e^{-\lambda}$, and $e^0 = 1$. An element of
$\mathbb{C}[P]$ is an exponential polynomial on $A_{\mathbb{C}}$ of the form $\sum_{\lambda \in P} c_\lambda e^\lambda$ where $c_\lambda \in \mathbb{C}$ and
$c_\lambda \neq 0$ for only finitely many $\lambda \in P$.

---

[1] Our multiplicity notation is different from the one used by Heckman and Opdam. The root
system R they use is related to our 2Σ, and the multiplicity function k in Heckman and Opdam's
work is related to our m by $k_{2\alpha} = \frac{1}{2}m_\alpha$.

3. The Heckman-Opdam Hypergeometric Functions

In this section we mainly review some facts of the theory of Heckman-Opdam hypergeometric functions which will be used later. The Harish-Chandra expansion was the basic tool in Heckman and Opdam's theory of (generalized) hypergeometric functions for arbitrary multiplicity functions. For a complete construction of the Heckman-Opdam hypergeometric functions, see [3, 4, 10, 11] or [5, Part I, Chapter 4].

Let $\{\xi_j\}_{j=1}^n$ be an orthonormal basis for $\mathfrak{a}$ and ∂_{ξ_j} the corresponding directional derivative with respect to ξ_j, i.e.

$$(\partial_{\xi_j}\phi)(X) = \frac{d}{dt}\phi(X + t\xi_j)\Big|_{t=0}.$$

We define a modified operator $ML(m) = L(m) + \langle \rho(m), \rho(m) \rangle$, where

$$L(m) := \sum_{j=1}^n \partial_{\xi_j}^2 + \sum_{\alpha \in \Sigma^+} m_\alpha \frac{1 + e^{-2\alpha}}{1 - e^{-2\alpha}} \partial_\alpha.$$

Let $\Xi := \{\sum_{j=1}^n n_j \alpha_j \mid n_j \in \mathbb{Z}^+, \alpha_j \in \Pi\}$ (here Π is the set of simple roots in Σ^+). The Harish-Chandra series corresponding to a multiplicity function m is defined by

$$\Phi(\lambda, m; a) = a^{\lambda - \rho} \sum_{\mu \in \Xi} \Gamma_\mu(\lambda, m) a^{-\mu}, \qquad a \in A^+$$

where $A^+ = \{a \in A \mid e^\alpha(a) > 1, \forall \alpha \in \Sigma^+\}$ and $\lambda \in \mathfrak{a}_\mathbb{C}^*$. The coefficients $\Gamma_\mu(\lambda, m) \in \mathbb{C}$ are uniquely determined by $\Gamma_0(\lambda, m) = 1$ and the recurrence relations (using the eigenvalue equation of $L(m)$)

$$\langle \mu, \mu - 2\lambda \rangle \Gamma_\mu(\lambda, m) = 2 \sum_{\alpha \in \Sigma^+} m_\alpha \sum_{\substack{k \in \mathbb{N} \\ \mu - 2k\alpha \in \Xi}} \Gamma_{\mu - 2k\alpha}(\lambda, m) \langle \mu + \rho - 2k\alpha - \lambda, \alpha \rangle$$

provided λ satisfies $\langle \mu, \mu - 2\lambda \rangle \neq 0$ for all $\mu \in \Xi \setminus \{0\}$.

DEFINITION 3.1. Define the meromorphic functions $\widetilde{c}, c : \mathfrak{a}_\mathbb{C}^* \times \mathcal{M} \to \mathbb{C}$ by

$$c(\lambda, m) = \frac{\widetilde{c}(\lambda, m)}{\widetilde{c}(\rho, m)}, \quad \widetilde{c}(\lambda, m) = \prod_{\alpha \in \Sigma^+} \frac{\Gamma(\lambda_\alpha + \frac{m_{\alpha/2}}{4})}{\Gamma(\lambda_\alpha + \frac{m_{\alpha/2}}{4} + \frac{m_\alpha}{2})}$$

where Γ is the Euler Gamma function given by $\Gamma(x) = \int_0^\infty t^{x-1} e^{-t}\, dt$. The function c is the well-known Harish-Chandra's c-function.

We note that

$$c(\lambda, m)c(-\lambda, m) = \prod_{\alpha \in \Sigma} \frac{\Gamma(\lambda_\alpha + \frac{m_{\alpha/2}}{4})}{\Gamma(\lambda_\alpha + \frac{m_{\alpha/2}}{4} + \frac{m_\alpha}{2})}$$

is clearly W-invariant in λ whereas $c(\lambda, m)$ is not.

DEFINITION 3.2. The function

$$(3.1) \qquad F(\lambda, m; a) = \sum_{w \in W} c(w\lambda, m)\Phi(w\lambda, m; a)$$

is called the *hypergeometric function* on A associated with the triple $(\mathfrak{a}, \Sigma, m)$.

REMARK 3.3. In the theory of spherical functions the equation (3.1) is the well known Harish-Chandra's asymptotic expansion for the spherical function [2]. In that case an explicit expression for Harish-Chandra's c-function was given by Gindikin and Karpelevič [1].

THEOREM 3.4. *Let $\mathcal{P} \subset \mathcal{M}$ be defined by*

$$\mathcal{P} = \{m \in \mathcal{M} \mid \widetilde{c}(\rho, m) = 0\}.$$

Then $F(\lambda, m; a)$ is holomorphic in $\mathfrak{a}_{\mathbb{C}}^* \times (\mathcal{M} \setminus \mathcal{P}) \times T$, where T is a W-invariant tubular neighborhood of A in $A_{\mathbb{C}}$, and $F(\lambda, m; a)$ is W-invariant on the same domain (in the variables λ and a).

PROOF. See Theorem 4.4.2 in [5]. □

PROPOSITION 3.5. Let $m \in \mathcal{M}^+$. There exists a constant C such that

$$|F(\lambda, m; \exp(X + iY))| \leq C \exp(\max_{w \in W} \mathrm{Re}\, w\lambda(X) - \min_{w \in W} \mathrm{Im}\, w\lambda(Y))$$

where $X, Y \in \mathfrak{b}$ with $|\alpha(X)| \leq \pi/2$ for all $\alpha \in \Sigma$ and $\lambda \in \mathfrak{a}_{\mathbb{C}}^*$.

PROOF. This follows from Proposition 6.1 in [13]. □

4. Hypergeometric Shift Operators

We will introduce the hypergeometric shift operators of Opdam and discuss an example of such a shift operator when the root system is of type BC_n. For a comprehensive information about these shift operators we recommend the reader to check [10–12] or [5, Part I, Chapter 3].

We keep the notations as in Section 2. For the moment denote by k a multiplicity in $\mathcal{M}$. The hypergeometric shift operators with shift k were constructed in [10, 11]. They are differential operators $D(k)$ on $A_{\mathbb{C}}$ (or $\mathfrak{a}_{\mathbb{C}}$) satisfying the commuting relation

$$D(k) \circ ML(m) = ML(m + k) \circ D(k), \quad \forall m \in \mathcal{M}.$$

In particular shift operators are W-invariant on $A_{\mathbb{C}}^{\mathrm{reg}}$ where $A_{\mathbb{C}}^{\mathrm{reg}} = \{a \in A_{\mathbb{C}} \mid wa \neq a, \forall w \in W, w \neq e\}$ (cf. [5, Corollary 3.1.4]).

DEFINITION 4.1. (cf. [5, Def 3.4.1]) Let $\Sigma = \cup_{i=1}^{\ell} \mathcal{O}_i$ be the disjoint union of W-orbits in Σ where $\ell \in \mathbb{Z}^+$ is the number of Weyl group orbits in Σ. Define $e_i \in \mathcal{M}$ by $e_i(\alpha) = \delta_{ij}$ (Kronecker's symbol) for all $\alpha \in \mathcal{O}_j$. Let $\mathcal{B} = \{b_1, \cdots, b_\ell\}$ be the basis of $\mathcal{M}$ given by

$$b_i = \begin{cases} 2e_i & \text{if } 2\mathcal{O}_i \cap \Sigma = \emptyset \\ 4e_i - 2e_j & \text{if } 2\mathcal{O}_i = \mathcal{O}_j \text{ for some } j. \end{cases}$$

Note that $m \in \mathcal{M}$ is integral if and only if $m \in \mathbb{Z}\mathcal{B}$. A shift operator with shift $k \in \mathbb{Z}\mathcal{B}$ is called a raising operator if $k \in \mathbb{Z}^+\mathcal{B}$ and a lowering operator if $k \in \mathbb{Z}^-\mathcal{B}$.

REMARK 4.2. The open set $\mathcal{M} \setminus \mathcal{P}$ contains the closed subset

$$\{m \in \mathcal{M} \mid \mathrm{Re}\,(m_{\alpha/2} + m_\alpha) \geq 0, \forall \alpha \in \Sigma_*\}$$

which in turn contains the set $\mathbb{C}^+\mathcal{B}$.

Denote by D^* the formal transpose of a differential operator D on $A_{\mathbb{C}}$ with respect to the Haar measure da on A:

$$\int_A (Df(a))\, g(a)\, da = \int_A f(a)\, (D^*g(a))\, da.$$

In a same way we define the transpose of a differential operator on B. Define the weight function δ by

$$\delta = \delta(m) = \prod_{\alpha \in \Sigma^+} (e^\alpha - e^{-\alpha})^{m_\alpha}.$$

THEOREM 4.3. *(Existence of shift operators).*

(1) *There exist nontrivial shift operators of shift k if and only if $k \in \mathbb{Z}\mathcal{B}$.*
(2) *Let $k \in \mathbb{Z}^+\mathcal{B}$ and $m \in \mathcal{M}$ with $m, m \pm k \notin \mathcal{P}$. Then there is a unique shift operator G_- of shift $-k$ (a lowering operator) such that*

(4.1) $$G_-(-k, m)F(\lambda, m;\, \cdot) = F(\lambda, m - k;\, \cdot).$$

Define

$$G_+(k, m) := \delta(-k - m) \circ G_-^*(-k, m + k) \circ \delta(m).$$

It is a raising operator with shift k.

PROOF. The existence of nontrivial shift operators is asserted by Theorem 3.6 in [**11**]. Also see Theorem 3.4.3 and Corollary 3.4.4 in [**5**]. So now there is a lowering operator $G_-(-k, m)$ which satisfies the formula (4.4.8) in [**5**]. Multiplying it by

$$\frac{\widetilde{c}(\rho(m), m)}{\widetilde{c}(\rho(m - k), m - k)},$$

which is a constant depending on m and k, and has no poles by the assumption that $m, m \pm k \notin \mathcal{P}$, the derived operator is again a lowering operator with shift $-k$ and satisfies (4.1).   □

EXAMPLE 4.4. Let $\{\varepsilon_j\}_{j=1}^n$ be an orthonormal basis of $\mathfrak{a}^*$. We consider an irreducible root system of type BC_n:

$$\Sigma = \pm\{\varepsilon_i, 2\varepsilon_i, \varepsilon_j \pm \varepsilon_k \mid 1 \le i \le n, 1 \le j < k \le n\}.$$

Then the ε_i are short, $2\varepsilon_i$ are long and the roots $\varepsilon_j \pm \varepsilon_k$ are medium. We have

$$\Sigma = \bigcup_{i=1}^3 \mathcal{O}_i = \mathcal{O}_s \bigcup \mathcal{O}_m \bigcup \mathcal{O}_l$$

where the three disjoint W-orbits correspond to short, medium, and long roots, respectively. Similarly, $(m_s, m_m, m_l) \in \mathcal{M}$ are multiplicities with respect to each of them. Note that some of the multiplicities are allowed to be zero. If $\alpha \in \mathcal{O}_s$, then $2\alpha \in \mathcal{O}_l$. So by Definition 4.1,

$$b_1 = 4e_1 + 0e_2 - 2e_3$$

which means shifting up the multiplicity of short roots by 4, shifting down on long roots by 2, and no change on medium roots, that is, $b_1 = (4, 0, -2)$. Similarly,

$$b_2 = 0 + 2e_2 + 0 = (0, 2, 0), \quad b_3 = 0 + 0 + 2e_3 = (0, 0, 2).$$

Thus, $\mathcal{B} = \{b_1, b_2, b_3\}$ forms a basis of $\mathcal{M}$ (associated to Σ).

As an example of shifts in the multiplicity parameters of the hypergeometric functions using shift operators we discuss here our fundamental example. Let $l \in \mathbb{Z}$. Let $k = |l| b_1 \in \mathbb{Z}^+ \mathcal{B}$. We start with a multiplicity parameter

$$m' = (m_s + 2|l|, m_m, m_l) \in \mathcal{M} \setminus \mathcal{P}.$$

It satisfies

$$
\begin{aligned}
m_+(l) &:= (m_s - 2|l|, m_m, m_l + 2|l|) \\
&= (m_s + 2|l|, m_m, m_l) - |l|(4, 0, -2) \\
&= m' - k.
\end{aligned}
$$

If $m' - k \notin \mathcal{P}$, then since $k \in \mathbb{Z}^+ \mathcal{B}$, by (4.1), there exists a shift operator G_- such that for all $\lambda \in \mathfrak{a}_{\mathbb{C}}^*$,

(4.2) $\qquad F(\lambda, m_+(l); \cdot) = G_-(-k, m')F(\lambda, m'; \cdot), \quad \text{with } k = |l| b_1.$

In short, applying $G_-(-k, m')$ to the hypergeometric function with parameter m' gives the hypergeometric function with parameter $m_+(l)$.

EXAMPLE 4.5. (cf. Example 1.3.2 and Proposition 3.3.1 in [5]) We work on the rank one root system of type BC_1, say $\Sigma = \{\pm\alpha, \pm 2\alpha\}$. Fix $\Sigma^+ = \{\alpha, 2\alpha\}$ with $\langle \alpha, \alpha \rangle = 1$. Set $k_1 = m_\alpha$ and $k_2 = m_{2\alpha}$. The Weyl group W associated to this Σ is just $\{\pm 1\}$. It is well-known that $\mathbb{C}[P] = \mathbb{C}[x, x^{-1}]$ with $x = e^{2\alpha}$, and $\mathbb{C}[x, x^{-1}]^W = \mathbb{C}[s]$ with $s = (x + x^{-1})/2$. We have the following shift operator with shift $-b_1 = (-4, 0, 2)$ (in terms of the coordinate x)

(4.3) $\qquad E_- = E_-(-b_1, m) = \dfrac{1 - x^{-1}}{1 + x^{-1}} x \dfrac{d}{dx} + C, \quad C = k_1 + k_2 - 1.$

For any $m \in \mathcal{M}$ and all $F, H \in \mathbb{C}[P]$, define an inner product

$$
\begin{aligned}
(F, H)_m &= \int_B F(x)\overline{H(x)}|\delta(m, x)| \, dx \\
&= |W| \int_{B^+} F(x)\overline{H(x)}\delta(m, x) \, dx
\end{aligned}
$$

provided that the integral exists. Here $|W|$ is the number of elements in W and B^+ is a positive Weyl chamber associated to a choice of positive roots. Thus B^+ is open, $W \cdot B^+$ is open, dense and of full measure in B and dx is the Haar measure on B normalized by $\int_B dx = 1$.

PROPOSITION 4.6. If $F, H \in \mathbb{C}[P]^W$, then

$$(G_-(-k, m)F, H)_{m-k} = (F, G_+(k, m-k)H)_m, \quad \forall k \in \mathbb{Z}^+ \mathcal{B}.$$

PROOF. We have

$$(G_-(-k, m)F, H)_{m-k}$$

$$= |W| \int_{B^+} [G_-(-k, m)F(x)]\overline{H(x)}\delta(m-k, x)\, dx$$

$$= (F, G_-^*(-k, m)H)_{m-k}$$

$$= |W| \int_{B^+} F(x)\overline{G_-^*(-k, m)\, H(x)}\delta(m-k, x)\, dx$$

$$= |W| \int_{B^+} F(x)\overline{G_-^*(-k, m) \circ \delta(m-k)H(x)}\, dx$$

$$= |W| \int_{B^+} F(x)\overline{\delta(-m) \circ G_-^*(-k, m) \circ \delta(m-k)H(x)}\delta(m, x)\, dx$$

$$= |W| \int_{B^+} F(x)\overline{G_+(k, m-k)H(x)}\delta(m, x)\, dx$$

$$= (F, G_+(k, m-k)\, H)_m,$$

where on B^+, $G_+(k, m-k) := \delta(-m) \circ G_-^*(-k, m) \circ \delta(m-k)$ is a differential operator (note: G_+ is a raising operator with shift k, the same as in Theorem 4.3).   □

5. An Application to the χ-spherical functions

We consider the situation when root multiplicities correspond to those of an irreducible Hermitian compact symmetric space. Hence Σ is of type BC_n and $m_l = 1$. Then we describe how to apply shift operators to the χ-spherical functions which are connected to the hypergeometric functions associated with root system Σ and certain nonpositive multiplicities (cf. Proposition 5.1).

Let us briefly review some necessary notations and facts about symmetric spaces. The book [6] is served as a good reference. We keep most of notations as in [8] and refer to it for more information.

Let G be a noncompact semisimple connected Lie group with Lie algebra $\mathfrak{g}$. Denote by θ a Cartan involution on G. Write $\mathfrak{g} = \mathfrak{k} \oplus \mathfrak{s}$ where $\mathfrak{k} = \mathfrak{g}^\theta$ and $\mathfrak{s} = \mathfrak{g}^{-\theta}$ for the decomposition of $\mathfrak{g}$ into θ-eigenspaces. Let $K = G^\theta$. The compact dual of G is denoted by U. The space G/K is a Riemannian symmetric space of the noncompact type, and its dual U/K is the one of compact type. Let $\mathfrak{a}$ be a maximal abelian subspace in $\mathfrak{s}$. Then $n = \dim \mathfrak{a}$ is the rank of G/K. Let $\Sigma \subset \mathfrak{a}^* \setminus \{0\}$ be a system of roots of $\mathfrak{a}$ in $\mathfrak{g}$. Choose a positive root system Σ^+ in Σ. Set $\mathfrak{b} = i\mathfrak{a}$. Let $A = \exp \mathfrak{a}$ and $B = \exp \mathfrak{b}$.

We assume, in addition, that G/K (resp. U/K) is an irreducible Hermitian symmetric space. Write $m = (m_s, m_m, m_l)$ for the multiplicities of roots in Σ as before. In our case, $m_l = 1$.

Choose a generator χ_1 for the set of one-dimensional characters of K. Then a (nontrivial) character of K is given by $\chi = \chi_l = \chi_1^l$ for some $l \in \mathbb{Z}$. The χ-spherical functions $\varphi_{\lambda,l}$ on G with spectral parameters $\lambda \in \mathfrak{a}_{\mathbb{C}}^*$ can be defined by integral or differential equations (cf. [8, Section 4]). These functions are closely related to the Heckman-Opdam hypergeometric functions in the sense which will be explained as follows. Let $\mathcal{O}_s^+$ be the set of short roots in Σ^+. Define

$$\eta_l^{\pm} = \prod_{\alpha \in \mathcal{O}_s^+} \left(\frac{e^{\alpha} + e^{-\alpha}}{2} \right)^{\pm 2|l|}.$$

Note that η_l^+ is holomorphic on $A_{\mathbb{C}}$.

PROPOSITION 5.1. In case G/K is an irreducible Hermitian symmetric space we have

$$\varphi_{\lambda,l}|_A = \eta_l^{\pm} F(\lambda, m_{\pm}(l); \cdot)$$

where $\lambda \in \mathfrak{a}_{\mathbb{C}}^*$, $m_{\pm}(l) \in \mathcal{M} \cong \mathbb{C}^3$ is given by

$$m_{\pm}(l) = (m_s \mp 2|l|, m_m, m_l \pm 2|l|),$$

and the $\pm$ sign indicates that both possibilities are valid.

PROOF. See [5, p.76, Theorem 5.2.2] for the proof of this proposition. □

Note that the space of smooth sections of homogeneous line bundles $\mathcal{L}_\chi$ over U/K is isomorphic to the space of all smooth functions $f : U \to \mathbb{C}$ such that

$$f(uk) = \chi(k)^{-1} f(u), \qquad \forall k \in K, u \in U.$$

We consider the subspace $C_r^{\infty}(U//K; \mathcal{L}_\chi)$ whose elements satisfy

$$f(k_1 u k_2) = \chi(k_1 k_2)^{-1} f(u), \qquad \forall k_1, k_2 \in K, u \in U$$

and the support of f is contained in a geodesic ball of radius r for some sufficient small $r > 0$. The χ-spherical Fourier transform is defined by

$$(5.1) \qquad \mathcal{S}(f)(\lambda) = \frac{1}{|W|} \int_U f(u) \varphi_{\lambda + \rho, l}(u^{-1}) du, \quad \lambda \in \mathfrak{a}_{\mathbb{C}}^*$$

where the normalization is chosen to simplify some formulas involving integration over B^+ (see below). In [8, Theorem 5.1], a Paley-Wiener theorem for the χ-spherical Fourier transform of these line bundles was proved. The noncompact case was treated in [14].

For the compact group U we have the Cartan decomposition $U = KBK$. But $B = \cup_{w \in W} w \overline{B^+}$ with B^+ a positive Weyl chamber. The χ-spherical Fourier transform (5.1) becomes (in view of Proposition 5.1), up to a constant,

$$(5.2) \qquad \mathcal{S}(f)(\lambda) = \int_{B^+} f(x) \eta_l^+(x) F(\lambda + \rho, m_+(l); x) \delta(x) \, dx.$$

Here, $\delta(x) = \delta(m)(x) = \delta(m, x)$. Our goal is to prove that (5.2) has at most an exponential growth of type r by using Proposition 3.5 and some shift operators. Therefore, assume that $f|_B$ (still call it f) is compactly supported in a geodesic ball of radius r with r small enough. Define

$$\Delta_s := \prod_{\alpha \in \mathcal{O}_s^+} |e^{\alpha} - e^{-\alpha}| = |\delta(1, 0, 0)|,$$

and similarly, $\Delta_m = |\delta(0, 1, 0)|$, $\Delta_l = |\delta(0, 0, 1)|$. Note that the absolute value is not needed on B^+. In the following we will for simplicity write $\varphi_{\lambda}(m; x)$ for $F(\lambda + \rho, m; x)$ and $d := (\frac{\Delta_s}{\Delta_l})^{2|l|}$.

Recall the notations from Example 4.4: $k = |l|b_1 \in \mathbb{Z}^+\mathcal{B}$, $m' = (m_s + 2|l|, m_m, m_l) \in \mathcal{M}^+$, and $m_+(l) = m' - k$. By (4.2) and Proposition 4.6, we have up to a constant,

$$(5.3)\,\mathcal{S}(f)(\lambda) \;=\; \int_{B^+} f(x)\eta_l^+(x)\varphi_\lambda(m_+(l);x)\delta(m,x)\,dx$$

$$\overset{(4.2)}{=} \int_{B^+} f(x)\eta_l^+(x)G_-(-k,m')\varphi_\lambda(m';x)\delta(m,x)\,dx$$

$$= \int_{B^+} f(x)\eta_l^+(x)G_-(-k,m')\varphi_\lambda(m';x)(\Delta_s^{m_s}\Delta_m^{m_m}\Delta_l)(x)\,dx$$

$$= \int_{B^+} d(x)f(x)\eta_l^+(x)G_-(-k,m')\varphi_\lambda(m';x)$$
$$\cdot\,(\Delta_s^{m_s-2|l|}\Delta_m^{m_m}\Delta_l^{1+2|l|})(x)\,dx$$

$$(5.4)\qquad = \int_{B^+} d(x)f(x)\eta_l^+(x)G_-(-k,m')\varphi_\lambda(m';x)\delta(m_+(l),x)\,dx$$

$$= \int_{B^+} G_+(k,m_+(l))\,\big[d(x)f(x)\eta_l^+(x)\big]\,\varphi_\lambda(m';x)\delta(m',x)\,dx\,.$$

Since $m = (m_s, m_m, 1) \in \mathcal{M}^+$, the term $\delta(m,x)$ in (5.3) will not blow up and so the integral (5.3) is finite. The subsequent four integrals are simply rewritings of (5.3), so they are finite too. Hence, it makes sense to apply Proposition 4.6 to (5.4). Moreover,

$$\frac{\Delta_s}{\Delta_l} = \prod_{\alpha\in\mathcal{O}_s^+}\left|\frac{e^\alpha - e^{-\alpha}}{e^{2\alpha} - e^{-2\alpha}}\right| = \prod_{\alpha\in\mathcal{O}_s^+}\left|\frac{1}{e^\alpha + e^{-\alpha}}\right|,$$

the denominator is never zero (because $|\alpha(\cdot)| \le \pi/4$ for $\alpha \in \mathcal{O}_s^+$), so this term will not blow up. It follows that d is a bounded function on B. Since $m' \in \mathcal{M}^+$, we can use the estimate given in Proposition 3.5. So there is a $C > 0$ such that

$$|\mathcal{S}(f)(\lambda)| \;\le\; \int_{B^+}\big|G_+(k,m_+(l))\,\big[d(x)f(x)\eta_l^+(x)\big]\big|\,|\varphi_\lambda(m';x)|\delta(m',x)\,dx$$

$$\le\; C\max_{x\in B^+}\{|d(x)f(x)\eta_l^+(x)|\}\exp(\max_{w\in W}\mathrm{Re}w(\lambda+\rho)(X))\delta(m',x),$$

where we write $x = e^X$ with X in a ball centered at 0 of radius r and

$$|\exp(\max_{w\in W}\mathrm{Re}w(\lambda+\rho)(X))| < e^{r\|\mathrm{Re}\lambda\|}.$$

On the other hand, since f has a compact support in a ball of radius r, and since η_l^+ and d are both bounded functions, applying a differential operator G_+ doesn't increase the radius of the ball, giving that

$$(5.5)\qquad \big|G_+(k,m_+(l))\,\big[d(x)f(x)\eta_l^+(x)\big]\big| \le \max_{x\in B^+}\{|d(x)f(x)\eta_l^+(x)|\}.$$

The right-hand side of (5.5) is actually a constant depending on r. Moreover, $|\delta(m',x)|$ is bounded by some constant. Hence, there is a constant $C_1 > 0$ (depending on r) such that

$$|\mathcal{S}(f)(\lambda)| \le C_1 e^{r\|\mathrm{Re}\lambda\|}.$$

The polynomial decay of $\mathcal{S}(f)(\lambda)$ is obtained by applying an invariant differential operator on G/K to $\varphi_{\lambda,l}$. Hence, $\mathcal{S}(f)(\lambda)$ has an exponential growth of type r.

6. The Rank One Case

We treat the rank one case as a simple example for the previous results in Section 5. In this case the χ-spherical Fourier transform is given by (5.2) with dx being the invariant measure on the torus $\mathbb{T}$. We give an explicit choice of G_- and show that how easily we can achieve an exponential estimate for $\mathcal{S}(f)(\lambda)$ in this case.

The rank one case corresponds to $n = \dim \mathfrak{a} = 1$. The only rank one Hermitian compact symmetric space U/K is

$$\mathrm{SU}(q+1)/S(\mathrm{U}(1) \times \mathrm{U}(q)), \quad q \geq 1.$$

This is the Grassmann manifold of one-dimensional linear subspaces of $\mathbb{C}^{q+1}$. The root system Σ is of type BC_1. Recall Example 4.5 for more information. In this case we have $k_1 = 2(q-1)$, $q \geq 1$, and $k_2 = 1$. That is, root multiplicities associated to this symmetric space are $(m_s, m_m, m_l) = (2(q-1), 0, 1)$. We identify $i\mathfrak{a}$ with $i\mathbb{R}$, and $\mathfrak{a}_{\mathbb{C}}$ with $\mathbb{C}$. So $B = \exp \mathfrak{b} \cong \mathbb{T}$. The Weyl group $W = \{\pm 1\}$ acting on $i\mathbb{R}$ and $\mathbb{C}$ by multiplication.

A good candidate for G_- is $E_-^{|l|}$ which means repeated applications of E_- the number of $|l|$ times (Here $l \in \mathbb{Z}$ and E_- is given by (4.3)). By (4.2),

$$\begin{aligned}
F(\lambda + \rho, m_+(l); \cdot) &= G_-(-|l|b_1, m')F(\lambda + \rho, m'; \cdot) \\
&= \underbrace{E_-(-b_1, m_+(l) + b_1) \cdots E_-(-b_1, m')}_{|l| \text{ copies}} F(\lambda + \rho, m'; \cdot)
\end{aligned}$$

We can write an element $x \in B$ as $x = e^{it}$ with $-\pi \leq t \leq \pi$. We also need to ensure $f|_B$ compactly supported in a geodesic ball which meets the conditions in Proposition 3.5, so it reduces the domain to $-\pi/2 \leq t \leq \pi/2$. Note that $f|_B$ is W-invariant, so it is a function of s with $s = (x + x^{-1})/2 = \cos t$. We know from Theorem 3.4 that the hypergeometric functions $F(\lambda, m; x)$ are W-invariant in x. A similar argument asserts that $F(\lambda, m; \cdot)$ can be viewed as a function of s. Since all shift operators are W-invariant, in particular, so is E_-. With change of variables $x = e^{it}$ and $s = \cos t$, the operator (4.3) is just a first order differential operator, having the form

$$E_- = (s-1)\frac{d}{ds} + C$$

where C is the same constant as in (4.3). In the rank one case, $\alpha \in \Sigma^+$ is the only short root, so

$$\eta_l^+(x) = \eta_l^+(e^{it}) = (\cos(t/2))^{2|l|} = \left(\frac{1+s}{2}\right)^{|l|}.$$

Next, the weight measure becomes

$$|\delta(m,x)|\, dx = 2^q (1-s)^{q-1}\, ds, \qquad q \geq 1.$$

We choose a positive Weyl chamber B^+ such that $0 \leq t \leq \pi/2$. Then $0 \leq s \leq 1$, and $ds = \sin t\, dt$ is the invariant measure. With change of the variable $s = (x + x^{-1})/2$, $0 \leq s \leq 1$, we have, up to a constant,

$$\begin{aligned}
\mathcal{S}(f)(\lambda) &= \int_{B^+} f(x)\eta_l^+(x)[(E_- \cdots E_-)\varphi_\lambda(m'; x)]\delta(m,x)\, dx \\
(6.1) \qquad &= \int_0^1 f(s)\left(\frac{1+s}{2}\right)^{|l|} 2^q (1-s)^{q-1}[((s-1)\frac{d}{ds} + C)^{|l|}\varphi_\lambda(m'; x)]\, ds.
\end{aligned}$$

Note that $f(s)$ is smooth compactly supported in a ball of radius r, and vanishes on the boundary. Also, the function

$$2^q(1-s)^{q-1}\left(\frac{1+s}{2}\right)^{|l|}$$

is smooth and bounded. Since each E_- is a first order differential operator, we can apply integration by parts ($|l|$ times) to (6.1) and then use Proposition 3.5 to get the desired r-type exponential growth of $\mathcal{S}(f)(\lambda)$.

We mention in the next remark an alternative method to prove $\mathcal{S}(f)(\lambda)$ has exponential growth of type r in some particular higher rank cases. It is a simple and nice generalization of the rank one case.

REMARK 6.1. Suppose U/K is a rank n (for $n > 1$) Hermitian compact symmetric space which associates with the multiplicity $(m_s, m_m, 1)$, where m_m is even. Let $\mathfrak{u}$ be the Lie algebra of U. Let $\mathfrak{u}_j$ be rank one Lie algebras, $j = 1, \ldots, n$, associated to $(2(q-1), 0, 1)$ with q chosen so that $2(q-1) = m_s$. Then $\mathfrak{b} = \mathfrak{b}_1 \oplus \cdots \oplus \mathfrak{b}_n$ is a maximal abelian subspace in $\mathfrak{u}$, where each $\mathfrak{b}_j$ is a maximal abelian subspace in $\mathfrak{u}_j$. Let $B = \exp \mathfrak{b}$. Then $B = B_1 \times \cdots \times B_n$ with $B_j = \exp \mathfrak{b}_j$. So $x = (x_1, \ldots, x_n) \in B$, $|\delta(x)| = \prod_{j=1}^n |\delta(x_j)|$, and

$$\begin{aligned} F(\lambda, m; x) &= F((\lambda_1, \ldots, \lambda_n), m; (x_1, \ldots, x_n)) \\ &= \prod_{j=1}^n F(\lambda_j, m; x_j) \end{aligned}$$

Let $f \in C_r^\infty(U//K; \mathcal{L}_\chi)$. We write $f(x) = f(x_1, \ldots, x_n)$, $x \in B$. We have (5.2) as the χ-spherical Fourier transform of f with B^+ a positive Weyl chamber in the higher rank Lie group.

Let G_+ be the fundamental shift operator with shift $b_2 = (0,2,0)$, repeatedly applied $m_m/2$ times. The existence of such a G_+ is asserted by Theorem 4.3. The hypergeometric function $F(\lambda, m_+(l); \cdot)$ can be realized by applying G_+ (multiplied by some constant factor) to $F(\lambda, (m_s - 2|l|, 0, 1 + 2|l|); \cdot)$. Let

$$G_- = 4^n(E_-^{|l|})^n = 4^n E_-^{|l|} \times \cdots \times E_-^{|l|}$$

where E_- is the same as in (4.3) and $E_-^{|l|}$ means applying E_- the number of $|l|$ times. The existence of G_- is also followed from Theorem 4.3 (and check [5, Theorem 3.4.3]). Therefore,

$$\begin{aligned} F(\lambda, (m_s - 2|l|, 0, 1+2|l|); x) &= G_- F(\lambda, m'; x) \\ &= 4^n \prod_{j=1}^n E_-^{|l|} F(\lambda_j, m'; x_j) \end{aligned}$$

where $m' = (m_s + 2|l|, 0, 1) \in \mathcal{M}^+$. Hence, by using the composition of G_+ and G_- we can write $\mathcal{S}(f)(\lambda)$ as a n-fold iterated integral of rank one integrals with which we have done. The desired r-type exponential growth of $\mathcal{S}(f)(\lambda)$ thus follows from the rank one case.

References

[1] S. G. Gindikin and F. I. Karpelevič, *Plancherel measure for symmetric Riemannian spaces of non-positive curvature* (Russian), Dokl. Akad. Nauk SSSR **145** (1962), 252–255. MR0150239 (27 #240)

[2] Harish-Chandra, *Spherical functions on a semisimple Lie group I-II*, Amer. J. Math. **80** (1958), 241–310 and 553–613.

[3] G. J. Heckman, *Root systems and hypergeometric functions. II*, Compositio Math. **64** (1987), no. 3, 353–373. MR918417 (89b:58192b)

[4] G. J. Heckman and E. M. Opdam, *Root systems and hypergeometric functions. I*, Compositio Math. **64** (1987), no. 3, 329–352. MR918416 (89b:58192a)

[5] Gerrit Heckman and Henrik Schlichtkrull, *Harmonic analysis and special functions on symmetric spaces*, Perspectives in Mathematics, vol. 16, Academic Press, Inc., San Diego, CA, 1994. MR1313912 (96j:22019)

[6] Sigurdur Helgason, *Differential geometry, Lie groups, and symmetric spaces*, Pure and Applied Mathematics, vol. 80, Academic Press, Inc. [Harcourt Brace Jovanovich, Publishers], New York-London, 1978. MR514561 (80k:53081)

[7] Sigurdur Helgason, *Groups and geometric analysis*, Mathematical Surveys and Monographs, vol. 83, American Mathematical Society, Providence, RI, 2000. Integral geometry, invariant differential operators, and spherical functions; Corrected reprint of the 1984 original. MR1790156 (2001h:22001)

[8] V.M. Ho and G. Ólafsson, *Paley-Wiener theorem for line bundles over compact symmetric spaces and new estimates for the Heckman-Opdam hypergeometric functions*, submitted (2014), arXiv:1407.1489.

[9] T.H. Koornwinder, *Orthogonal polynomials in two variables which are eigenfunctions of two algebraically independent partial differential operators I-IV*, Indag. Math. **36** (1974), 48–66 and 357–381.

[10] E. M. Opdam, *Root systems and hypergeometric functions. III*, Compositio Math. **67** (1988), no. 1, 21–49. MR949270 (90k:17021)

[11] E. M. Opdam, *Root systems and hypergeometric functions. IV*, Compositio Math. **67** (1988), no. 2, 191–209. MR951750 (90c:58079)

[12] E. M. Opdam, *Some applications of hypergeometric shift operators*, Invent. Math. **98** (1989), no. 1, 1–18, DOI 10.1007/BF01388841. MR1010152 (91h:33024)

[13] Eric M. Opdam, *Harmonic analysis for certain representations of graded Hecke algebras*, Acta Math. **175** (1995), no. 1, 75–121, DOI 10.1007/BF02392487. MR1353018 (98f:33025)

[14] Nobukazu Shimeno, *The Plancherel formula for spherical functions with a one-dimensional K-type on a simply connected simple Lie group of Hermitian type*, J. Funct. Anal. **121** (1994), no. 2, 330–388, DOI 10.1006/jfan.1994.1052. MR1272131 (95b:22036)

DEPARTMENT OF MATHEMATICS, LOUISIANA STATE UNIVERSITY, BATON ROUGE, LOUISIANA 70803

E-mail address: vivian@math.lsu.edu

DEPARTMENT OF MATHEMATICS, LOUISIANA STATE UNIVERSITY, BATON ROUGE, LOUISIANA 70803

E-mail address: olafsson@math.lsu.edu

Contemporary Mathematics
Volume **650**, 2015
http://dx.doi.org/10.1090/conm/650/13009

Harmonic analysis of a class of reproducing kernel Hilbert spaces arising from groups

Palle E. T. Jorgensen, Steen Pedersen, and Feng Tian

ABSTRACT. We study two extension problems, and their interconnections: (i) extension of positive definite continuous functions defined on subsets in locally compact groups G; and (ii) (in case of Lie groups G) representations of the associated Lie algebras $La\,(G)$, i.e., representations of $La\,(G)$ by unbounded skew-Hermitian operators acting in a reproducing kernel Hilbert space $\mathscr{H}_F$ (RKHS). Our analysis is non-trivial even if $G = \mathbb{R}^n$, and even if $n = 1$. If $G = \mathbb{R}^n$, (ii), we are concerned with finding systems of strongly commuting selfadjoint operators $\{T_i\}$ extending a system of commuting Hermitian operators with common dense domain in $\mathscr{H}_F$.

Our general results include non-compact and non-Abelian Lie groups, where the study of unitary representations in $\mathscr{H}_F$ is subtle.

CONTENTS

2010 *Mathematics Subject Classification.* Primary 47L60, 46N30, 46N50, 42C15, 65R10; Secondary 46N20, 22E70, 31A15, 58J65, 81S25.

Key words and phrases. Unbounded operators, deficiency-indices, Hilbert space, reproducing kernels, boundary values, unitary one-parameter group, convex, harmonic decompositions, stochastic processes, representations of Lie groups, potentials, quantum measurement, renormalization, partial differential operators, rank-one perturbation, Friedrichs extension, Green's function.

The co-authors thank the following for enlightening discussions: Professors Sergii Bezuglyi, Dorin Dutkay, Paul Muhly, Myung-Sin Song, Wayne Polyzou, Gestur Olafsson, Robert Niedzialomski, and members in the Math Physics seminar at the University of Iowa.

1. Introduction

The theory of positive definite functions has a large number of applications in a host of areas; for example, in harmonic analysis, in representation theory (of both algebras and groups), in physics, and in the study of probability models, such as stochastic processes. One reason for this is the theorem of Bochner which links continuous positive definite functions on locally compact abelian groups G to measures on the corresponding dual group. Analogous uses of positive definite functions exist for classes for non-abelian groups. Even the seemingly modest case of $G = \mathbb{R}$ is of importance in the study of spectral theory for Schrödinger equations. And further, counting the study of Gaussian stochastic processes, there are even generalizations (Gelfand-Minlos) to the case of continuous positive definite functions on Fréchet spaces of test functions which make up part of a Gelfand triple. These cases will be explored below, but with the following important change in the starting point of the analysis; – we focus on the case when the given positive definite function is only partially defined, i.e., is only known on a proper subset of the ambient group, or space. How much of duality theory carries over when only partial information is available?

In detail; the purpose of the present paper is to explore what can be said when the given continuous positive definite function is only given on a subset of the ambient group (which is part of the application setting sketched above.) For this problem of partial information, even the case of positive definite functions defined only on bounded subsets of $G = \mathbb{R}$ (say an interval), or on bounded subsets of $G = \mathbb{R}^n$, is of substantial interest.

In this paper, we study two extension problems, and their interconnections. The first class of extension problems concerns (i) positive definite continuous functions on Lie groups G, and the second deals with (ii) Lie algebras of unbounded skew-Hermitian operators in a certain family of reproducing kernel Hilbert spaces (RKHS). Our analysis is non-trivial even if $G = \mathbb{R}^n$, and even if $n = 1$. If $G = \mathbb{R}^n$, we are concerned in (ii) with the study of systems of n skew-Hermitian operators $\{S_i\}$ on a common dense domain in Hilbert space, and in deciding whether it is possible to find a corresponding system of strongly commuting selfadjoint operators $\{T_i\}$ such that, for each value of i, the operator T_i extends S_i.

The version of this for non-commutative Lie groups G will be stated in the language of unitary representations of G, and corresponding representations of the Lie algebra $La(G)$ by skew-Hermitian unbounded operators.

In summary, for (i) we are concerned with partially defined positive definite continuous functions F on a Lie group; i.e., at the outset, such a function F will only be defined on a connected proper subset in G. From this partially defined p.d. function F we then build a reproducing kernel Hilbert space $\mathscr{H}_F$, and the operator extension problem (ii) is concerned with operators acting on $\mathscr{H}_F$, as well as with unitary representations of G acting on $\mathscr{H}_F$. If the Lie group G is not simply connected, this adds a complication, and we are then making use of the associated simply connected covering group.

1.1. Two Extension Problems.

While each of the two extension problems has received a considerable amount of attention in the literature, our emphasis here will be the interplay between the two problems: Our aim is a duality theory; and, in the case $G = \mathbb{R}^n$, and $G = \mathbb{T}^n = \mathbb{R}^n/\mathbb{Z}^n$, we will state our theorems in the language of Fourier duality of abelian groups: With the time frequency duality formulation of Fourier duality for $G = \mathbb{R}^n$ we have that both the time domain and the frequency domain constitute a copy of $\mathbb{R}^n$. We then arrive at a setup such that our extension questions (i) are in time domain, and extensions from (ii) are in frequency domain. Moreover we show that each of the extensions from (i) has a variant in (ii). Specializing to $n = 1$, we arrive of a spectral theoretic characterization of all skew-Hermitian operators with dense domain in a separable Hilbert space, having deficiency-indices $(1, 1)$.

A systematic study of densely defined Hermitian operators with deficiency indices $(1, 1)$, and later (d, d), was initiated by M. Krein [35], and is also part of de Branges' model theory; see [12, 13]. The direct connection between this theme and the problem of extending continuous positive definite functions F when they are only defined on a fixed open subset to $\mathbb{R}^n$ was one of our motivations. One desires continuous p.d. extensions to $\mathbb{R}^n$.

If F is given, we denote the set of such extensions $Ext\,(F)$. If $n = 1$, $Ext\,(F)$ is always non-empty, but for $n = 2$, Rudin gave examples in [43, 44] when $Ext\,(F)$ may be empty. Here we extend these results, and we also cover a number of classes of positive definite functions on locally compact groups in general; so cases when $\mathbb{R}^n$ is replaced with other groups, both Abelian and non-abelian.

Our results in the framework of locally compact Abelian groups are more complete than their counterparts for non-Abelian Lie groups, one reason is the availability of Bochner's duality theorem for locally compact Abelian groups; – not available for non-Abelian Lie groups.

1.2. Earlier Papers.

Below we mention some earlier papers dealing with one or the other of the two extension problems (i) or (ii). To begin with, there is a rich literature on (i), a little on (ii), but comparatively much less is known about their interconnections.

As for positive definite functions, their use and applications are extensive and includes such areas as stochastic processes, see e.g., [6, 7, 23, 34]; harmonic analysis (see [11, 32, 33]) , and the references there); potential theory [19, 36]; operators in Hilbert space [2, 4, 5]; and spectral theory [10, 14, 15, 40]. We stress that the literature is vast, and the above list is only a small sample.

Extensions of positive definite continuous functions defined on subsets of Lie groups G was studied in [29]. In our present analysis of connections between extensions of positive definite continuous functions and extension questions for associated operators in Hilbert space, we will be making use of tools from spectral theory, and from the theory of reproducing kernel-Hilbert spaces, such as can be found in e.g., [3, 8, 24, 38].

There is a different kind of notion of positivity involving reflections, restrictions, and extensions. It comes up in physics and in stochastic processes [18, 21, 22, 31], and is somewhat related to our present theme. While they have several names, "refection positivity" is a popular term [**32, 33**].

In broad terms, the issue is about realizing geometric reflections as "conjugations" in Hilbert space. When the program is successful, for a given unitary representation U of a Lie group G, for example $G = \mathbb{R}$, it is possible to renormalize the Hilbert space on which U is acting.

Now the Bochner transform F of a probability measure (e.g., the distribution of a stochastic process) which further satisfies refection positivity, has two positivity properties: one (i) because F is the transform of a positive measure, so F is positive definite; and in addition the other, (ii) because of refection symmetry. We have not followed up below with structural characterizations of this family of positive definite functions, but readers interested in the theme, will find details in [**9, 32, 33, 41**], and in the references given there.

1.3. Preliminaries. In our theorems and proofs, we shall make use of the particular reproducing kernel Hilbert spaces (RKHSs) which allow us to give explicit formulas for our solutions. The general framework of RKHSs were pioneered by Aronszajn in the 1950s [**8**]; and subsequently they have been used in a host of applications; e.g., [**49, 50**].

The RKHS $\mathscr{H}_F$. For simplicity we focus on the case $G = \mathbb{R}$. Modifications for other groups will be described in the body of paper.

DEFINITION 1.1. Fix $0 < a$, let $\Omega = (0, a)$. A function $F : \Omega - \Omega \to \mathbb{C}$ is *positive definite* if

$$(1.1) \qquad \sum_i \sum_j \overline{c_i} c_j F(x_i - x_j) \geq 0$$

for all finite sums with $c_i \in \mathbb{C}$, and all $x_i \in \Omega$. We assume that all the p.d. functions are continuous and bounded.

LEMMA 1.2. *F is p.d. iff*

$$\int_\Omega \int_\Omega \overline{\varphi(x)} \varphi(y) F(x - y) dx dy \geq 0 \text{ for all } \varphi \in C_c^\infty(\Omega).$$

PROOF. Standard.   □

Consider a continuous positive definite function so F is defined on $\Omega - \Omega$. Set

$$(1.2) \qquad F_y(x) := F(x - y), \ \forall x, y \in \Omega.$$

Let $\mathscr{H}_F$ be the *reproducing kernel Hilbert space (RKHS)*, which is the completion of

$$(1.3) \qquad \left\{ \sum_{\text{finite}} c_j F_{x_j} \mid x_j \in \Omega, \ c_j \in \mathbb{C} \right\}$$

with respect to the inner product

$$(1.4) \qquad \left\langle \sum_i c_i F_{x_i}, \sum_j d_j F_{y_j} \right\rangle_{\mathscr{H}_F} := \sum_i \sum_j \overline{c_i} d_j F(x_i - y_j);$$

modulo the subspace of functions of $\|\cdot\|_{\mathscr{H}_F}$-norm zero.

Throughout, we use the convention that the inner product is conjugate linear in the first variable, and linear in the second variable. When more than one inner product is used, subscripts will make reference to the Hilbert space.

Below, we introduce an equivalent characterization of the RKHS $\mathscr{H}_F$, which we will be working with in the rest of the paper.

LEMMA 1.3. *Fix* $\Omega = (0, \alpha)$. *Let* $\varphi_{n,x}(t) = n\varphi(n(t-x))$, *for all* $t \in \Omega$; *where* φ *satisfies*

(1) supp $(\varphi) \subset (-\alpha, \alpha)$;
(2) $\varphi \in C_c^\infty$, $\varphi \geq 0$;
(3) $\int \varphi(t)\, dt = 1$. *Note that* $\lim_{n \to \infty} \varphi_{n,x} = \delta_x$, *the Dirac measure at* x.

LEMMA 1.4. *The RKHS,* $\mathscr{H}_F$, *is the Hilbert completion of the functions*

$$(1.5) \qquad F_\varphi(x) = \int_\Omega \varphi(y) F(x-y)\, dy, \ \forall \varphi \in C_c^\infty(\Omega), x \in \Omega$$

with respect to the inner product

$$(1.6) \qquad \langle F_\varphi, F_\psi \rangle_{\mathscr{H}_F} = \int_\Omega \int_\Omega \overline{\varphi(x)} \psi(y) F(x-y)\, dx dy, \ \forall \varphi, \psi \in C_c^\infty(\Omega).$$

In particular,

$$(1.7) \qquad \|F_\varphi\|_{\mathscr{H}_F}^2 = \int_\Omega \int_\Omega \overline{\varphi(x)} \varphi(y) F(x-y)\, dx dy, \ \forall \varphi \in C_c^\infty(\Omega)$$

and

$$(1.8) \qquad \langle F_\varphi, F_\psi \rangle_{\mathscr{H}_F} = \int_\Omega \overline{\varphi(x)} F_\psi(x)\, dx, \ \forall \phi, \psi \in C_c^\infty(\Omega).$$

PROOF. Indeed, by Lemma 1.4, we have

$$(1.9) \qquad \left\| F_{\varphi_{n,x}} - F(\cdot - x) \right\|_{\mathscr{H}_F} \to 0, \text{ as } n \to \infty.$$

Hence $\{F_\varphi\}_{\varphi \in C_c^\infty(\Omega)}$ spans a dense subspace in $\mathscr{H}_F$.

For more details, see [**25, 26, 28**]. □

These two conditions $(1.10)(\Leftrightarrow(1.11))$ below will be used to characterize elements in the Hilbert space $\mathscr{H}_F$.

THEOREM 1.5. *A continuous function* $\xi : \Omega \to \mathbb{C}$ *is in* $\mathscr{H}_F$ *if and only if there exists* $A_0 > 0$, *such that*

$$(1.10) \qquad \sum_i \sum_j \overline{c_i} c_j \overline{\xi(x_i)} \xi(x_j) \leq A_0 \sum_i \sum_j \overline{c_i} c_j F(x_i - x_j)$$

for all finite system $\{c_i\} \subset \mathbb{C}$ *and* $\{x_i\} \subset \Omega$.
Equivalently, for all $\psi \in C_c^\infty(\Omega)$,

$$(1.11) \qquad \left| \int_\Omega \psi(y) \overline{\xi(y)} dy \right|^2 \leq A_0 \int_\Omega \int_\Omega \overline{\psi(x)} \psi(y) F(x-y)\, dx dy$$

Note that, if $\xi \in \mathscr{H}_F$, *then the LHS of* (1.11) *is* $|\langle F_\psi, \xi \rangle_{\mathscr{H}_F}|^2$. *Indeed,*

$$\left| \langle F_\psi, \xi \rangle_{\mathscr{H}_F} \right|^2 = \left| \left\langle \int_\Omega \psi(y) F_y\, dy, \xi \right\rangle_{\mathscr{H}_F} \right|^2$$

$$= \left| \int_\Omega \psi(y) \langle F_y, \xi \rangle_{\mathscr{H}_F}\, dy \right|^2$$

$$= \left| \int_\Omega \psi(y) \xi(y)\, dy \right|^2 \ \textit{(by the reproducing property of } F_\psi\textit{)}$$

The Operator $D^{(F)}$. Fix $0 < a$ and a continuous positive definite function F defined on $\Omega - \Omega$, where $\Omega = (0, a)$ as in Definition 1.1. Let $\mathscr{H}_F$ be the corresponding RKHS as in (1.3).

DEFINITION 1.6. Define $D^{(F)}$ on the dense domain $C_c^\infty (\Omega) * F$ by $D^{(F)} F_\psi = F_{\psi'}$, where $\psi' = \frac{d\psi}{dt}$.

One shows that $D^{(F)}$ is well defined by using Schwarz' inequality to prove that $F_\psi = 0$ implies that $F_{\psi'} = 0$.

LEMMA 1.7. $D^{(F)}$ *is a well-defined operator with dense domain in $\mathscr{H}_F$. Moreover, it is skew-symmetric and densely defined in $\mathscr{H}_F$.*

PROOF. By Lemma 1.4 dom $\left(D^{(F)} \right)$ is dense in $\mathscr{H}_F$. If $\psi \in C_c^\infty (0, a)$ and $|t| < \text{dist} \left(\text{supp} (\psi), \text{endpoints} \right)$, then

$$(1.12) \qquad \left\| F_{\psi(\cdot + t)} \right\|_{\mathscr{H}_F}^2 = \left\| F_\psi \right\|_{\mathscr{H}_F}^2 = \int_0^a \int_0^a \overline{\psi (x)} \psi (y) F (x - y) \, dx dy$$

see (1.7), so

$$\frac{d}{dt} \left\| F_{\psi(\cdot + t)} \right\|_{\mathscr{H}_F}^2 = 0$$

which is equivalent to

$$(1.13) \qquad \left\langle D^{(F)} F_\psi, F_\psi \right\rangle_{\mathscr{H}_F} + \left\langle F_\psi, D^{(F)} F_\psi \right\rangle_{\mathscr{H}_F} = 0.$$

It follows that $D^{(F)}$ is skew-symmetric.

To show that $D^{(F)} F_\psi = F_{\psi'}$ is a well-defined operator on is dense domain in $\mathscr{H}_F$, we proceed as follows:

LEMMA 1.8. *The following implication holds:*

$$(1.14) \qquad [\psi \in C_c^\infty (\Omega), \ F_\psi = 0 \ in \ \mathscr{H}_F]$$

$$\Downarrow$$

$$(1.15) \qquad [F_{\psi'} = 0 \ in \ \mathscr{H}_F]$$

PROOF. Substituting (1.14) into

$$\left\langle F_\varphi, F_{\psi'} \right\rangle_{\mathscr{H}_F} + \left\langle F_{\varphi'}, F_\psi \right\rangle_{\mathscr{H}_F} = 0$$

we get

$$\left\langle F_\varphi, F_{\psi'} \right\rangle_{\mathscr{H}_F} = 0, \ \forall \varphi \in C_c^\infty (\Omega).$$

Taking $\varphi = \psi'$, yields

$$\left\langle F_{\psi'}, F_{\psi'} \right\rangle = \left\| F_{\psi'} \right\|_{\mathscr{H}_F}^2 = 0$$

which is the desired conclusion (1.15). □

This finishes the proof of Lemma 1.7. □

LEMMA 1.9. *Let $\Omega = (\alpha, \beta)$. Suppose F is a real valued positive definite function defined on $\Omega - \Omega$. The operator J on $\mathscr{H}_F$ determined by*

$$J F_\varphi = \overline{F_{\varphi(\alpha + \beta - x)}}, \varphi \in C_c^\infty (\Omega)$$

is a conjugation, i.e., J is conjugate-linear, J^2 is the identity operator, and

$$(1.16) \qquad \left\langle J F_\phi, J F_\psi \right\rangle_{\mathscr{H}_F} = \left\langle F_\psi, F_\phi, \right\rangle_{\mathscr{H}_F}.$$

Moreover,

$$(1.17) \qquad\qquad D^{(F)} J = -J D^{(F)}.$$

PROOF. Let $a := \alpha + \beta$ and $\phi \in C_c^\infty (\Omega)$. Since F is real valued

$$J F_\phi(x) = \int_\alpha^\beta \overline{\phi(a-y)}\, \overline{F(x-y)} dy = \int_\alpha^\beta \psi(y)\, F(x-y) dy$$

where $\psi(y) := \overline{\phi(a-y)}$ is in $C_c^\infty(\Omega)$. It follows that J maps the domain dom $\left(D^{(F)}\right)$ of $D^{(F)}$ onto itself. For $\phi, \psi \in C_c^\infty(\Omega)$,

$$\begin{aligned}
\langle J F_\phi, F_\psi \rangle_{\mathscr{H}_F} &= \int_\alpha^\beta F_{\phi(a-\cdot)}(x) \psi(x) dx \\
&= \int_\alpha^\beta \int_\alpha^\beta \phi(a-y) F(x-y) \psi(x) dy dx.
\end{aligned}$$

Making the change of variables $(x,y) \to (a-x, a-y)$ and interchanging the order of integration we see that

$$\begin{aligned}
\langle J F_\phi, F_\psi \rangle_{\mathscr{H}_F} &= \int_\alpha^\beta \int_\alpha^\beta \phi(y) F(y-x) \psi(a-x) dy dx \\
&= \int_\alpha^\beta \phi(y) F_{\psi(a-\cdot)}(y) dy \\
&= \langle J F_\psi, F_\phi \rangle_{\mathscr{H}_F},
\end{aligned}$$

establishing (1.16). For $\phi \in C_c^\infty(\Omega)$,

$$J D^{(F)} F_\phi = \overline{F_{\phi'(a-\cdot)}} = -\overline{F_{\frac{d}{dx}(\phi(a-\cdot))}} = -D^{(F)} J F_\phi,$$

hence (1.17) holds. $\qquad\square$

REMARK 1.10. Note that

$$\overline{F_\psi}(x) = \int \overline{\psi(y)} F(y-x)\, dy = F_{\frac{\vee}{\psi}}(x)$$

where $F^\vee(x) = F(-x)$ is defined on $(-\beta, -\alpha)$.

DEFINITION 1.11. Let $\left(D^{(F)}\right)^*$ be the adjoint of $D^{(F)}$. The deficiency spaces $DEF^\pm$ consists of $\xi_\pm \in \mathscr{H}_F$, such that $\left(D^{(F)}\right)^* \xi_\pm = \pm \xi_\pm$. That is,

$$DEF^\pm = \left\{ \xi_\pm \in \mathscr{H}_F : \langle F_{\psi'}, \xi_\pm \rangle_{\mathscr{H}_F} = \langle F_\psi, \pm\xi_\pm \rangle_{\mathscr{H}_F}, \forall \psi \in C_c^\infty(\Omega) \right\}.$$

COROLLARY 1.12. *If F is real valued, then DEF^+ and DEF^- have the same dimension.*

PROOF. This follows from Lemma 1.9, see e.g, [**1**] or [**16**]. $\qquad\square$

LEMMA 1.13. *If $\xi \in DEF^\pm$ then $\xi(y) = $ constant $e^{\mp y}$.*

PROOF. Specifically, $\xi \in DEF^+$ if and only if

$$\int_0^a \psi'(y) \xi(y)\, dy = \int_0^a \psi(y) \xi(y)\, dy, \ \forall \psi \in C_c^\infty(0,a).$$

Equivalently, $y \mapsto \xi(y)$ is a weak solution to $-\xi' = \xi$, i.e., a strong solution in C^1. Thus, $\xi(y) = $ constant e^{-y}. The DEF^- case is similar. $\qquad\square$

COROLLARY 1.14. *Suppose F is real valued. Let $\xi_{\pm}(y) := e^{\mp y}$, for $y \in \Omega$. Then $\xi_+ \in \mathscr{H}_F$ iff $\xi_- \in \mathscr{H}_F$. In the affirmative case $\|\xi_-\|_{\mathscr{H}_F} = e^a \|\xi_+\|_{\mathscr{H}_F}$.*

PROOF. Let J be the conjugation from Lemma 1.9. A short calculation:

$$\langle J\xi, F_\phi \rangle_{\mathscr{H}_\mathscr{F}} = \left\langle F_{\overline{\phi(a-\cdot)}}, \xi \right\rangle_{\mathscr{H}_\mathscr{F}} = \int \phi(a-x)\xi(x)dx$$

$$= \int \phi(x)\xi(a-x)dx = \left\langle \overline{\xi(a-\cdot)}, F_\phi \right\rangle_{\mathscr{H}_\mathscr{F}}$$

shows that $(J\xi)(x) = \overline{\xi(a-x)}$, for $\xi \in \mathscr{H}_F$. In particular, $J\xi_- = e^a \xi_+$. Since, $\|J\xi_-\|_{\mathscr{H}_F} = \|\xi_-\|_{\mathscr{H}_F}$, the proof is easily completed. $\qquad\square$

COROLLARY 1.15. *The deficiency indices of $D^{(F)}$, with its dense domain in $\mathscr{H}_F$ are $(0,0)$, $(0,1)$, $(1,0)$, or $(1,1)$.*

The second case in the above corollary happens precisely when $y \mapsto e^{-y} \in \mathscr{H}_F$. We can decide this with the use of $(1.10)(\Leftrightarrow(1.11))$.

The Extension of F. By Corollary 1.15, we conclude that there exists skew-adjoint extension $A^{(F)} \supset D^{(F)}$ in $\mathscr{H}_F$. That is, $\left(A^{(F)}\right)^* = -A^{(F)}$, and

$$\{F_\psi\}_{\psi \in C_c^\infty(0,a)} \subset \text{dom}(A^{(F)}) \subset \mathscr{H}_F.$$

Hence, set $U(t) = e^{tA^{(F)}} : \mathscr{H}_F \to \mathscr{H}_F$, and get the unitary one-parameter group

$$\{U(t) : t \in \mathbb{R}\}, \; U(s+t) = U(s)U(t), \; \forall s,t \in \mathbb{R};$$

and if

$$\xi \in dom\left(A^{(F)}\right) = \left\{ \xi \in \mathscr{H}_F : \text{ s.t. } \lim_{t \to 0} \frac{U(t)\xi - \xi}{t} \text{ exists} \right\}$$

then

(1.18) $$A^{(F)}\xi = \lim_{t \to 0} \frac{U(t)\xi - \xi}{t}.$$

Now use $F_x(\cdot) = F(\cdot - x)$ defined in $(0,a)$; and set

(1.19) $$F_A(t) := \langle F_0, U(t) F_0 \rangle_{\mathscr{H}_F}, \; \forall t \in \mathbb{R}$$

then using (1.9), we see that F_A is a continuous positive definite extension of F on $(-a,a)$, i.e., a continuous positive definite function on $\mathbb{R}$, and if $x \in (0,a)$, then we get the following conclusion:

LEMMA 1.16. *F_A is a continuous bounded positive definite function of $\mathbb{R}$ and*

(1.20) $$F_A(t) = F(t).$$

for $t \in (-a,a)$.

PROOF. But $\mathbb{R} \ni t \mapsto F_A(t)$ is bounded and continuous, since $\{U(t)\}$ is a strongly continuous unitary group acting on $\mathscr{H}_F$, and

$$|F_A(t)| = |\langle F_0, U(t) F_0 \rangle| \leq \|F_0\|_{\mathscr{H}_F} \|U(t) F_0\|_{\mathscr{H}_F} = \|F_0\|_{\mathscr{H}_F}^2$$

where $|\langle F_0, U(t) F_0 \rangle| \leq \|F_0\|_{\mathscr{H}_F}^2 = F(0)$. See the proof of Theorem 2.2 and [**27, 28, 30**] for the remaining details. $\qquad\square$

REMARK 1.17. F can be normalized by $F(0) = 1$. Recall that F is defined on $(-a,a) = \Omega - \Omega$ if $\Omega = (0,a)$.

Consider the spectral representation:

$$(1.21) \qquad U(t) = \int_{-\infty}^{\infty} e_t(\lambda) P(d\lambda)$$

where $e_t(\lambda) = e^{i2\pi\lambda t}$; and $P(\cdot)$ is a projection-valued measure on $\mathbb{R}$, $P(B) : \mathscr{H}_F \to \mathscr{H}_F$, $\forall B \in Borel(\mathbb{R})$. Then

$$d\mu(\lambda) = \|P(d\lambda) F_0\|_{\mathscr{H}_F}^2$$

satisfies

$$(1.22) \qquad F_A(t) = \int_{-\infty}^{\infty} e_t(\lambda) d\mu(\lambda), \ \forall t \in \mathbb{R}.$$

Conclusion. The extension F_A from (1.19) has nice transform properties, and via (1.22) we get

$$\mathscr{H}_{F_A} \simeq L^2(\mathbb{R}, \mu)$$

where $\mathscr{H}_{F_A}$ is the RKHS of F_A.

2. Extensions of Continuous Positive Definite Functions

Our main theme is the interconnection between (i) the study of extensions of locally defined continuous and positive definite functions F on groups on the one hand, and, on the other, (ii) the question of extensions for an associated system of unbounded Hermitian operators with dense domain in a reproducing kernel Hilbert space (RKHS) $\mathscr{H}_F$ associated to F.

Because of the role of positive definite functions in harmonic analysis, in statistics, and in physics, the connections in both directions is of interest, i.e., from (i) to (ii), and vice versa. This means that the notion of "extension" for question (ii) must be inclusive enough in order to encompass all the extensions encountered in (i). For this reason enlargement of the initial Hilbert space $\mathscr{H}_F$ are needed. In other words, it is necessary to consider also operator extensions which are realized in a dilation-Hilbert space; a new Hilbert space containing $\mathscr{H}_F$ isometrically, and with the isometry intertwining the respective operators.

2.1. Enlarging the Hilbert Space. Below we describe the dilation-Hilbert space in detail, and prove some lemmas which will then be used in the following sections.

Let $F : \Omega - \Omega \to \mathbb{C}$ be a continuous p.d. function. Let $\mathscr{H}_F$ be the corresponding RKHS and $\xi_x := F(\cdot - x) \in \mathscr{H}_F$.

$$(2.1) \qquad \langle \xi_x, \xi_y \rangle_{\mathscr{H}_F} = F(x - y), \forall x, y \in \Omega.$$

As usual $F_\varphi = \varphi * F$, $\varphi \in C_c^\infty(\Omega)$. Then

$$\langle F_\varphi, F_\psi \rangle_{\mathscr{H}_F} = \langle \xi_0, \pi(\varphi^\# * \psi) \xi_0 \rangle = \langle \pi(\phi)\xi_0, \pi(\psi)\xi_0 \rangle$$

where $\pi(\varphi)\xi_0 = F_\varphi$. The following lemma also holds in $\mathbb{R}^n$ with $n > 1$, but we state it for $n = 1$ to illustrate the "enlargement" of $\mathscr{H}_F$ question.

REMARK 2.1. We have introduced three equivalent notations:

$$F_\varphi = \varphi * F = \pi(\varphi) F.$$

Recall that if π is a representation of a locally compact topological group G on a Banach space $\mathscr{B}$, then one defines

$$\pi(\varphi)v = \int_G \varphi(x)\pi(x)v\,dx,\ \forall v \in \mathscr{B}$$

where dx denotes the Haar measure on G. Using the left regular representation, one gets $\pi(\varphi)F = \varphi * F$, or

$$\pi(\varphi)F(x) = \int \varphi(y)F(x-y)\,dy = \pi(\varphi)\xi_0(x).$$

THEOREM 2.2. *The following two conditions are equivalent:*

(i) *F is extendable to a continuous p.d. function $\widetilde{F}$ defined on $\mathbb{R}$, i.e., $\widetilde{F}$ is a continuous p.d. function defined on $\mathbb{R}$ and $F(x) = \widetilde{F}(x)$ for all x in $\Omega - \Omega$.*

(ii) *There is a Hilbert space $\mathscr{K}$, an isometry $W : \mathscr{H}_F \to \mathscr{K}$, and a strongly continuous unitary group $U_t : \mathscr{K} \to \mathscr{K}$, $t \in \mathbb{R}$ such that, if A is the skew-adjoint generator of U_t, i.e.,*

(2.2) $$\tfrac{1}{t}(U_t k - k) \to Ak, \forall k \in \mathrm{dom}(A),$$

then

(2.3) $$WF_\varphi \in \mathrm{domain}(A), \forall \varphi \in C_c^\infty(\Omega)$$

and

(2.4) $$AWF_\varphi = WF_{\varphi'}, \forall \varphi \in C_c^\infty(\Omega).$$

PROOF. $\Uparrow$: First, assume there exists $\mathscr{K}$, W, U_t, and A as in (ii). Set

(2.5) $$\widetilde{F}(t) = \langle W\xi_0, U_t W\xi_0 \rangle,\ t \in \mathbb{R}.$$

Then, if $U_t = \int_\mathbb{R} e_t(\lambda)P(d\lambda)$, set

$$d\mu(\lambda) = \|P(d\lambda)W\xi_0\|_\mathscr{K}^2 = \langle W\xi_0, P(d\lambda)W\xi_0 \rangle_\mathscr{K}$$

and $\widetilde{F} = \widehat{d\mu}$ is the Bochner transform.

LEMMA 2.3. *If $s,t \in \Omega$, and $\varphi \in C_c^\infty(\Omega)$, then*

(2.6) $$\langle WF_\varphi, U_t W F_\varphi \rangle = \langle \xi_0, \pi(\varphi^\# * \varphi_t)\xi_0 \rangle_{\mathscr{H}_F},$$

where $\varphi_t(\cdot) = \varphi(t - \cdot)$.

Suppose first (2.6) has been checked. Let ϕ_ϵ be an approximate identity at $x = 0$. Then

(2.7) $$\begin{aligned} \widetilde{F}(t) &= \langle W\xi_0, U_t W\xi_0 \rangle \\ &= \lim_{\epsilon \to 0} \langle \xi_0, \pi((\phi_\epsilon)_t)\xi_0 \rangle_{\mathscr{H}_F} \\ &= \langle \xi_0, \xi_t \rangle = F(t) \end{aligned}$$

by (2.5), (2.6), and (2.1).

PROOF OF LEMMA 2.3. Now (2.6) follows from

(2.8) $$U_t W F_\varphi = W F_{\varphi_t},$$

$\varphi \in C_c^\infty(\Omega), t \in \Omega$. Consider now

$$(2.9) \qquad U_{t-s}WF_{\varphi_s} = \begin{cases} U_t W F_\varphi & \text{at } s = 0 \\ W F_{\varphi_t} & \text{at } s = t \end{cases}$$

and

$$(2.10) \qquad \int_0^t \frac{d}{ds} U_{t-s} W F_{\varphi_s} ds = W F_{\varphi_t} - U_t W F_\varphi.$$

We claim that the left hand side of (2.10) equals zero. By (2.2) and (2.3)

$$\frac{d}{ds} [U_{t-s} W F_{\varphi_s}] = -U_{t-s} A W F_{\varphi_s} + U_{t-s} W F_{\varphi_s'}.$$

But, by (2.4) applied to φ_s, we get

$$(2.11) \qquad A W F_{\varphi_s} = W F_{\varphi_s'}$$

and the desired conclusion (2.8) follows. $\qquad\qquad\qquad\qquad\qquad\qquad\square$

$\Downarrow$: Assume (i), let $\widetilde{F} = \widehat{d\mu}$ be a p.d. extension and Bochner transform. Then $\mathcal{H}_{\widetilde{F}} \simeq L^2(\mu)$; and for $\varphi \in C_c^\infty(\Omega)$, set

$$(2.12) \qquad W F_\varphi = \widetilde{F}_\varphi,$$

then $W : \mathcal{H}_F \to \mathcal{H}_{\widetilde{F}}$ is an isometry.

PROOF THAT (2.12) IS AN ISOMETRY . Let $\varphi \in C_c^\infty(\Omega)$. Then

$$\left\| \widetilde{F}_\varphi \right\|_{\mathcal{H}_{\widetilde{F}}}^2 = \int \int \overline{\varphi(s)} \varphi(t) \widetilde{F}(t-s) ds dt$$
$$= \| F_\varphi \|_{\mathcal{H}_F}^2 = \int_{\mathbb{R}} |\widehat{\varphi}(\lambda)|^2 d\mu(\lambda)$$

since $\widetilde{F}$ is an extension of F. $\qquad\qquad\qquad\qquad\qquad\qquad\qquad\square$

Now set $U_t : L^2(\mu) \to L^2(\mu)$,

$$(U_t f)(\lambda) = e_t(\lambda) f(\lambda),$$

a unitary group acting in $\mathcal{H}_{\widetilde{F}} \simeq L^2(\mu)$. Using (2.1), we get

$$(2.13) \qquad (W F_\varphi)(x) = \int e_x(\lambda) \widehat{\varphi}(\lambda) d\mu(\lambda), \forall x \in \Omega, \forall \varphi \in C_c^\infty(\Omega).$$

And therefore (ii) follows. By (2.13)

$$(W F_{\varphi'})(x) = \int e_x(\lambda) i\lambda \widehat{\varphi}(\lambda) d\mu(\lambda)$$
$$= \frac{d}{dt}\Big|_{t=0} U_t W F_\varphi = A W F_\varphi$$

as claimed. $\qquad\qquad\qquad\qquad\qquad\qquad\qquad\qquad\qquad\qquad\qquad\square$

2.2. $Ext_1(F)$ and $Ext_2(F)$.

DEFINITION 2.4. Let G be a locally compact group, and let Ω be an open connected subset of G. Let $F : \Omega^{-1} \cdot \Omega \to \mathbb{C}$ be a continuous positive definite function.

Consider a strongly continuous unitary representation U of G acting in some Hilbert space $\mathscr{K}$, containing the RKHS $\mathscr{H}_F$. We say that $(U, \mathscr{K}) \in Ext(F)$ iff there is a vector $k_0 \in \mathscr{K}$ such that

$$(2.14) \qquad F(g) = \langle k_0, U(g) k_0 \rangle_{\mathscr{K}}, \ \forall g \in \Omega^{-1} \cdot \Omega.$$

I. The subset of $Ext(F)$ consisting of $(U, \mathscr{H}_F, k_0 = F_e)$ with

$$(2.15) \qquad F(g) = \langle F_e, U(g) F_e \rangle_{\mathscr{H}_F}, \ \forall g \in \Omega^{-1} \cdot \Omega$$

is denoted $Ext_1(F)$; and we set

$$Ext_2(F) := Ext(F) \backslash Ext_1(F);$$

i.e., $Ext_2(F)$ consists of the solutions to problem (2.14) for which $\mathscr{K} \supsetneqq \mathscr{H}_F$, i.e., unitary representations realized in an enlargement Hilbert space.

(We write $F_e \in \mathscr{H}_F$ for the vector satisfying $\langle F_e, \xi \rangle_{\mathscr{H}_F} = \xi(e)$, $\forall \xi \in \mathscr{H}_F$, where e is the neutral (unit) element in G, i.e., $e g = g$, $\forall g \in G$.)

II. In the special case, where $G = \mathbb{R}^n$, and $\Omega \subset \mathbb{R}^n$ is open and connected, we consider

$$F : \Omega - \Omega \to \mathbb{C}$$

continuous and positive definite. In this case,

$$(2.16) \qquad Ext(F) = \left\{ \mu \in \mathscr{M}_+(\mathbb{R}^n) \ \Big| \ \widehat{\mu}(x) = \int_{\mathbb{R}^n} e^{i\lambda \cdot x} d\mu(\lambda) \right.$$

$$\left. \text{is a p.d. extension of } F \right\}.$$

REMARK 2.5. Note that (2.16) is consistent with (2.14): For if $(U, \mathscr{K}, k_0)$ is a unitary representation of $G = \mathbb{R}^n$, such that (2.14) holds; then, by a theorem of Stone, there is a projection-valued measure (PVM) $P_U(\cdot)$, defined on the Borel subsets of $\mathbb{R}^n$ s.t.

$$(2.17) \qquad U(x) = \int_{\mathbb{R}^n} e^{i\lambda \cdot x} P_U(d\lambda), \ x \in \mathbb{R}^n.$$

Setting

$$(2.18) \qquad d\mu(\lambda) := \| P_U(d\lambda) k_0 \|_{\mathscr{K}}^2,$$

it is then immediate that we have: $\mu \in \mathscr{M}_+(\mathbb{R}^n)$, and that the finite measure μ satisfies

$$(2.19) \qquad \widehat{\mu}(x) = F(x), \ \forall x \in \Omega - \Omega.$$

2.3. The Case of $G = \mathbb{R}$. Start with a local p.d. continuous function F on $\mathbb{R}$, and let $\mathscr{H}_F$ be the corresponding RKHS. Let $Ext(F)$ be the compact convex set of probability measures on $\mathbb{R}$ defining extensions of F.

The deficiency indices (computed in $\mathscr{H}_F$) for the canonical skew-Hermitian operator $D^{(F)}$ in $\mathscr{H}_F$ (see Definition 1.6) can be only $(0,0)$ or $(1,1)$, and we have:

THEOREM 2.6. *The deficiency indices computed in $\mathscr{H}_F$ are $(0,0)$ if and only if $Ext_1(F)$ is a singleton. (Note that even if $Ext_1(F)$ is a singleton, we can still have non-empty $Ext_2(F)$.)*

Now let $Ext_1(F)$ be the subset of $Ext(F)$ corresponding to extensions when the unitary representation $U(t)$ acts in $\mathscr{H}_F$ (Remark 2.5), and $Ext_2(F)$ denote the part of $Ext(F)$ associated to unitary representations $U(t)$ acting in a proper enlargement Hilbert space (Section 2.1) $\mathscr{K}$ (if any), i.e., acting in a Hilbert space $\mathscr{K}$ corresponding to a proper dilation. The Polya extensions (sect. 3.1) account for a part of $Ext_2(F)$.

In Section 3, we include a host of examples, including one with a Polya extension where $\mathscr{K}$ is infinite dimensional, while $\mathscr{H}_F$ is 2 dimensional. (If $\mathscr{H}_F$ is 2 dimensional, then obviously we must have deficiency indices $(0,0)$.) In other examples we have $\mathscr{H}_F$ infinite dimensional, non-trivial Polya extensions and deficiency indices $(0,0)$.

2.4. The Case of Locally Compact Abelian Groups. We are concerned with extensions of locally defined continuous and positive definite functions F on Lie groups, but some results apply to locally compact groups as well. In the case of locally compact groups, we have stronger theorems, due to the powerful Fourier analysis theory.

We must fix notations:

- G: a given locally compact abelian group, write the operation in G additively;
- dx: denotes the Haar measure of G (unique up to a scalar multiple.)
- $\widehat{G}$: the dual group, i.e., $\widehat{G}$ consists of all continuous homomorphisms: $\lambda : G \to \mathbb{T}$, $\lambda(x+y) = \lambda(x)\lambda(y)$, $\forall x, y \in G$; $\lambda(-x) = \overline{\lambda(x)}$, $\forall x \in G$. Occasionally, we shall write $\langle \lambda, x \rangle$ for $\lambda(x)$. Note that $\widehat{G}$ also has its Haar measure.

THEOREM 2.7 (Pontryagin [**46**]). $\widehat{\widehat{G}} \simeq G$, and we have the following:

$$[G \text{ is compact}] \Longleftrightarrow [\widehat{G} \text{ is discrete}]$$

Let $\emptyset \neq \Omega \subset G$ be an open connected subset, and let $F : \Omega - \Omega \to \mathbb{C}$ be a fixed continuous positive definite function. We choose the normalization $F(0) = 1$; and introduce the corresponding reproducing kernel Hilbert space (RKHS):

LEMMA 2.8. *For $\varphi \in C_c(\Omega)$, set*

(2.20) $$F_\varphi(\cdot) = \int_\Omega \varphi(y) F(\cdot - y)\, dy,$$

then $\mathscr{H}_F$ is the Hilbert completion of

$$\{F_\varphi \mid \varphi \in C_c(\Omega)\} / \{F_\varphi \mid ||F_\varphi||_{\mathscr{H}_F} = 0\}$$

with respect to the inner product:

(2.21) $$\langle F_\varphi, F_\psi \rangle_{\mathscr{H}_F} = \int_\Omega \int_\Omega \overline{\varphi(x)}\psi(y) F(x-y)\, dxdy.$$

Here $C_c(\Omega) :=$ all continuous compactly supported functions in Ω.

LEMMA 2.9. *The Hilbert space $\mathscr{H}_F$ is also a Hilbert space of continuous functions on Ω as follows:*

If $\xi : \Omega \to \mathbb{C}$ is a fixed continuous function, then $\xi \in \mathscr{H}_F$ if and only if $\exists$ $K = K_\xi < \infty$ such that

$$(2.22) \qquad \left| \int_\Omega \overline{\xi(x)} \varphi(x) \, dx \right|^2 \leq K \int_\Omega \int_\Omega \overline{\varphi(y_1)} \varphi(y_2) \, F(y_1 - y_2) \, dy_1 dy_2.$$

When (2.22) holds, then

$$\langle \xi, F_\varphi \rangle_{\mathscr{H}_F} = \int_\Omega \overline{\xi(x)} \varphi(x) \, dx, \text{ for all } \varphi \in C_c(\Omega).$$

PROOF. We refer to the basics on the theory of RKHSs; e.g., [**8**]. $\qquad \square$

LEMMA 2.10. *There is a bijective correspondence between all continuous p.d. extensions $\tilde{F}$ to G of the given p.d. function F on $\Omega - \Omega$, on the one hand; and all Borel probability measures μ on $\widehat{G}$, on the other, i.e., all $\mu \in \mathscr{M}(\widehat{G})$ s.t.*

$$(2.23) \qquad F(x) = \widehat{\mu}(x), \; \forall x \in \Omega - \Omega$$

where

$$\widehat{\mu}(x) = \int_{\widehat{G}} \lambda(x) \, d\mu(\lambda) = \int_{\widehat{G}} \langle \lambda, x \rangle \, d\mu(\lambda), \; \forall x \in G.$$

PROOF. This is an immediate application of Bochner's characterization of the continuous positive definite functions on locally compact abelian groups. $\qquad \square$

DEFINITION 2.11. Set

$$Ext(F) = \left\{ \mu \in \mathscr{M}(\widehat{G}) \mid s.t. \text{ (2.23) holds} \right\}.$$

LEMMA 2.12. *$Ext(F)$ is weak $*$-compact and convex.*

PROOF. Left to the reader; see e.g., [**45**]. $\qquad \square$

THEOREM 2.13.
(1) *Let F and $\mathscr{H}_F$ be as above; and let $\mu \in \mathscr{M}(\widehat{G})$; then there is a positive Borel function h on $\widehat{G}$ s.t. $h^{-1} \in L^\infty(\widehat{G})$, and $h d\mu \in Ext(F)$, if and only if $\exists K_\mu < \infty$ such that*

$$(2.24) \qquad \int_{\widehat{G}} |\widehat{\varphi}(\lambda)|^2 \, d\mu(\lambda) \leq K_\mu \int_\Omega \int_\Omega \overline{\varphi(y_1)} \varphi(y_2) \, F(y_1 - y_2) \, dy_1 dy_2.$$

(2) *Assume $\mu \in Ext(F)$, then*

$$(2.25) \qquad (f d\mu)^\vee \in \mathscr{H}_F, \; \forall f \in L^2(\widehat{G}, \mu).$$

PROOF. The assertion in (2.24) is immediate from Lemma 2.9.
Our conventions for the two transforms used in (2.24) and (2.25) are as follows:

$$(2.26) \qquad \widehat{\varphi}(\lambda) = \int_G \overline{\langle \lambda, x \rangle} \varphi(x) \, dx.$$

The transform in (2.25) is:

$$(2.27) \qquad (f d\mu)^\vee = \int_{\widehat{G}} \langle \lambda, x \rangle \, f(\lambda) \, d\mu(\lambda).$$

The remaining computations are left to the reader. $\qquad \square$

COROLLARY 2.14.

(1) *Let F be as above; then $\mu \in Ext(F)$ iff the following operator*

$$T\left(F_\varphi\right) = \widehat{\varphi}, \ \varphi \in C_c\left(\Omega\right)$$

is well-defined on $\mathscr{H}_F$, and bounded as follows: $T : \mathscr{H}_F \to L^2(\widehat{G}, \mu)$.

(2) *In this case, the adjoint operator $T^* : L^2(\widehat{G}, \mu) \to \mathscr{H}_F$ is given by*

(2.28) $$T^*\left(f\right) = \left(f d\mu\right)^\vee, \ \forall f \in L^2(\widehat{G}, \mu).$$

PROOF. If $\mu \in Ext(F)$, then for all $\varphi \in C_c\left(\Omega\right)$, and $x \in \Omega$, we have (see (2.20))

$$
\begin{aligned}
F_\varphi\left(x\right) &= \int_\Omega \varphi\left(y\right) F\left(x - y\right) dy \\
&= \int_\Omega \varphi\left(y\right) \widehat{\mu}\left(x - y\right) dy \\
&= \int_\Omega \varphi\left(y\right) \langle \lambda, x - y \rangle \, d\mu\left(\lambda\right) dy \\
&\overset{\text{(Fubini)}}{=} \int_{\widehat{G}} \langle \lambda, x \rangle \, \widehat{\varphi}\left(\lambda\right) d\mu\left(\lambda\right).
\end{aligned}
$$

By Lemma 2.9, we note that $\left(\widehat{\varphi} d\mu\right)^\vee \in \mathscr{H}_F$, see (2.27). Hence $\exists K < \infty$ such that the estimate (2.24) holds. To see that $T\left(F_\varphi\right) = \widehat{\varphi}$ is well-defined on $\mathscr{H}_F$, we must check the implication:

$$\left(F_\varphi = 0 \text{ in } \mathscr{H}_F\right) \implies \left(\widehat{\varphi} = 0 \text{ in } L^2(\widehat{G}, \mu)\right)$$

but this now follows from estimate (2.24).

Using the definition of the respective inner products in $\mathscr{H}_F$ and in $L^2(\widehat{G}, \mu)$, we check directly that, if $\varphi \in C_c\left(\Omega\right)$, and $f \in L^2(\widehat{G}, \mu)$ then we have:

(2.29) $$\langle \widehat{\varphi}, f \rangle_{L^2(\mu)} = \left\langle F_\varphi, \left(f d\mu\right)^\vee \right\rangle_{\mathscr{H}_F}.$$

On the RHS in (2.29), we note that, when $\mu \in Ext(F)$, then $\widehat{f d\mu} \in \mathscr{H}_F$. This last conclusion is a consequence of Lemma 2.9. Indeed, since μ is finite, $L^2(\widehat{G}, \mu) \subset L^1(\widehat{G}, \mu)$, so $\widehat{f d\mu}$ in (2.27) is continuous on G by Riemann-Lebesgue; and so is its restriction to Ω. If μ is further assumed absolutely continuous, then $\widehat{f d\mu} \to 0$ at ∞.

With a direct calculation, using the reproducing property in $\mathscr{H}_F$, and Fubini's theorem, we check directly that the following estimate holds:

$$\left| \int_\Omega \overline{\varphi\left(x\right)} \left(f d\mu\right)^\vee\left(x\right) dx \right|^2 \leq \left(\int_\Omega \int_\Omega \overline{\varphi\left(y_1\right)} \varphi\left(y_2\right) F\left(y_1 - y_2\right) dy_1 dy_2 \right) \|f\|^2_{L^2(\mu)}$$

and so Lemma 2.9 applies; we get $\left(f d\mu\right)^\vee \in \mathscr{H}_F$.

It remains to verify the formula (2.29) for all $\varphi \in C_c\left(\Omega\right)$ and all $f \in L^2(\widehat{G}, \mu)$; but this now follows from the reproducing property in $\mathscr{H}_F$, and Fubini.

Once we have this, both assertions in (1) and (2) in the Corollary follow directly from the definition of the adjoint operator T^* with respect to the two Hilbert spaces in $\mathscr{H}_F \overset{T}{\longrightarrow} L^2(\widehat{G}, \mu)$. Indeed then (2.28) follows.   □

We recall a general result on continuity of positive definite functions on any locally compact Lie group:

THEOREM 2.15. *If F is p.d. function on a locally compact group G, assumed continuous only in a neighborhood of $e \in G$; then it is automatically continuous everywhere on G.*

PROOF. Since F is positive definite, we may apply the Gelfand-Naimark-Segal (GNS) theorem to get a cyclic unitary representation $(U, \mathscr{H}, v)$, v denoting the cyclic vector, such that $F(g) = \langle v, U(g)v \rangle$, $g \in G$. The stated assertion about continuity for unitary representations is easy to verify; and so it follows for F. □

QUESTION. *Suppose $Ext(F) \neq \emptyset$, then what are its extreme points? Equivalently, characterize $ext(Ext(F))$.*

Let $\Omega \subset G$, $\Omega \neq \emptyset$, Ω open and connected, and let

$$K_\Omega(\lambda) = \widehat{\chi_\Omega(\lambda)}, \ \forall \lambda \in \widehat{G}.$$

THEOREM 2.16. *Let $F : \Omega - \Omega \to \mathbb{C}$ be continuous, and positive definite on $\Omega - \Omega$; and assume $Ext(F) \neq \emptyset$. Let $\mu \in Ext(F)$, and let $T_\mu(F_\phi) := \widehat{\varphi}$, defined initially only for $\varphi \in C_c(\Omega)$, be the isometry $T_\mu : \mathscr{H}_F \to L^2(\mu) = L^2(\widehat{G}, \mu)$. Then $Q_\mu := T_\mu T_\mu^*$ is a projection in $L^2(\mu)$ with $K_\Omega(\cdot)$ as kernel:*

$$(2.30) \qquad (Q_\mu f)(\lambda) = \int_{\widehat{G}} K_\Omega(\lambda - \xi) f(\xi) \, d\mu(\xi), \ \forall f \in L^2(\widehat{G}, \mu), \forall \lambda \in \widehat{G}.$$

PROOF. We showed in Theorem 2.13 that $T_\mu : \mathscr{H}_F \to L^2(\mu)$ is isometric, and so $Q_\mu := T_\mu T_\mu^*$ is the projection in $L^2(\mu)$. For $f \in L^2(\mu)$, $\lambda \in \widehat{G}$, we have the following computation, where the interchanging of integrals is justified by Fubini's theorem:

$$\begin{aligned} (Q_\mu f)(\lambda) &= \int_\Omega (f d\mu)^\vee(x) \langle \lambda, x \rangle \, dx \\ &= \int_\Omega \langle \lambda, x \rangle \left(\int_{\widehat{G}} f(\xi) \overline{\langle \xi, x \rangle} d\mu(\xi) \right) dx \\ &\overset{\text{Fubini}}{=} \int_{\widehat{G}} K_\Omega(\lambda - \xi) f(\xi) \, d\mu(\xi) \end{aligned}$$

which is the desired conclusion (2.30). Here, dx denotes the Haar measure on G. □

2.5. The Case of $G = \mathbb{R}^n$. As a special case of the setting of locally compact Abelian groups from above, the results available for $\mathbb{R}^n$ are more refined. This is also the setting of the more classical studies of extension questions.

Let $\Omega \subset \mathbb{R}^n$ be a fixed open and connected subset; and let $F : \Omega - \Omega \to \mathbb{C}$ be a given continuous and positive definite function defined on

$$(2.31) \qquad \Omega - \Omega := \left\{ x - y \in \mathbb{R}^n \mid x, y \in \Omega \right\}.$$

Let $\mathscr{H}_F$ be the corresponding reproducing kernel Hilbert space (RKHS). We showed that $Ext(F) \neq \emptyset$ if and only if there is a strongly continuous unitary representation $\{U(t)\}_{t \in \mathbb{R}^n}$ acting on $\mathscr{H}_F$ such that

$$(2.32) \qquad \mathbb{R}^n \ni t \mapsto \langle F_0, U(t) F_0 \rangle_{\mathscr{H}_F}$$

is a p.d. extension of F, extending from (2.31) to $\mathbb{R}^n$. Finally, if U is a unitary representation of $G = \mathbb{R}^n$ we denote by $P_U(\cdot)$ the associated projection valued measure (PVM) on $\mathscr{B}(\mathbb{R}^n)$ (= the sigma-algebra of all Borel subsets in $\mathbb{R}^n$).

We have

(2.33) $$U(t) = \int_{\mathbb{R}^n} e^{it\cdot\lambda} P_U(d\lambda), \ \forall t \in \mathbb{R}^n;$$

where $t = (t_1, \ldots, t_n)$, $\lambda = (\lambda_1, \ldots, \lambda_n)$, and $t \cdot \lambda = \sum_{j=1}^n t_j \lambda_j$. Recall, setting

(2.34) $$d\mu(\cdot) = \|P_U(\cdot) F_0\|_{\mathscr{H}_F}^2,$$

then the p.d. function on RHS in (2.32) satisfies

(2.35) $$\text{RHS}_{(2.32)} = \int_{\mathbb{R}^n} e^{it\cdot\lambda} d\mu(\lambda), \ \forall t \in \mathbb{R}^n.$$

The purpose of the next theorem is to give an orthogonal splitting of the RKHS $\mathscr{H}_F$ associated to a fixed (Ω, F) when it is assumed that $Ext(F)$ is non-empty. This orthogonal splitting of $\mathscr{H}_F$ depends on a choice of $\mu \in Ext(F)$, and the splitting is into three orthogonal subspaces of $\mathscr{H}_F$, correspond a splitting of spectral types into atomic, absolutely continuous (with respect to Lebesgue measure), and singular.

THEOREM 2.17. *Let $\Omega \subset \mathbb{R}^n$ be given, $\Omega \neq \emptyset$, open and connected. Suppose F is given p.d. and continuous on $\Omega - \Omega$, and assume $Ext(F) \neq \emptyset$. Let U be the corresponding unitary representations of $G = \mathbb{R}^n$, and let $P_U(\cdot)$ be its associated PVM acting on $\mathscr{H}_F$ (= the RKHS of F.)*

(1) Then $\mathscr{H}_F$ splits up as an orthogonal sum of three closed and $U(\cdot)$ invariant subspaces

(2.36) $$\mathscr{H}_F = \mathscr{H}_F^{(atom)} \oplus \mathscr{H}_F^{(ac)} \oplus \mathscr{H}_F^{(sing)}$$

with these subspaces characterized as follows:

(a) The PVM $P_U(\cdot)$ restricted to $\mathscr{H}_F^{(atom)}$ is purely atomic;

(b) $P_U(\cdot)$ restricted to $\mathscr{H}_F^{(ac)}$ is absolutely continuous with respect to the Lebesgue measure $d\lambda = d\lambda_1 \cdots d\lambda_n$ on $\mathbb{R}^n$; and

(c) $P_U(\cdot)$ is continuous, purely singular, when restricted to $\mathscr{H}_F^{(sing)}$.

(2) Case $\mathscr{H}_F^{(atom)}$. If $\lambda \in \mathbb{R}^n$ is an atom in $P_U(\cdot)$, i.e., $P_U(\{\lambda\}) \neq 0$, where $\{\lambda\}$ denotes the singleton with λ fixed; then $P_U(\{\lambda\})\mathscr{H}_F$ is one-dimensional, and the function $e_\lambda(x) := e^{i\lambda\cdot x}$, (complex exponential) restricted to Ω, is in $\mathscr{H}_F$. We have:

(2.37) $$P_U(\{\lambda\}) \mathscr{H}_F = \mathbb{C} e_\lambda \big|_\Omega.$$

Case $\mathscr{H}_F^{(ac)}$. If $\xi \in \mathscr{H}_F^{(ac)}$, then it is represented as a continuous function on Ω, and

(2.38) $$\langle \xi, F_\varphi \rangle_{\mathscr{H}_F} = \int_\Omega \overline{\xi(x)} \varphi(x) \, dx_{(Lebesgue\ meas.)}, \ \forall \varphi \in C_c(\Omega).$$

Moreover, there is a $f \in L^2(\mathbb{R}^n, \mu)$ (where μ is given in (2.34)) such that

(2.39) $$\int_\Omega \overline{\xi(x)} \varphi(x) \, dx = \int_{\mathbb{R}^n} \overline{f(\lambda)} \widehat{\varphi}(\lambda) \, d\mu(\lambda), \ \forall \varphi \in C_c(\Omega);$$

and

(2.40) $$\xi = (f d\mu)^\vee \big|_\Omega.$$

174   PALLE E. T. JORGENSEN, STEEN PEDERSEN, AND FENG TIAN

(We say that $(f d\mu)^{\vee}$ is the μ-extension of ξ.)

<u>Conclusion.</u> *Every μ-extension of ξ is continuos on $\mathbb{R}^n$, and goes to 0 at infinity (in $\mathbb{R}^n$,); so the μ-extension $\tilde{\xi}$ satisfies $\lim_{|x|\to\infty} \tilde{\xi}(x) = 0$.*

Case $\mathscr{H}_F^{(sing)}$. *Vectors $\xi \in \mathscr{H}_F^{(sing)}$ are characterized by the following property:*
The measure

$$(2.41) \qquad d\mu_\xi(\cdot) := \|P_U(\cdot)\xi\|_{\mathscr{H}_F}^2$$

is continuous and purely singular.

PROOF. Most of the proof details are contained in the previous discussion.

For (2), Case $\mathscr{H}_F^{(atom)}$; suppose $\lambda \in (\mathbb{R}^n)$ is an atom, and that $\xi \in \mathscr{H}_F \backslash \{0\}$ satisfies

$$(2.42) \qquad P_U(\{\lambda\})\xi = \xi;$$

then

$$(2.43) \qquad U(t)\xi = e^{it\cdot\lambda}\xi, \ \forall t \in \mathbb{R}^n.$$

Using now (2.32)-(2.33), we conclude that ξ (as a continuous function on $\mathbb{R}^n$) is a weak solution to the following elliptic system

$$(2.44) \qquad \frac{\partial}{\partial x_j}\xi = \sqrt{-1}\lambda_j\xi \ (\text{on } \Omega), \ 1 \leq j \leq n.$$

Hence $\xi = \text{const} \cdot e_\lambda\big|_\Omega$ as asserted in (2).

Case (2), $\mathscr{H}_F^{(ac)}$ follows from (2.40) and the Riemann-Lebesgue theorem applied to $\mathbb{R}^n$; and case $\mathscr{H}_F^{(sing)}$ is immediate. $\qquad \square$

EXAMPLE 2.18. Consider the following continuous positive definite function F on $\mathbb{R}$, or on some bounded interval $(-a, a)$, $a > 0$.

$$(2.45) \qquad F(x) = \frac{1}{3}\left(e^{-ix} + \prod_{n=1}^{\infty}\cos\left(\frac{2\pi x}{3^n}\right) + e^{i3x/2}\frac{\sin(x/2)}{(x/2)}\right).$$

   (1) This is the decomposition (2.36) of the corresponding RKHSs $\mathscr{H}_F$, all three subspaces $\mathscr{H}_F^{(atom)}$, $\mathscr{H}_F^{(ac)}$, and $\mathscr{H}_F^{(sing)}$ are non-zero; the first one is one-dimensional, and the other two are infinite-dimensional.
   (2) The operator

$$(2.46) \qquad D^{(F)}(F_\varphi) := F_{\varphi'} \text{ on } dom(D^{(F)}) = \{F_\varphi \mid \varphi \in C_c^\infty(0, a)\}$$

      is bounded, and so extends by <u>closure</u> to a skew-adjoint operator, i.e., $\overline{D^{(F)}} = -(D^{(F)})^*$.

PROOF. Using infinite convolutions of operators, and results from [**17**], we conclude that F defined in (2.45) is entire analytic, and $F = \widehat{d\mu}$ (Bochner-transform) where

$$(2.47) \qquad d\mu(\lambda) = \frac{1}{3}\left(\delta_{-1} + \mu_{Cantor} + \chi_{[1,2]}(\lambda)\,d\lambda\right).$$

The measures on the RHS in (2.47) are as follows: $\qquad\qquad\qquad\qquad \square$

- δ_{-1} is the Dirac mass at -1, i.e., $\delta(\lambda + 1)$.
- μ_{Cantor} = the middle-third Cantor measure μ_c determined as the unique solution in $\mathscr{M}_+^{prob}(\mathbb{R})$ to

$$\int f(\lambda)\,d\mu_c(\lambda) = \frac{1}{2}\left(\int f\left(\frac{\lambda+1}{3}\right)d\mu_c(\lambda) + \int f\left(\frac{\lambda-1}{3}\right)d\mu_c\right)$$

  for all $f \in C_c(\mathbb{R})$; and the last term
- $\chi_{[1,2]}(\lambda)\,d\lambda$ is restriction to the closed interval $[1,2]$ of the Lebesgue measure.

It follows from the literature (e.g. [17]) that μ_c is supported in $\left[-\frac{1}{2}, \frac{1}{2}\right]$; and so the three measures on the RHS in (2.47) have disjoint compact support, with the three supports positively separately.

The conclusions asserted in the example follow from this, in particular the properties for $D^{(F)}$, in fact

$$(2.48) \qquad \text{spectrum}\left(D^{(F)}\right) \subseteq \{-1\} \cup \left[-\tfrac{1}{2}, \tfrac{1}{2}\right] \cup [1,2]$$

2.6. Lie Groups.

DEFINITION 2.19. Let G be a Lie group. We consider the extension problem for continuous positive definite functions

$$(2.49) \qquad F : \Omega^{-1}\Omega \to \mathbb{C}$$

where $\Omega \neq \emptyset$, is a connected and open subset in G, i.e., it is assumed that

$$(2.50) \qquad \sum_i \sum_j \overline{c_i} c_j F\left(x_j^{-1} x_i\right) \geq 0,$$

for all finite systems $\{c_i\} \subset \mathbb{C}$, and points $\{x_i\} \subset \Omega$. Equivalent,

$$(2.51) \qquad \int_\Omega \overline{\varphi(x)}\varphi(y) F\left(y^{-1}x\right) dx dy \geq 0,$$

for all $\varphi \in C_c(\Omega)$; where dx denotes a choice of $\underline{\text{left}}$-invariant Haar measure on G.

LEMMA 2.20. *Let F be defined as in (2.49)-(2.50); and for all $X \in La(G) =$ the Lie algebra of G, set*

$$(2.52) \qquad (\tilde{X}\varphi)(g) := \frac{d}{dt}\varphi\left(\exp_G(-tX)g\right)$$

for all $\varphi \in C_c^\infty(\Omega)$. Set

$$(2.53) \qquad F_\varphi(x) := \int_\Omega \varphi(y) F\left(y^{-1}x\right) dy;$$

then

$$(2.54) \qquad S_X^{(F)}(F_\varphi) := F_{\tilde{X}\varphi}, \; \varphi \in C_c^\infty(\Omega)$$

defines a representation of the Lie algebra $La(G)$ by skew-Hermitian operators in the RKHS $\mathscr{H}_F$, with the operator in (2.54) defined on the common dense domain $\left\{F_\varphi \mid \varphi \in C_c^\infty(\Omega)\right\} \subset \mathscr{H}_F$.

PROOF. The arguments here follow those of the proof of 1.7 *mutatis mutandis*.
$\square$

DEFINITION 2.21.

(1) We say that a continuous p.d. function $F : \Omega^{-1}\Omega \to \mathbb{C}$ is extendable iff there is a continuous p.d. function $F_{ex} : G \to G$ such that

$$(2.55) \qquad F_{ex}\big|_{\Omega^{-1}\Omega} = F.$$

(2) Let $U \in Rep(G, \mathscr{K})$ be a strongly continuous unitary representation of G acting in some Hilbert space $\mathscr{K}$. We say that $U \in Ext(F)$ iff (Def.) there is an isometry $\mathscr{H}_F \hookrightarrow \mathscr{K}$ such that the function

$$(2.56) \qquad G \ni g \mapsto \langle JF_e, U(g) JF_e \rangle_{\mathscr{K}}$$

satisfies the condition in (1).

THEOREM 2.22. *Every extension of some continuous p.d. function F on $\Omega^{-1}\Omega$ as in (1) arises from a unitary representation of G as specified in (2).*

PROOF. First assume some unitary representation U of G satisfies (2), then (2.56) is an extension of F. This follows from the argument in our proof of 1.7.

For the converse; assume some continuous p.d. function F_{ex} on G satisfies (2.55). Now apply the GNS-theorem to F_{ex}; and, as a result, we get a cyclic representation $(U, \mathscr{K}, v_0)$ where

- $\mathscr{K}$ is a Hilbert space;
- U is a strongly continuous unitary representation of G acting on $\mathscr{K}$; and
- $v_0 \in \mathscr{K}$ is a cyclic vector, $\|v_0\| = 1$; and

$$(2.57) \qquad F_{ex}(g) = \langle v_0, U(g) v_0 \rangle, \; g \in G.$$

Defining now $J : \mathscr{H}_F \to \mathscr{K}$ as follows,

$$J(F(\cdot g)) := U(g^{-1})v_0, \; \forall g \in \Omega;$$

and extension by limit, we check that J is isometric and satisfies the condition from (2) in Definition 2.21. We omit details as they parallel arguments already contained in section 1.3. $\square$

THEOREM 2.23. *Let Ω, G, $La(G)$, and $F : \Omega^{-1}\Omega \to \mathbb{C}$ be as in Definition 2.19. Let $\tilde{G}$ be the simply connected universal covering group for G. Then F has an extension to a p.d. continuous function on $\tilde{G}$ iff there is a unitary representation U of $\tilde{G}$ and an isometry $\mathscr{H}_F \xrightarrow{J} \mathscr{K}$ such that*

$$(2.58) \qquad JS_X^{(F)} = dU(X) J$$

holds on $\left\{ F_\varphi \mid \varphi \in C_c^\infty(\Omega) \right\}$, for all $X \in La(G)$; where

$$dU(X) U(\varphi) v_0 \;=\; U(\tilde{X}\varphi)v_0, \; and$$

$$U(\varphi) \;=\; \int_{\tilde{G}} \varphi(g) U(g^{-1}) \, dg.$$

PROOF. Details are contained in sections 2.1 and 3. $\square$

Assume G is connected. Note that on $C_c^\infty(\Omega)$, the Lie group G acts locally, i.e., by $\varphi \mapsto \varphi_g$ where φ_g denotes translation of φ by some element $g \in G$, such that φ_g is also supported in Ω. Then

$$(2.59) \qquad \|F_\varphi\|_{\mathscr{H}_F} = \|F_{\varphi_g}\|_{\mathscr{H}_F};$$

but only for elements $g \in G$ in a neighborhood of $e \in G$, and with the neighborhood depending on φ.

COROLLARY 2.24. *Given*

$$(2.60) \qquad\qquad F : \Omega^{-1} \cdot \Omega \to \mathbb{C}$$

continuous and positive definite, then set

$$(2.61) \qquad\qquad L_g\left(F_\varphi\right) := F_{\varphi_g}, \ \varphi \in C_c^\infty\left(\Omega\right),$$

*defining a local representation of G in $\mathscr{H}_E$, see [**25, 26**].*

COROLLARY 2.25. *Given F, positive definite and continuous, as in (2.60), and let L be the corresponding local representation of G acting on $\mathscr{H}_F$. Then $Ext\left(F\right) \neq \emptyset$ if and only if the local representation (2.61) extends to a global unitary representation acting in some Hilbert space $\mathscr{K}$, containing $\mathscr{H}_F$ isometrically.*

PROOF. We refer to [**25, 26**] for details, as well as 3 below.   □

3. Type I v.s. Type II Extensions

In this section, we identify extensions of the initially give p.d. function F which are associated with operator extensions in the RKHS $\mathscr{H}_F$ itself (Type I), and those which require an enlargement of $\mathscr{H}_F$ (Type II). In the case of $G = \mathbb{R}$ (the real line) some of these continuous p.d. extensions arising from the second construction involve a spline-procedure, and a theorem of G. Polya [**42**], which leads to p.d. extensions of F that are symmetric, compactly supported in an interval around $x = 0$, and convex on the left and right half-lines. For splines and positive definite functions, we refer to [**20, 48**].

3.1. Polya Extensions. Part of this is the construction of Polya extensions as follow: Starting with a *convex* p.d. F on $(-a, a)$; we create a new F_{ex} on $\mathbb{R}$, such that $F_{ex}\big|_{\mathbb{R}_+}$ is convex, and $F_{ex}\left(-x\right) = F_{ex}\left(x\right)$. Polya's theorem [**42**] states that F_{ex} is positive definite.

In Figure 3.1, the slope of L_+ is chosen to be $F'\left(a\right)$; and we take the slope of L_- to be $F'\left(-a\right) = -F'\left(a\right)$. Recall that F is defined initially only on some fixed interval $(-a, a)$. It then follows by Polya's theorem that each of these spline extensions is continuous and positive definite.

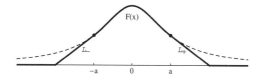

FIGURE 3.1. Spline extension of $F : (-a, a) \to \mathbb{R}$

After extending F from $(-a, a)$ by adding one or more line-segments over $\mathbb{R}_+$, and using symmetry by $x = 0$; the lines in the extensions will be such that there is a c, $0 < a < c$, and the extension F_{ex} satisfies $F_{ex}\left(x\right) = 0$ for all $|x| \geq c$. See Figure 3.2 below.

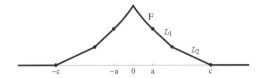

FIGURE 3.2. An example of Polya extension of F on $(-a, a)$.

PROPOSITION 3.1. *Given $F : (-a, a) \to \mathbb{C}$, and assume F has a Polya extension F_{ex}, then the corresponding measure $\mu_{ex} \in Ext(F)$ has the following form*

$$d\mu_{ex}(\lambda) = \Phi_{ex}(\lambda)\, d\lambda$$

where

$$\Phi_{ex}(\lambda) = \frac{1}{2\pi} \int_{-c}^{c} e^{-i\lambda y} F_{ex}(y)\, dy$$

is entire analytic in λ.

PROOF. An application of Fourier inversion, and the Paley-Wiener theorem. □

EXAMPLE 3.2 (Cauchy distribution). $F_1(x) = \dfrac{1}{1+x^2}$; $|x| < 1$. F_1 is concave near $x = 0$.

FIGURE 3.3. Extension of $F_1(x) = \frac{1}{1+x^2}$; $\Omega = (0, 1)$

EXAMPLE 3.3. $F_2(x) = 1 - |x|$; $|x| < \frac{1}{2}$. Consider the following Polya extension:

$$F(x) = \begin{cases} 1 - |x| & \text{if } |x| < \frac{1}{2} \\ \dfrac{1}{3}(2 - |x|) & \text{if } \frac{1}{2} \le |x| < 2 \\ 0 & \text{if } |x| \ge 2 \end{cases}$$

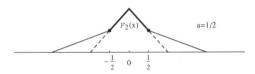

FIGURE 3.4. Extension of $F_2(x) = 1 - |x|$; $\Omega = \left(0, \frac{1}{2}\right)$

This is a p.d. spline extension which is convex on $\mathbb{R}_+$. The corresponding measure $\mu \in Ext(F)$ has the following form $d\mu(\lambda) = \Phi(\lambda)\, d\lambda$, where $d\lambda =$ Lebesgue measure on $\mathbb{R}$, and where

$$\Phi(\lambda) = \begin{cases} \dfrac{3}{4\pi} & \text{if } \lambda = 0 \\ \dfrac{1}{3\pi\lambda^2}(3 - 2\cos(\lambda/2) - \cos(2\lambda)) & \text{if } \lambda \ne 0. \end{cases}$$

This solution is in $Ext_2(F)$. By contrast, the measure μ_2 in Table 4 satisfies $\mu_2 \in Ext_1(F)$. See 2.2.

EXAMPLE 3.4 (Ornstein-Uhlenbeck). $F_3(x) = e^{-|x|}$; $|x| < 1$. A p.d. spline extension which is convex on $\mathbb{R}_+$.

FIGURE 3.5. Extension of $F_3(x) = e^{-|x|}$; $\Omega = (0, 1)$

EXAMPLE 3.5 (Shannon). $F_4(x) = \left(\dfrac{\sin \pi x}{\pi x}\right)^2$; $|x| < \frac{1}{2}$. F_4 is concave near $x = 0$.

FIGURE 3.6. Extension of $F_4(x) = \left(\frac{\sin \pi x}{\pi x}\right)^2$; $\Omega = \left(0, \frac{1}{2}\right)$

EXAMPLE 3.6 (Gaussian distribution). $F_5(x) = e^{-x^2/2}$; $|x| < 1$. F_5 is concave in $-1 < x < 1$.

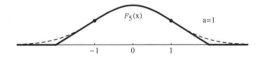

FIGURE 3.7. Extension $F_5(x) = e^{-x^2/2}$; $\Omega = (0, 1)$

REMARK 3.7. Some spline extensions may not be positive definite. In order for Polya's theorem to be applicable, the spline extended function F_{ex} to $\mathbb{R}$ must be convex on $\mathbb{R}_+$. By construction, our extension to $\mathbb{R}$ is by mirror symmetry around $x = 0$. If we start with a symmetric p.d. function F in $(-a, a)$ which is concave near $x = 0$, then the spline extension does not satisfy the premise in Polya's theorem.

For example, it is easy to check that the two partially defined functions F in Figures 3.3 and 3.6 are concave near $x = 0$ (just calculate the double derivative F''). The corresponding spline extensions are *not* p.d..

REMARK 3.8. Polya's theorem only applies when convexity holds on $\mathbb{R}_+$. In that case, the spline extensions will be p.d.. And so Polya's theorem only accounts for those spline extensions F_{ex} which are convex when restricted to $\mathbb{R}_+$. Now there may be p.d. spline extensions that are not convex when restricted to $\mathbb{R}_+$, and we leave open the discovery of those. See Figures 3.1-3.8.

Of the p.d. functions in Table 2, we note that F_1, F_4, F_5, and F_6 satisfy this: there is a $c > 0$ such that the function in question is concave in the interval $[0, c]$, the value of c varies from one to the next. So these four cases do not yield spline extensions F_{ex} which are convex when restricted to $\mathbb{R}_+$.

We get the nicest spline extensions if we make the derivative $F' = \frac{dF}{dx}$ a spline at the break-points. In Example 3.2-3.6, we compute $F'(a)$, see Table 1. We then use symmetry for the left-hand-side of the figure.

TABLE 1. Spline extension at break-points.

$F_1'(1) = -1/2$	$F_4'(1/2) = -16\pi^{-2}$
$F_2'(1/2) = -1$	$F_5'(1) = -e^{-1/2}$
$F_3'(1) = -e^{-1}$	

For each locally defined p.d. function F_i, we then get a deficiency index-problem in the RKHSs $\mathscr{H}_{F_i}$, $i = 1, \ldots, 5$, for the operator $D^{(F_i)}F_\varphi^{(i)} = F_{\varphi'}^{(i)}$, $\forall \varphi \in C_c^\infty(0, a)$. And all the five skew-Hermitian operators in $\mathscr{H}_{F_i}$ will have deficiency indices $(1, 1)$.

Following is an example with deficiency indices $(0, 0)$

EXAMPLE 3.9. $F_6(x) = \cos(x)$; $|x| < \frac{\pi}{4}$

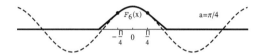

FIGURE 3.8. Extension of $F_6(x) = \cos(x)$; $\Omega = \left(0, \frac{\pi}{4}\right)$

LEMMA 3.10. $\mathscr{H}_{F_6}$ is 2-dimensional.

PROOF. Easy. □

Thus, in all five examples above, $\mathscr{H}_{F_i}$ $(i = 1, \ldots, 5)$ is infinite-dimensional; but $\mathscr{H}_{F_6}$ is 2-dimensional.

In the given five examples, we have p.d. continuous extensions to $\mathbb{R}$ of the following form, $\widehat{d\mu_i}(\cdot)$, $i = 1, \ldots, 5$, where these measures are given in Table 5; also see Figure 4.1.

COROLLARY 3.11. *In all five examples above, we get isometries as follows*

$$T^{(i)} : \mathscr{H}_{F_i} \to L^2(\mathbb{R}, \mu_i)$$

$$T^{(i)}\left(F_\varphi^{(i)}\right) = \widehat{\varphi}$$

for all $\varphi \in C_c^\infty(\Omega)$, where we note that $\widehat{\varphi} \in L^2(\mathbb{R}, \mu_i)$, $i = 1, \ldots, 5$; and

$$\left\|F_\varphi^{(i)}\right\|_{\mathscr{H}_{F_i}}^2 = \|\widehat{\varphi}\|_{L^2(\mu)}^2 = \int_{\mathbb{R}} |\widehat{\varphi}|^2 \, d\mu_i, \ i = 1, \ldots, 5;$$

but note that $T^{(i)}$ is only isometric into $L^2(\mu_i)$.

For the adjoint operator:

$$\left(T^{(i)}\right)^* : L^2(\mathbb{R}, \mu_i) \to \mathscr{H}_{F_i}$$

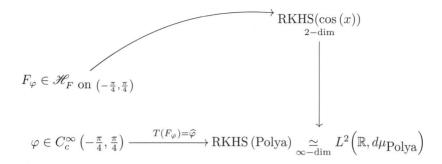

FIGURE 3.9. $\dim\left(\mathrm{RKHS}\left(\cos x \text{ on } \mathbb{R}\right)\right) = 2$; but RKHS(Polya ext. to $\mathbb{R}$) is ∞-dimensional.

we have

$$\left(T^{(i)}\right)^* f = \left(fd\mu_i\right)^\vee, \ \forall f \in L^2\left(\mathbb{R}, \mu_i\right).$$

Here is an infinite-dimensional example as a version of F_6. Fix some positive p, $0 < p < 1$, and set

$$\prod_{n=1}^{\infty} \cos\left(2\pi p^n x\right) = F_p\left(x\right)$$

then this is a continuous positive definite function on $\mathbb{R}$, and the law is the corresponding Bernoulli measure μ_p satisfying $F_p = \widehat{d\mu_p}$. Note that some of those measures μ_p are fractal measures.

For fixed $p \in (0, 1)$, the measure μ_p is the law of the following random power series

$$X_p\left(w\right) := \sum_{n=1}^{\infty} \left(\pm\right) p^n$$

where $w \in \prod_1^\infty \{\pm 1\}$ (= infinite Cartesian product) and where the distribution of each factor is $\left\{-\frac{1}{2}, \frac{1}{2}\right\}$, and statically independent. For relevant references on random power series, see [37, 39].

The extensions we generate with the application of Polya's theorem are realized in a bigger Hilbert space. The deficiency indices are computed for the RKHS $\mathscr{H}_F$, i.e., for the "small" p.d. function $F : \Omega - \Omega \to \mathbb{C}$.

EXAMPLE 3.12. F_6 on $\left(-\frac{\pi}{4}, \frac{\pi}{4}\right)$ has the obvious extension $\mathbb{R} \ni x \to \cos x$, with a 2-dimensional Hilbert space; but the other p.d. extensions (from Polya) will be in infinite-dimensional Hilbert spaces. See Figure 3.9.

We must make a distinction between two classes of p.d. extensions of F : $\Omega - \Omega \to \mathbb{C}$ to continuous p.d. functions on $\mathbb{R}$.

Case 1. There exists a unitary representation $U\left(t\right) : \mathscr{H}_F \to \mathscr{H}_F$ such that

$$(3.1) \qquad F\left(t\right) = \left\langle \xi_0, U\left(t\right) \xi_0 \right\rangle_{\mathscr{H}_F}, \ t \in \Omega - \Omega$$

Case 2. (e.g., Polya extension) There exist a Hilbert space $\mathscr{K}$, and an isometry $J : \mathscr{H}_F \to \mathscr{K}$, and a unitary representation $U\left(t\right) : \mathscr{K} \to \mathscr{K}$, such that

$$(3.2) \qquad F\left(t\right) = \left\langle J\xi_0, U\left(t\right) J\xi_0 \right\rangle_{\mathscr{K}}, \ t \in \Omega - \Omega$$

In both cases, $\xi_0 = F\left(0 - \cdot\right) \in \mathscr{H}_F$.

In case 1, the unitary representation is realized in $\mathcal{H}_{(F,\Omega-\Omega)}$, while, in case 2, the unitary representation $U(t)$ lives in the expanded Hilbert space $\mathcal{K}$.

Note that the RHS in both (3.1) and (3.2) is defined for all $t \in \mathbb{R}$.

LEMMA 3.13. *Let F_{ex} be one of the Polya extensions if any. Then by the Galfand-Naimark-Segal (GNS) construction applied to $F_{ext} : \mathbb{R} \to \mathbb{R}$, there is a Hilbert space $\mathcal{K}$ and a vector $v_0 \in \mathcal{K}$ and a unitary representation $\{U(t)\}_{t\in\mathbb{R}}$; $U(t) : \mathcal{K} \to \mathcal{K}$, such that*

$$(3.3) \qquad F_{ex}(t) = \langle v_0, U(t) v_0 \rangle_{\mathcal{K}}, \ \forall t \in \mathbb{R}.$$

Setting $J : \mathcal{H}_F \to \mathcal{K}$, $J\xi_0 = v_0$, then J defines (by extension) an isometry such that

$$(3.4) \qquad U(t) J\xi_0 = J \ (local\ translation\ in\ \Omega)$$

holds locally (i.e., for t sufficiently close to 0.)

Moreover, the function

$$(3.5) \qquad \mathbb{R} \ni t \mapsto U(t) J\xi_0 = U(t) v_0$$

is compactly supported.

PROOF. The existence of $\mathcal{K}$, v_0, and $\{U(t)\}_{t\in\mathbb{R}}$ follows from the GNS-construction.

The conclusions in (3.4) and (3.5) follow from the given data, i.e., $F : \Omega - \Omega \to \mathbb{R}$, and the fact that F_{ex} is a spline-extension, i.e., it is of compact support; but by (3.3), this means that (3.5) is also compactly supported. □

Example 3.3 gives a p.d. F in $\left(-\frac{1}{2}, \frac{1}{2}\right)$ with $D^{(F)}$ of index $(1,1)$ and explicit measures in $Ext_1(F)$ and in $Ext_2(F)$.

We have the following:

Deficiency $(0,0)$: The p.d. extension of type 1 is unique; see (3.1); but there may still be p.d. extensions of type 2; see (3.2).

Deficiency $(1,1)$: This is a one-parameter family of extensions of type 1; and some more p.d. extensions are type 2.

So we now divide

$$Ext(F) = \left\{\mu \in \text{Prob}(\mathbb{R}) \mid \widehat{d\mu} \text{ is an extension of } F\right\}$$

up in subsets

$$Ext(F) = Ext_{type1}(F) \cup Ext_{type2}(F);$$

where $Ext_{type2}(F)$ corresponds to the Polya extensions.

Return to a continuous p.d. function $F : (-a, a) \to \mathbb{C}$, we take for the RKHS $\mathcal{H}_F$, and the skew-Hermitian operator

$$D(F_\varphi) = F_{\varphi'}, \ \varphi' = \frac{d\varphi}{dx}$$

If $D \subseteq A$, $A^* = -A$ in $\mathcal{H}_F$ then there exists an isometry $J : \mathcal{H}_F \to L^2(\mathbb{R}, \mu)$, where $d\mu(\cdot) = \|P_U(\cdot)\xi_0\|^2$,

$$U_A(t) = e^{tA} = \int_{\mathbb{R}} e^{it\lambda} P_U(d\lambda),$$

$\xi_0 = F(\cdot - 0) \in \mathcal{H}_F$, $J\xi_0 = 1 \in L^2(\mu)$.

TABLE 2. The deficiency indices of $D^{(F)} : F_\varphi \mapsto F_{\varphi'}$ in examples 3.2-3.9

$F : (-a, a) \to \mathbb{C}$	Indices	The Operator $D^{(F)}$
$F_1(x) = \frac{1}{1+x^2}$, $\|x\| < 1$	$(0,0)$	$D^{(F)}$ unbounded, skew-adjoint
$F_2(x) = 1 - \|x\|$, $\|x\| < \frac{1}{2}$	$(1,1)$	$D^{(F)}$ has unbounded sk. adj. extensions
$F_3(x) = e^{-\|x\|}$, $\|x\| < 1$	$(1,1)$	$D^{(F)}$ has unbounded sk. adj. extensions
$F_4(x) = \left(\frac{\sin \pi x}{\pi x}\right)^2$, $\|x\| < \frac{1}{2}$	$(0,0)$	$D^{(F)}$ bounded, skew-adjoint
$F_5(x) = e^{-x^2/2}$, $\|x\| < 1$	$(0,0)$	$D^{(F)}$ unbounded, skew-adjoint
$F_6(x) = \cos x$, $\|x\| < \frac{\pi}{4}$	$(0,0)$	$D^{(F)}$ is rank-one, $\dim(\mathscr{H}_{F_6}) = 2$

TABLE 3. Type II extensions. Six cases of p.d. continuous functions F_i defined on a finite interval $(-a, a)$.

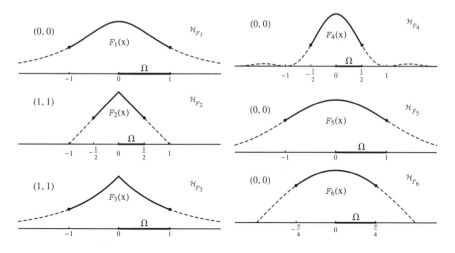

TABLE 4. The canonical isometric embeddings: $\mathscr{H}_{F_i} \hookrightarrow L^2(\mathbb{R}, d\mu_i)$, $i = 1, \ldots, 6$.

$d\mu_1(\lambda) = \frac{1}{2}e^{-\|\lambda\|}d\lambda$	$d\mu_4(\lambda) = \chi_{(-1,1)}(\lambda)(1 - \|\lambda\|) d\lambda$, cpt. support
$d\mu_2(\lambda) = \left(\frac{\sin \pi\lambda}{\pi\lambda}\right)^2 d\lambda$, Shannon	$d\mu_5(\lambda) = \frac{1}{\sqrt{2\pi}}e^{-\lambda^2/2}d\lambda$, Gaussian
$d\mu_3(\lambda) = \frac{d\lambda}{\pi(1+\lambda^2)}$, Cauchy	$d\mu_6(\lambda) = \frac{1}{2}(\delta_1 + \delta_{-1})$, atomic; two Dirac masses

4. Models for Operator Extensions

A special case of our extension question for continuous positive definite functions on a fixed finite interval $|x| < a$ in $\mathbb{R}$ is the following: It offers a spectral model representation for ALL Hermitian operators with dense domain in Hilbert space and with deficiency indices $(1,1)$.

Specifically, on $\mathbb{R}$, all the partially defined continuous p.d. functions extend, and we can make a translation of our p.d. problem into the problem of finding all $(1,1)$ restrictions selfadjoint operators.

By the Spectral theorem, every selfadjoint operator with simple spectrum has a representation as a multiplication operator M_λ in some $L^2(\mathbb{R}, \mu)$ for some probability measure μ on $\mathbb{R}$. So this accounts for all Hermitian restrictions operators with deficiency indices $(1,1)$.

4.1. Model for Restrictions of Continuous p.d. Functions on $\mathbb{R}$.
Let $\mathscr{H}$ be a Hilbert space, A a skew-adjoint operator, $A^* = -A$, which is *unbounded*; let $v_0 \in \mathscr{H}$ satisfying $\|v_0\|_{\mathscr{H}} = 1$. Then we get an associated p.d. continuous function F_A defined on $\mathbb{R}$ as follows:

$$(4.1) \qquad F_A(t) := \langle v_0, e^{tA} v_0 \rangle = \langle v_0, U_A(t) v_0 \rangle, \ t \in \mathbb{R},$$

where $U_A(t) = e^{tA}$ is a unitary representation of $\mathbb{R}$. Note that $U_A(t)$ is defined by the Spectral Theorem, and (4.1) holds for all $t \in \mathbb{R}$.

Let $P_U(\cdot)$ be the projection-valued measure (PVM) of A, then

$$(4.2) \qquad U(t) = \int_{-\infty}^{\infty} e^{i\lambda t} P_U(d\lambda), \ \forall t \in \mathbb{R}.$$

LEMMA 4.1.

(i) Setting $d\mu = \|P_U(d\lambda) v_0\|^2$, we then get

$$(4.3) \qquad F_A(t) = \widehat{d\mu}(t), \ \forall t \in \mathbb{R}$$

Moreover, every probability measure μ on $\mathbb{R}$ arises this way.

(ii) For Borel functions f on $\mathbb{R}$, let

$$(4.4) \qquad f(A) = \int_{\mathbb{R}} f(\lambda) P_U(d\lambda)$$

be given by functional calculus. We note that

$$(4.5) \qquad v_0 \in dom(f(A)) \Longleftrightarrow f \in L^2(\mu)$$

where μ is the measure in part (i). Then

$$(4.6) \qquad \|f(A) v_0\|^2 = \int_{\mathbb{R}} |f|^2 \, d\mu.$$

PROOF. (i) A direct computation using (4.1). (ii) This is an application of the Spectral Theorem. □

Now we consider restriction of F_A to, say $(-1,1)$, i.e.,

$$(4.7) \qquad F(\cdot) = F_A\big|_{(-1,1)}(\cdot)$$

LEMMA 4.2. *Let $\mathcal{H}_F$ be the RKHS computed for F in (4.3); and for $\varphi \in C_c(0,1)$, set $F_\varphi =$ the generating vectors in $\mathcal{H}_F$, as usual. Set*

$$(4.8) \qquad U(\varphi) := \int_0^1 \varphi(y) U(-y) \, dy$$

where $dy =$ Lebesgue measure on $(0,1)$; then

$$(4.9) \qquad F_\varphi(x) = \langle v_0, U(x) U(\varphi) v_0 \rangle, \; \forall x \in (0,1).$$

PROOF. We have

$$F_\varphi(x) \quad = \quad \int_0^1 \varphi(y) F(x-y) \, dy$$

$$\overset{\text{(by (4.1))}}{=} \quad \int_0^1 \varphi(y) \langle v_0, U_A(x-y) v_0 \rangle \, dy$$

$$= \quad \left\langle v_0, U_A(x) \int_0^1 \varphi(y) U_A(-y) v_0 \, dy \right\rangle$$

$$\overset{\text{(by (4.8))}}{=} \quad \langle v_0, U_A(x) U(\varphi) v_0 \rangle$$

$$= \quad \langle v_0, U(\varphi) U_A(x) v_0 \rangle$$

for all $x \in (0,1)$, and all $\varphi \in C_c(0,1)$. $\qquad\square$

COROLLARY 4.3. *Let A, $U(t) = e^{tA}$, $v_0 \in \mathcal{H}$, $\varphi \in C_c(0,1)$, and F p.d. on $(0,1)$ be as above; let $\mathcal{H}_F$ be the RKHS of F; then, for the inner product in $\mathcal{H}_F$, we have*

$$(4.10) \qquad \langle F_\varphi, F_\psi \rangle_{\mathcal{H}_F} = \langle U(\varphi) v_0, U(\psi) v_0 \rangle_{\mathcal{H}}, \; \forall \varphi, \psi \in C_c(0,1).$$

PROOF. Note that

$$\langle F_\varphi, F_\psi \rangle_{\mathcal{H}_F} \quad = \quad \int_0^1 \int_0^1 \overline{\varphi(x)} \psi(y) F(x-y) \, dx \, dy$$

$$\overset{\text{(by (4.7))}}{=} \quad \int_0^1 \int_0^1 \overline{\varphi(x)} \psi(y) \langle v_0, U_A(x-y) v_0 \rangle_{\mathcal{H}} \, dx \, dy$$

$$= \quad \int_0^1 \int_0^1 \langle \varphi(x) U_A(-x) v_0, \psi(y) U_A(-y) v_0 \rangle_{\mathcal{H}} \, dx \, dy$$

$$\overset{\text{(by (4.8))}}{=} \quad \langle U(\varphi) v_0, U(\psi) v_0 \rangle_{\mathcal{H}}$$

$\qquad\square$

COROLLARY 4.4. *Set $\varphi^\#(x) = \overline{\varphi(-x)}$, $x \in \mathbb{R}$, $\varphi \in C_c(\mathbb{R})$, or in this case, $\varphi \in C_c(0,1)$; then we have:*

$$(4.11) \qquad \langle F_\varphi, F_\psi \rangle_{\mathcal{H}_F} = \langle v_0, U(\varphi^\# * \psi) v_0 \rangle_{\mathcal{H}}, \; \forall \varphi, \psi \in C_c(0,1).$$

PROOF. Immediate from (4.10) and Fubini. $\qquad\square$

COROLLARY 4.5. *Let F and $\varphi \in C_c(0,1)$ be as above; then in the RKHS $\mathcal{H}_F$ we have:*

$$(4.12) \qquad \|F_\varphi\|_{\mathcal{H}_F}^2 = \|U(\varphi) v_0\|_{\mathcal{H}}^2 = \int |\widehat{\varphi}|^2 \, d\mu$$

where μ is the measure in part (i) of Lemma 4.1. $\widehat{\varphi} =$ Fourier transform: $\widehat{\varphi}(\lambda) = \int_0^1 e^{-i\lambda x} \varphi(x) \, dx$, $\lambda \in \mathbb{R}$.

PROOF. Immediate from (4.11); indeed:

$$
\begin{aligned}
\|F_\varphi\|^2_{\mathscr{H}_F} &= \int_0^1 \int_0^1 \overline{\varphi(x)}\varphi(y) \int_{\mathbb{R}} e_\lambda(x-y)\, d\mu(\lambda) \\
&= \int_{\mathbb{R}} |\widehat{\varphi}(\lambda)|^2\, d\mu(\lambda), \ \forall \varphi \in C_c(0,1).
\end{aligned}
$$

$\square$

COROLLARY 4.6. *Every Borel probability measure μ on $\mathbb{R}$ arises this way.*

PROOF. We shall need to following:

LEMMA 4.7. *Let A, $\mathscr{H}$, $\{U_A(t)\}_{t \in \mathbb{R}}$, $v_0 \in \mathscr{H}$ be as above; and set*

$$(4.13) \qquad d\mu = d\mu_A(\cdot) = \|P_U(\cdot)\, v_0\|^2$$

as in Lemma 4.1. Assume v_0 is cyclic; then $W_\mu f(A)\, v_0 = f$ defines a unitary isomorphism $W_\mu : \mathscr{H} \to L^2(\mu)$; and

$$(4.14) \qquad W_\mu U_A(t) = e^{it\cdot} W_\mu$$

where $e^{it\cdot}$ is seen as a multiplication operator in $L^2(\mu)$. More precisely:

$$(4.15) \qquad (W_\mu U(t)\xi)(\lambda) = e^{it\lambda}(W_\mu \xi)(\lambda), \ \forall t, \lambda \in \mathbb{R}, \forall \xi \in \mathscr{H}.$$

(We say that the isometry W_μ __intertwines__ the two unitary one-parameter groups.)

PROOF. Since v_0 is cyclic, it is enough to consider $\xi \in \mathscr{H}$ of the following form: $\xi = f(A)\, v_0$, with $f \in L^2(\mu)$, see (4.5) in Lemma 4.1. Then

$$(4.16) \qquad \|\xi\|^2_{\mathscr{H}} = \int_{\mathbb{R}} |f(\lambda)|^2\, d\mu(\lambda), \text{ so}$$

$$\|W_\mu \xi\|_{L^2(\mu)} = \|\xi\|_{\mathscr{H}} \ (\Longleftrightarrow (4.16))$$

For the adjoint operator $W_\mu^* : L^2(\mathbb{R}, \mu) \to \mathscr{H}$, we have

$$W_\mu^* f = f(A)\, v_0,$$

see (4.4)-(4.6). Note that $f(A)\, v_0 \in \mathscr{H}$ is well-defined for all $f \in L^2(\mu)$. Also $W_\mu^* W_\mu = I_{\mathscr{H}}$, $W_\mu W_\mu^* = I_{L^2(\mu)}$.

Proof of (4.15). Take $\xi = f(A)\, v_0$, $f \in L^2(\mu)$, and apply the previous lemma, we have

$$W_\mu U(t)\xi = W_\mu U(t) f(A)_0 = W_\mu \left(e^{it\cdot} f(\cdot)\right)(A)\, v_0 = e^{it\cdot} f(\cdot) = e^{it\cdot} W_\mu \xi;$$

or written differently:

$$W_\mu U(t) = M_{e^{it\cdot}} W_\mu, \ \forall t \in \mathbb{R}$$

where $M_{e^{it\cdot}}$ is the multiplication operator by $e^{it\cdot}$. $\square$

$\square$

REMARK 4.8. Deficiency indices $(1,1)$ occur for probability measures μ on $\mathbb{R}$ such that

$$(4.17) \qquad \int_{\mathbb{R}} |\lambda|^2\, d\mu(\lambda) = \infty.$$

See examples below.

TABLE 5. Application of Theorem 4.9 to Table 2.

measure	condition (4.17)	deficiency indices
μ_1	$\int_{\mathbb{R}} \lvert\lambda\rvert^2 e^{-\lvert\lambda\rvert} d\lambda < \infty$	$(0,0)$
μ_2	$\int_{\mathbb{R}} \lvert\lambda\rvert^2 \left(\frac{\sin \pi\lambda}{\pi\lambda}\right)^2 d\lambda = \infty$	$(1,1)$
μ_3	$\int_{\mathbb{R}} \lvert\lambda\rvert^2 \frac{d\lambda}{\pi(1+\lambda^2)} = \infty$	$(1,1)$
μ_4	$\int_{\mathbb{R}} \lvert\lambda\rvert^2 \chi_{(-1,1)}(\lambda)(1-\lvert\lambda\rvert) d\lambda < \infty$	$(0,0)$
μ_5	$\int_{\mathbb{R}} \lvert\lambda\rvert^2 \frac{1}{\sqrt{2\pi}}e^{-\lambda^2/2} d\lambda = 1 < \infty$	$(0,0)$

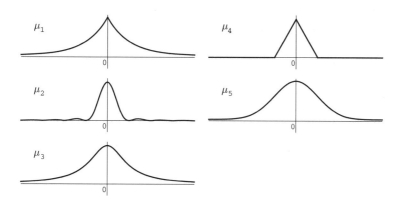

FIGURE 4.1. The measures $\mu_i \in Ext(F_i)$ extending p.d. functions F_i in Table 2, $i = 1, 2, \ldots 5$.

SUMMARY. Restrictions with deficiency indices $(1,1)$.

THEOREM 4.9. *If μ is a fixed probability measure on $\mathbb{R}$, then the following two conditions are equivalent:*

(1) $\int_{\mathbb{R}} \lambda^2 d\mu(\lambda) = \infty$;
(2) The set

$$\left\{ f \in L^2(\mu) \ \middle| \ \lambda f \in L^2(\mu), \int_{\mathbb{R}} (\lambda + i) f(\lambda) d\mu(\lambda) = 0 \right\}$$

is the $\underline{dense}$ domain of a restriction operator $S \subset M_\lambda$ with deficiency indices $\overline{(1,1)}$, and the deficiency space $DEF_+ = \mathbb{C}\mathbf{1}$, $(\mathbf{1} = $ the constant function 1 in $L^2(\mu)$.)

4.2. A Model of ALL Deficiency Index-$(1,1)$ Operators.

LEMMA 4.10. *Let μ be a Borel probability measure on $\mathbb{R}$, and denote $L^2(\mathbb{R}, d\mu)$ by $L^2(\mu)$. Then we have TFAE:*

(1) $\int_{\mathbb{R}} \lvert\lambda\rvert^2 d\mu(\lambda) = \infty$;

(2) *The following two subspaces in $L^2(\mu)$ are dense (in the $L^2(\mu)$-norm):*

$$(4.18) \qquad \left\{ f \in L^2(\mu) \,\Big|\, [(\lambda \pm i)\, f(\lambda)] \in L^2(\mu) \ \text{and} \ \int (\lambda \pm i)\, f(\lambda)\, d\mu(\lambda) = 0 \right\}$$

where $i = \sqrt{-1}$.

PROOF. See [**24**].   □

REMARK 4.11. If (1) holds, then the two dense subspaces $\mathscr{D}_\pm \subset L^2(\mu)$ in (4.18) form the dense domain of a restriction S of M_λ in $L^2(\mu)$; and this restriction has deficiency indices $(1,1)$. Moreover, all Hermitian operators having deficiency indices $(1,1)$ arise this way.

LEMMA 4.12. *With $i = \sqrt{-1}$, set*

$$(4.19) \qquad dom(S) = \left\{ f \in L^2(\mu) \,\Big|\, \lambda f \in L^2(\mu) \ \text{and} \ \int (\lambda + i)\, f(\lambda)\, d\mu(\lambda) = 0 \right\}$$

then $S \subset M_\lambda \subset S^*$; and the deficiency subspaces $DEF_\pm$ are as follow:

$$(4.20) \qquad\qquad DEF_+ \;=\; \text{the constant function in } L^2(\mu) = \mathbb{C}1$$

$$(4.21) \qquad\qquad DEF_- \;=\; span\left\{ \frac{\lambda - i}{\lambda + i} \right\}_{\lambda \in \mathbb{R}} \subseteq L^2(\mu)$$

where DEF_- is also a 1-dimensional subspace in $L^2(\mu)$.

PROOF. Let $f \in dom(S)$, then, by definition,

$$\int_{\mathbb{R}} (\lambda + i)\, f(\lambda)\, d\mu(\lambda) = 0 \text{ and so}$$

$$(4.22) \qquad\qquad \langle 1, (S + iI)\, f \rangle_{L^2(\mu)} = \int_{\mathbb{R}} (\lambda + i)\, f(\lambda)\, d\mu(\lambda) = 0$$

hence (4.20) follows.

Note we have formula (4.19) for $dom(S)$. Moreover $dom(S)$ is dense in $L^2(\mu)$ because of (4.18) in Lemma 4.10.

Now to (4.21): Let $f \in dom(S)$; then

$$\left\langle \frac{\lambda - i}{\lambda + i}, (S - iI)\, f \right\rangle_{L^2(\mu)} \;=\; \int_{\mathbb{R}} \left(\frac{\lambda + i}{\lambda - i} \right) (\lambda - i)\, f(\lambda)\, d\mu(\lambda)$$

$$\;=\; \int_{\mathbb{R}} (\lambda + i)\, f(\lambda)\, d\mu(\lambda) = 0$$

again using the definition of $dom(S)$ in (4.19).   □

We have established a representation for <u>all</u> Hermitian operators with dense domain in a Hilbert space, and having deficiency indices $(1,1)$. In particular, we have justified the answers in Table 2 for F_i, $i = 1, \dots, 5$.

To further emphasize to the result we need about deficiency indices $(1,1)$, we have the following:

THEOREM 4.13. *Let $\mathscr{H}$ be a separable Hilbert space, and let S be a Hermitian operator with dense domain in $\mathscr{H}$. Suppose the deficiency indices of S are (d,d); and suppose one of the selfadjoint extensions of S has simple spectrum.*

Then the following two conditions are equivalent:

(1) $d = 1$;

(2) for each of the selfadjoint extensions T of S, we have a unitary equivalence between $(S, \mathscr{H})$ on the one hand, and a system $\left(S_\mu, L^2(\mathbb{R}, \mu)\right)$ on the other, where μ is a Borel probability measure on $\mathbb{R}$. Moreover,

$$(4.23) \qquad (S_\mu f)(\lambda) = \lambda f(\lambda), \ \forall f \in dom(S_\mu), \forall \lambda \in \mathbb{R}, \ where$$

$$(4.24) \qquad dom(S_\mu) = \left\{ f \in L^2(\mu) \middle| \lambda f \in L^2(\mu), \ \int_{\mathbb{R}} (\lambda + i) f(\lambda) \, d\mu(\lambda) = 0 \right\}.$$

In case μ satisfies condition (4.23), then the constant function $\mathbf{1}$ (in $L^2(\mathbb{R}, \mu)$) is in the domain of S_μ^*, and

$$(4.25) \qquad\qquad\qquad S_\mu^* \mathbf{1} = i\mathbf{1}$$

i.e., $\left(S_\mu^* \mathbf{1}\right)(\lambda) = i$, a.a. λ w.r.t. $d\mu$.

PROOF. For the implication (2)$\Rightarrow$(1), see Lemma 4.12.

(1)$\Rightarrow$(2). Assume that the operator S, acting in $\mathscr{H}$ is Hermitian with deficiency indices $(1, 1)$. This means that each of the two subspaces $DEF_\pm \subset \mathscr{H}$ is one-dimensional, where

$$(4.26) \qquad DEF_\pm = \left\{ h_\pm \in dom(S^*) \middle| S^* h_\pm = \pm i h_\pm \right\}.$$

Now pick a selfadjoint extension, say T, extending S. We have

$$(4.27) \qquad\qquad\qquad S \subseteq T = T^* \subseteq S^*$$

where "$\subseteq$" in (4.27) means "containment of the respective graphs."

Now set $U(t) = e^{itT}$, $t \in \mathbb{R}$, and let $P_U(\cdot)$ be the corresponding projection-valued measure, i.e., we have:

$$(4.28) \qquad\qquad U(t) = \int_{\mathbb{R}} e^{it\lambda} P_U(d\lambda), \ \forall t \in \mathbb{R}.$$

Using the assumption (1), and (4.26), it follows that there is a vector $h_+ \in \mathscr{H}$ such that $\|h_+\|_{\mathscr{H}} = 1$, $h_+ \in dom(S^*)$, and $S^* h_+ = i h_+$. Now set

$$(4.29) \qquad\qquad d\mu(\lambda) := \|P_U(d\lambda) h_+\|_{\mathscr{H}}^2.$$

Using (4.28), we then verify that there is a unitary (and isometric) isomorphism of $L^2(\mu) \overset{W}{\longrightarrow} \mathscr{H}$ given by

$$(4.30) \qquad\qquad Wf = f(T) h_+, \ \forall f \in L^2(\mu);$$

where $f(T) = \int_{\mathbb{R}} f(T) P_U(d\lambda)$ is the functional calculus applied to the selfadjoint operator T. Hence

$$
\begin{aligned}
\|Wf\|_{\mathscr{H}}^2 &= \|f(T) h_+\|_{\mathscr{H}}^2 \\
&= \int_{\mathbb{R}} |f(\lambda)|^2 \|P_U(d\lambda) h_+\|^2 \\
&\overset{\text{(by 4.29)}}{=} \int_{\mathbb{R}} |f(\lambda)|^2 \, d\mu(\lambda) = \|f\|_{L^2(\mu)}^2.
\end{aligned}
$$

To see that W in (4.30) is an isometric isomorphism of $L^2(\mu)$ onto $\mathscr{H}$, we use the assumption that T has simple spectrum.

Now set

$$(4.31) \qquad\qquad S_\mu \quad := \quad W^* S W$$

$$(4.32) \qquad\qquad T_\mu \quad := \quad W^* T W.$$

We note that T_μ is then the multiplication operator M in $L^2(\mathbb{R}, \mu)$, given by

$$(4.33) \qquad\qquad (Mf)(\lambda) = \lambda f(\lambda), \ \forall f \in L^2(\mu)$$

such that $\lambda f \in L^2(\mu)$. This assertion is immediate from (4.30) and (4.29).

To finish the proof, we compute the integral in (4.24) in the theorem, and we use the intertwining properties of the isomorphism W from (4.30). Indeed, we have

$$
\begin{aligned}
\int_\mathbb{R} (\lambda + i) f(\lambda) \, d\mu(\lambda) \quad &= \quad \langle \mathbf{1}, (M + iI) f \rangle_{L^2(\mu)} \\
&= \quad \langle W\mathbf{1}, W(M + iI) f \rangle_{\mathscr{H}} \\
(4.34) \qquad\qquad &\overset{(4.29)}{=} \quad \langle h_+, (T + iI) W f \rangle_{\mathscr{H}}.
\end{aligned}
$$

Hence $Wf \in dom(S) \iff f \in dom(S_\mu)$, by (4.31); and, so for $Wf \in dom(S)$, the RHS in (4.34) yields $\langle (S^* - iI) h_+, Wf \rangle_{\mathscr{H}} = 0$; and the assertion (2) in the theorem follows. $\qquad\square$

4.3. The Case of Indices (d, d) **where** $d > 1$. Let μ be a Borel probability measure on $\mathbb{R}$, and let

$$(4.35) \qquad\qquad L^2(\mu) := L^2(\mathbb{R}, \mathscr{B}, \mu).$$

The notation $\mathrm{Prob}(\mathbb{R})$ will be used for these measures.

We saw that the restriction/extension problem for continuous positive definite functions F on $\mathbb{R}$ may be translated into a spectral theoretic model in some $L^2(\mu)$ for suitable $\mu \in \mathrm{Prob}(\mathbb{R})$. We saw that extension from a finite open ($\neq \emptyset$) interval leads to spectral representation in $L^2(\mu)$, and restrictions of

$$(4.36) \qquad\qquad (M_\mu f)(\lambda) = \lambda f(\lambda), \ f \in L^2(\mu)$$

having deficiency-indices $(1, 1)$; hence the case $d = 1$.

THEOREM 4.14. *Fix $\mu \in \mathrm{Prob}(\mathbb{R})$. There is a 1-1 bijective correspondence between the following:*

(1) certain closed subspaces $\mathscr{L} \subset L^2(\mu)$

(2) Hermitian restrictions $S_\mathscr{L}$ of M_μ (see (4.36)) such that

$$(4.37) \qquad\qquad DEF_+ (S_\mathscr{L}) = \mathscr{L}.$$

The closed subspaces in (1) are specified as follows:

(i) $\dim(\mathscr{L}) = d < \infty$

(ii) the following implication holds:

$$(4.38) \qquad g \neq 0, \ and \ g \in \mathscr{L} \implies ([\lambda \mapsto \lambda g(\lambda)] \notin L^2(\mu))$$

Then set

$$(4.39) \quad dom(S_\mathscr{L}) := \left\{ f \in dom(M_\mu) \ \Big| \ \int \overline{g(\lambda)} (\lambda + i) f(\lambda) \, d\mu(\lambda), \forall g \in \mathscr{L} \right\}$$

and set

$$(4.40) \qquad\qquad S_\mathscr{L} := M_\mu \Big|_{dom(S_\mathscr{L})}$$

where dom $(S_{\mathscr{L}})$ is specified as in (4.39).

PROOF. Note that the case $d = 1$ is contained in the previous theorem.

Proof of (1) $\Rightarrow$ (2). We will be using an idea from [**24**]. With assumptions (i)-(ii), in particular (4.38), one checks that $dom\,(S_{\mathscr{L}})$as specified in (4.39) is dense in $L^2(\mu)$. In fact, the converse implication is also true.

Now setting $S_{\mathscr{L}}$ to be the restriction in (4.40), we conclude that

(4.41) $$S_{\mathscr{L}} \subseteq M_\mu \subseteq S_{\mathscr{L}}^*, \text{ where}$$

$dom\,(S_{\mathscr{L}}^*)$ consists of $h \in L^2(\mu)$ s.t. $\exists C < \infty$, and

$$\left| \int_{\mathbb{R}} \overline{h\,(\lambda)} \lambda f\,(\lambda)\, d\mu\,(\lambda) \right|^2 \leq C \int_{\mathbb{R}} |f\,(\lambda)|^2\, d\mu\,(\lambda), \ \forall f \in dom\,(S_{\mathscr{L}}).$$

The assertions in (2) now follow from this.

Proof of (2) $\Rightarrow$ (1). Assume that S is a densely defined restriction of M_μ, and let $DEF_+(S) = $ the $(+)$ deficiency space, i.e.,

(4.42) $$DEF_+(S) = \{g \in dom\,(S^*) \mid S^*g = ig\}$$

Assume $\dim\,(DEF_+(S)) = d$, and $1 \leq d < \infty$. Then set $\mathscr{L} := DEF_+(S)$. Using [**24**], one checks that (1) then holds for this closed subspace in $L^2(\mu)$.

The fact that (4.38) holds for this subspace $\mathscr{L}$ follows from the observation:

$$DEF_+(S) \cap dom\,(M_\mu) = \{0\}$$

for every densely defined restriction S of M_μ. $\qquad\square$

4.4. Spectral Representation of Index $(1,1)$ Hermitian Operators. In this section, we give an explicit answer to the question: How to go from any index $(1,1)$ Hermitian operator to a $(\mathscr{H}_F, D^{(F)})$ model; i.e., from a given index $(1,1)$ Hermitian operator with dense domain in a separable Hilbert space $\mathscr{H}$, we build a p.d. continuous function F on $\Omega - \Omega$, where $\Omega = (0,a)$, $a > 0$.

So far, we have been concentrating on building transforms going in the other direction. But recall that, for a given continuous p.d. function F on $\Omega - \Omega$, it is often difficult to answer the question of whether the corresponding operator $D^{(F)}$ in the RKHS $\mathscr{H}_F$ has deficiency indices $(1,1)$ or $(0,0)$.

Now this question answers itself once we have an explicit transform going in the opposite direction. Specifically, given any index $(1,1)$ Hermitian operator S in a separable Hilbert space $\mathscr{H}$, we then to find a pair (F, Ω), p.d. function and interval, with the desired properties. There are two steps:

Step 1, writing down explicitly, a p.d. continuous function F on $\Omega - \Omega$, and the associated RKHS $\mathscr{H}_F$ with operator $D^{(F)}$.

Step 2, constructing an intertwining isomorphism $W : \mathscr{H} \to \mathscr{H}_F$, having the following properties: W is an isometric isomorphism, intertwining the pair $(\mathscr{H}, S)$ with $(\mathscr{H}_F, D^{(F)})$, i.e., satisfying $WS = D^{(F)}W$; and also intertwining the respective domains and deficiency spaces, in $\mathscr{H}$ and $\mathscr{H}_F$.

Moreover, starting with any $(1,1)$ Hermitian operator, we can even arrange a normalization for the p.d. function F such that $\Omega = (0,1)$ will do the job.

Details. We will have three pairs $(\mathscr{H}, S)$, $(L^2(\mathbb{R}, \mu)$, restriction of $M_\mu)$, and $(\mathscr{H}_F, D^{(F)})$, where:

(i) S is a fixed Hermitian operator with dense domain $dom\,(S)$ in a separable Hilbert space $\mathscr{H}$, and with deficiency indices $(1,1)$.

(ii) From (i), we will construct a finite Borel measure μ on $\mathbb{R}$ such that an index-$(1,1)$ restriction of $M_\mu : f \mapsto \lambda f(\lambda)$ in $L^2(\mathbb{R}, \mu)$, is equivalent to $(\mathscr{H}, S)$.

(iii) Here $F : (-1, 1) \to \mathbb{C}$ will be a p.d. continuous function, $\mathscr{H}_F$ the corresponding RKHS; and $D^{(F)}$ the usual operator with dense domain

$$(4.43) \qquad \{F_\varphi \mid \varphi \in C_c^\infty(0, 1)\}, \quad \text{and} \quad D^{(F)}(F_\varphi) = \frac{1}{i} F_{\varphi'}, \; \varphi' = \frac{d\varphi}{dx}.$$

We will accomplish the stated goal with the following system of intertwining operators: See Figure 4.2.

But we stress that, at the outset, only (i) is given; the rest (μ, F and $\mathscr{H}_F$) will be constructed. Further, the solutions (μ, F) in Figure 4.2 are not unique; rather they depend on choice of selfadjoint extension in (i): Different selfadjoint extensions of S in (i) yield different solutions (μ, F). But the selfadjoint extensions of S in $\mathscr{H}$ are parameterized by von Neumann's theory; see e.g., [**45**].

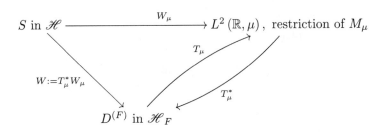

FIGURE 4.2. A system of intertwining operators.

REMARK 4.15. In our analysis of (i)-(iii), we may without loss of generality assume that the following normalizations hold:

(z_1) $\mu(\mathbb{R}) = 1$, so μ is a probability measure;

(z_2) $F(0) = 1$, and the p.d. continuous solution;

(z_3) $F : (-1, 1) \to \mathbb{C}$ is defined on $(-1, 1)$; so $\Omega := (0, 1)$.

Further, we may assume that the operator S in $\mathscr{H}$ from (i) has simple spectrum.

THEOREM 4.16. *Starting with $(\mathscr{H}, S)$ as in (i), there are solutions (μ, F) to (ii)-(iii), and intertwining operators W_μ, T_μ as in Figure 4.2, such that*

$$(4.44) \qquad\qquad W := T_\mu^* W_\mu$$

satisfies the intertwining properties for $(\mathscr{H}, S)$ and $(\mathscr{H}_F, D^{(F)})$.

PROOF. Since S has indices $(1,1)$, $\dim DEF_\pm(S) = 1$, and S has selfadjoint extensions indexed by partial isometries $DEF_+ \xrightarrow{v} DEF_-$; see [**16, 45**]. We now pick $g_+ \in DEF_+$, $\|g_+\| = 1$, and partial isometry v with selfadjoint extension S_v, i.e.,

$$(4.45) \qquad\qquad S \subset S_v \subset S_v^* \subset S^*.$$

Hence $\{U_v(t) : t \in \mathbb{R}\}$ is a strongly continuous unitary representation of $\mathbb{R}$, acting in $\mathscr{H}$, $U_v(t) := e^{itS_v}$, $t \in \mathbb{R}$. Let $P_{S_v}(\cdot)$ be the corresponding projection valued measure (PVM) on $\mathscr{B}(\mathbb{R})$, i.e., we have

$$(4.46) \qquad\qquad U_v(t) = \int_{\mathbb{R}} e^{it\lambda} P_{S_v}(d\lambda);$$

and set

$$(4.47) \qquad d\mu(\lambda) := d\mu_v(\lambda) = \|P_{S_v}(d\lambda)\, g_+\|_{\mathscr{H}}^2.$$

For $f \in L^2(\mathbb{R}, \mu_v)$, set

$$(4.48) \qquad W_{\mu_v}(f(S_v)\, g_+) = f;$$

then $W_{\mu_v} : \mathscr{H} \to L^2(\mathbb{R}, \mu_v)$ is isometric onto; and

$$(4.49) \qquad W_{\mu_v}^*(f) = f(S_v)\, g_+,$$

where

$$(4.50) \qquad f(S_v)\, g_+ = \int_{\mathbb{R}} f(\lambda)\, P_{S_v}(d\lambda)\, g_+.$$

For justification of these assertions, see e.g., [**38**]. Moreover, W_μ has the intertwining properties sketched in Figure 4.2.

Returning to (4.46) and (iii) in the theorem, we now set $F := F_\mu|_{(-1,1)}$, where

$$(4.51) \qquad F_\mu(t) \quad := \quad \langle g_+, U_v(t)\, g_+ \rangle$$

$$\overset{(4.46)}{=} \quad \left\langle g_+, \int_{\mathbb{R}} e^{it\lambda} P_{S_v}(d\lambda)\, g_+ \right\rangle$$

$$= \quad \int_{\mathbb{R}} e^{it\lambda} \|P_{S_v}(d\lambda)\, g_+\|^2$$

$$\overset{(4.47)}{=} \quad \int_{\mathbb{R}} e^{it\lambda} d\mu_v(\lambda) = \widehat{d\mu_v}(t),\ \forall t \in \mathbb{R}.$$

We now show that $F\left(:= F_\mu|_{(-1,1)}\right)$ has the desired properties.

From Corollary 2.14, we have the isometry $T_\mu(F_\varphi) = \widehat{\varphi}$, $\varphi \in C_c(0,1)$, with adjoint $T_\mu^*(f) = (f d\mu)^\vee$, see Fig 4.2.

The following properties are easily checked:

$$(4.52) \qquad W_\mu(g_+) = \mathbf{1} \in L^2(\mathbb{R}, \mu),\ \text{and}$$

$$(4.53) \qquad T_\mu^*(\mathbf{1}) = F_0 = F(\cdot - 0) \in \mathscr{H}_F,$$

as well as the intertwining properties stated in the theorem; see Fig. 4.2 for a summary.

<u>Proof of (4.52)</u> We will show instead that $W_\mu^*(\mathbf{1}) = g_+$. From (4.50) we note that if $f \in L^2(\mathbb{R}, \mu)$ satisfies $f = \mathbf{1}$, then $f(S_v) = I_{\mathscr{H}}$. Hence

$$W_\mu^*(\mathbf{1}) \overset{(4.49)}{=} \mathbf{1}(S_v)\, g_+ = g_+,$$

which is (4.52).

<u>Proof of (4.53)</u> For $\varphi \in C_c(0,1)$ we have $\widehat{\varphi} \in L^2(\mathbb{R}, \mu)$, and

$$T_\mu^* T_\mu(F_\varphi) = T_\mu^*(\widehat{\varphi}) \overset{(2.28)}{=} (\widehat{\varphi} d\mu)^\vee = F_\varphi.$$

Taking an approximation $(\varphi_n) \subset C_c(0,1)$ to the Dirac unit mass δ_0, we get (4.53). $\square$

COROLLARY 4.17. *The deficiency indices of $D^{(F)}$ in $\mathscr{H}_F$ for $F(x) = e^{-|x|}$, $|x| < 1$, are $(1,1)$.*

PROOF. Let $\mathscr{H} = L^2(\mathbb{R})$ w.r.t. the Lebesgue measure. Take $g_+ := ((\lambda + i)^{-1})^{\vee}(x)$, $x \in \mathbb{R}$; then $g_+ \in \mathscr{H}$ since

$$\int_{\mathbb{R}} |g_+(x)|^2 \, dx \overset{\text{Parseval}}{=} \int_{\mathbb{R}} \left| \frac{1}{\lambda + i} \right|^2 d\lambda = \int_{\mathbb{R}} \frac{1}{1 + \lambda^2} d\lambda = \pi.$$

Now for S and S_v in Theorem 4.16, we take

(4.54) $\quad S_v h = \dfrac{1}{i} \dfrac{d}{dx} h, \quad dom(S_v) := \{ h \in L^2(\mathbb{R}) \mid h' \in L^2(\mathbb{R}) \}$, and

(4.55) $\quad S = S_v$ restricted to $\{ h \in dom(S_v) \mid h(0) = 0 \}$;

then by [24], we know that S has index $(1, 1)$, and that $g_+ \in DEF_+(S)$. The corresponding p.d. continuous function F is the restriction to $|t| < 1$ of the p.d. function:

$$\langle g_+, U_v(t) g_+ \rangle_{\mathscr{H}} = \int_{\mathbb{R}} \frac{1}{\lambda - i} \frac{e^{it\lambda}}{\lambda + i} d\lambda = \left(\frac{1}{1 + \lambda^2} \right)^{\vee}(t) = \pi e^{-|t|}.$$

$\square$

EXAMPLE 4.18 (Lévy-measures (see e.g., [47])). Let $0 < \alpha \leq 2$, $-1 < \beta < 1$, $v > 0$; then the Lévy-measures μ on $\mathbb{R}$ are indexed by (α, β, ν), so $\mu = \mu_{(\alpha,\beta,\nu)}$. They are absolutely continuous with respect to the Lebesgue measure $d\lambda$ on $\mathbb{R}$; and for $\alpha = 1$,

(4.56) $$F_{(\alpha,\beta,\nu)}(x) = \widehat{\mu_{(\alpha,\beta,\nu)}}(x), \quad x \in \mathbb{R},$$

satisfies

(4.57) $$F_{(\alpha,\beta,\nu)}(x) = \exp\left(-\nu |x| \cdot \left(1 + \frac{2i\beta}{\pi} - \text{sgn}(x) \ln |x| \right) \right).$$

The case $\alpha = 2$, $\beta = 0$, reduces to the Gaussian distribution.

The measures $\mu_{(1,\beta,\nu)}$ have infinite variance, i.e.,

$$\int_{\mathbb{R}} \lambda^2 d\mu_{(1,\beta,\nu)} = \infty.$$

As a Corollary of Theorem 4.16, we therefore conclude that, for the restrictions (see (4.56)-(4.57)),

$$F_{(1,\beta,\nu)}^{(res)}(x) = F_{(1,\beta,\nu)}(x), \quad x \in (-1, 1)$$

the associated Hermitian operator $D^{F^{(res)}}$ all have deficiency indices $(1, 1)$.

References

[1] N. I. Akhiezer and I. M. Glazman, *Theory of linear operators in Hilbert space*, Dover Publications, Inc., New York, 1993. Translated from the Russian and with a preface by Merlynd Nestell; Reprint of the 1961 and 1963 translations; Two volumes bound as one. MR1255973 (94i:47001)

[2] Daniel Alpay, *On linear combinations of positive functions, associated reproducing kernel spaces and a non-Hermitian Schur algorithm*, Arch. Math. (Basel) **58** (1992), no. 2, 174–182, DOI 10.1007/BF01191883. MR1143167 (92m:46039)

[3] Daniel Alpay, Vladimir Bolotnikov, Aad Dijksma, and Henk de Snoo, *On some operator colligations and associated reproducing kernel Hilbert spaces*, Operator extensions, interpolation of functions and related topics (Timişoara, 1992), Oper. Theory Adv. Appl., vol. 61, Birkhäuser, Basel, 1993, pp. 1–27. MR1246577 (94i:47018)

[4] Daniel Alpay, Aad Dijksma, Heinz Langer, Simeon Reich, and David Shoikhet, *Boundary interpolation and rigidity for generalized Nevanlinna functions*, Math. Nachr. **283** (2010), no. 3, 335–364, DOI 10.1002/mana.200910135. MR2643017 (2011c:30088)

[5] Daniel Alpay and Harry Dym, *On applications of reproducing kernel spaces to the Schur algorithm and rational J unitary factorization*, I. Schur methods in operator theory and signal processing, Oper. Theory Adv. Appl., vol. 18, Birkhäuser, Basel, 1986, pp. 89–159, DOI 10.1007/978-3-0348-5483-2_5. MR902603 (89g:46051)

[6] Daniel Alpay, Palle Jorgensen, Ron Seager, and Dan Volok, *On discrete analytic functions: products, rational functions and reproducing kernels*, J. Appl. Math. Comput. **41** (2013), no. 1-2, 393–426, DOI 10.1007/s12190-012-0608-2. MR3017129

[7] Daniel Alpay and Palle E. T. Jorgensen, *Stochastic processes induced by singular operators*, Numer. Funct. Anal. Optim. **33** (2012), no. 7-9, 708–735, DOI 10.1080/01630563.2012.682132. MR2966130

[8] N. Aronszajn, *Theory of reproducing kernels*, Trans. Amer. Math. Soc. **68** (1950), 337–404. MR0051437 (14,479c)

[9] William Arveson, *Markov operators and OS-positive processes*, J. Funct. Anal. **66** (1986), no. 2, 173–234, DOI 10.1016/0022-1236(86)90071-6. MR832989 (87i:46139)

[10] Giles Auchmuty and Qi Han, *Spectral representations of solutions of linear elliptic equations on exterior regions*, J. Math. Anal. Appl. **398** (2013), no. 1, 1–10, DOI 10.1016/j.jmaa.2012.07.023. MR2984310

[11] Christian Berg, Jens Peter Reus Christensen, and Paul Ressel, *Harmonic analysis on semigroups*, Graduate Texts in Mathematics, vol. 100, Springer-Verlag, New York, 1984. Theory of positive definite and related functions. MR747302 (86b:43001)

[12] Louis de Branges, *Hilbert spaces of entire functions*, Prentice-Hall, Inc., Englewood Cliffs, N.J., 1968. MR0229011 (37 #4590)

[13] Louis de Branges and James Rovnyak, *Canonical models in quantum scattering theory*, Perturbation Theory and its Applications in Quantum Mechanics (Proc. Adv. Sem. Math. Res. Center, U.S. Army, Theoret. Chem. Inst., Univ. of Wisconsin, Madison, Wis., 1965), Wiley, New York, 1966, pp. 295–392. MR0244795 (39 #6109)

[14] Allen Devinatz, *On the extensions of positive definite functions*, Acta Math. **102** (1959), 109–134. MR0109992 (22 #875)

[15] Allen Devinatz, *The deficiency index of a certain class of ordinary self-adjoint differential operators*, Advances in Math. **8** (1972), 434–473. MR0298102 (45 #7154)

[16] Nelson Dunford and Jacob T. Schwartz, *Linear operators. Part II*, Wiley Classics Library, John Wiley & Sons, Inc., New York, 1988. Spectral theory. Selfadjoint operators in Hilbert space; With the assistance of William G. Bade and Robert G. Bartle; Reprint of the 1963 original; A Wiley-Interscience Publication. MR1009163 (90g:47001b)

[17] Dorin Ervin Dutkay and Palle E. T. Jorgensen, *Fourier duality for fractal measures with affine scales*, Math. Comp. **81** (2012), no. 280, 2253–2273, DOI 10.1090/S0025-5718-2012-02580-4. MR2945155

[18] B. J. Falkowski, *Factorizable and infinitely divisible PUA representations of locally compact groups*, J. Mathematical Phys. **15** (1974), 1060–1066. MR0372115 (51 #8332)

[19] Bent Fuglede, *Boundary minimum principles in potential theory*, Math. Ann. **210** (1974), 213–226. MR0357827 (50 #10293b)

[20] T. N. E. Greville, I. J. Schoenberg, and A. Sharma, *The behavior of the exponential Euler spline $S_n(x; t)$ as $n \to \infty$ for negative values of the base t*, Second Edmonton conference on approximation theory (Edmonton, Alta., 1982), CMS Conf. Proc., vol. 3, Amer. Math. Soc., Providence, RI, 1983, pp. 185–198. MR729330 (85c:41017)

[21] Takeyuki Hida, *Brownian motion*, Applications of Mathematics, vol. 11, Springer-Verlag, New York-Berlin, 1980. Translated from the Japanese by the author and T. P. Speed. MR562914 (81a:60089)

[22] Nobuyuki Ikeda and Shinzo Watanabe, *Stochastic differential equations and diffusion processes*, 2nd ed., North-Holland Mathematical Library, vol. 24, North-Holland Publishing Co., Amsterdam; Kodansha, Ltd., Tokyo, 1989. MR1011252 (90m:60069)

[23] P. E. T. Jorgensen and A. M. Paolucci, *q-frames and Bessel functions*, Numer. Funct. Anal. Optim. **33** (2012), no. 7-9, 1063–1069, DOI 10.1080/01630563.2012.682139. MR2966144

[24] Palle E. T. Jørgensen, *A uniqueness theorem for the Heisenberg-Weyl commutation relations with nonselfadjoint position operator*, Amer. J. Math. **103** (1981), no. 2, 273–287, DOI 10.2307/2374217. MR610477 (82g:81033)

[25] Palle E. T. Jorgensen, *Analytic continuation of local representations of Lie groups*, Pacific J. Math. **125** (1986), no. 2, 397–408. MR863534 (88m:22030)

[26] Palle E. T. Jorgensen, *Analytic continuation of local representations of symmetric spaces*, J. Funct. Anal. **70** (1987), no. 2, 304–322, DOI 10.1016/0022-1236(87)90115-7. MR874059 (88d:22021)

[27] Palle E. T. Jorgensen, *Positive definite functions on the Heisenberg group*, Math. Z. **201** (1989), no. 4, 455–476, DOI 10.1007/BF01215151. MR1004167 (90m:22024)

[28] Palle E. T. Jorgensen, *Extensions of positive definite integral kernels on the Heisenberg group*, J. Funct. Anal. **92** (1990), no. 2, 474–508, DOI 10.1016/0022-1236(90)90060-X. MR1069255 (91m:22013)

[29] Palle E. T. Jorgensen, *Integral representations for locally defined positive definite functions on Lie groups*, Internat. J. Math. **2** (1991), no. 3, 257–286, DOI 10.1142/S0129167X91000168. MR1104120 (92h:43017)

[30] Palle E. T. Jorgensen, *Integral representations for locally defined positive definite functions on Lie groups*, Internat. J. Math. **2** (1991), no. 3, 257–286, DOI 10.1142/S0129167X91000168. MR1104120 (92h:43017)

[31] Palle E. T. Jorgensen, *The measure of a measurement*, J. Math. Phys. **48** (2007), no. 10, 103506, 15, DOI 10.1063/1.2794561. MR2362879 (2008i:81011)

[32] Palle E. T. Jorgensen and Gestur Ólafsson, *Unitary representations of Lie groups with reflection symmetry*, J. Funct. Anal. **158** (1998), no. 1, 26–88, DOI 10.1006/jfan.1998.3285. MR1641554 (99m:22013)

[33] Palle E. T. Jorgensen and Gestur Ólafsson, *Unitary representations and Osterwalder-Schrader duality*, The mathematical legacy of Harish-Chandra (Baltimore, MD, 1998), Proc. Sympos. Pure Math., vol. 68, Amer. Math. Soc., Providence, RI, 2000, pp. 333–401. MR1767902 (2001f:22036)

[34] Palle E. T. Jorgensen and Erin P. J. Pearse, *A discrete Gauss-Green identity for unbounded Laplace operators, and the transience of random walks*, Israel J. Math. **196** (2013), no. 1, 113–160, DOI 10.1007/s11856-012-0165-2. MR3096586

[35] M. Krein, *Concerning the resolvents of an Hermitian operator with the deficiency-index (m, m)*, C. R. (Doklady) Acad. Sci. URSS (N.S.) **52** (1946), 651–654. MR0018341 (8,277a)

[36] Mark G. Krein and Heinz Langer, *Continuation of Hermitian positive definite functions and related questions*, Integral Equations Operator Theory **78** (2014), no. 1, 1–69, DOI 10.1007/s00020-013-2091-z. MR3147401

[37] A. A. Litvinyuk, *On types of distributions for sums of a class of random power series with independent identically distributed coefficients* (Ukrainian, with English and Ukrainian summaries), Ukraïn. Mat. Zh. **51** (1999), no. 1, 128–132, DOI 10.1007/BF02487417; English transl., Ukrainian Math. J. **51** (1999), no. 1, 140–145. MR1712765 (2000h:60022)

[38] Edward Nelson, *Topics in dynamics. I: Flows*, Mathematical Notes, Princeton University Press, Princeton, N.J.; University of Tokyo Press, Tokyo, 1969. MR0282379 (43 #8091)

[39] J. Neunhäuserer, *Absolutely continuous random power series in reciprocals of Pisot numbers*, Statist. Probab. Lett. **83** (2013), no. 2, 431–435, DOI 10.1016/j.spl.2012.10.024. MR3006973

[40] A. Edward Nussbaum, *Extension of positive definite functions and representation of functions in terms of spherical functions in symmetric spaces of noncompact type of rank 1*, Math. Ann. **215** (1975), 97–116. MR0385473 (52 #6334)

[41] Konrad Osterwalder and Robert Schrader, *Axioms for Euclidean Green's functions*, Comm. Math. Phys. **31** (1973), 83–112. MR0329492 (48 #7834)

[42] G. Pólya, *Remarks on characteristic functions*, Proceedings of the Berkeley Symposium on Mathematical Statistics and Probability, 1945, 1946, University of California Press, Berkeley and Los Angeles, 1949, pp. 115–123. MR0028541 (10,463c)

[43] Walter Rudin, *The extension problem for positive-definite functions*, Illinois J. Math. **7** (1963), 532–539. MR0151796 (27 #1779)

[44] Walter Rudin, *An extension theorem for positive-definite functions*, Duke Math. J. **37** (1970), 49–53. MR0254514 (40 #7722)

[45] Walter Rudin, *Functional analysis*, McGraw-Hill Book Co., New York-Düsseldorf-Johannesburg, 1973. McGraw-Hill Series in Higher Mathematics. MR0365062 (51 #1315)

[46] Walter Rudin, *Fourier analysis on groups*, Wiley Classics Library, John Wiley & Sons, Inc., New York, 1990. Reprint of the 1962 original; A Wiley-Interscience Publication. MR1038803 (91b:43002)

[47] Gennady Samorodnitsky and Murad S. Taqqu, *Lévy measures of infinitely divisible random vectors and Slepian inequalities*, Ann. Probab. **22** (1994), no. 4, 1930–1956. MR1331211 (96j:60023)

[48] I. J. Schoenberg, *Interpolating splines as limits of polynomials*, Linear Algebra Appl. **52/53** (1983), 617–628, DOI 10.1016/0024-3795(83)80039-1. MR709376 (84j:41019)

[49] Steve Smale and Ding-Xuan Zhou, *Learning theory estimates via integral operators and their approximations*, Constr. Approx. **26** (2007), no. 2, 153–172, DOI 10.1007/s00365-006-0659-y. MR2327597 (2009b:68184)

[50] Steve Smale and Ding-Xuan Zhou, *Geometry on probability spaces*, Constr. Approx. **30** (2009), no. 3, 311–323, DOI 10.1007/s00365-009-9070-2. MR2558684 (2011c:60006)

DEPARTMENT OF MATHEMATICS, THE UNIVERSITY OF IOWA, IOWA CITY, IOWA 52242-1419
E-mail address: palle-jorgensen@uiowa.edu
URL: http://www.math.uiowa.edu/~jorgen/

DEPARTMENT OF MATHEMATICS, WRIGHT STATE UNIVERSITY, DAYTON, OHIO 45435
E-mail address: steen@math.wright.edu
URL: http://www.wright.edu/~steen.pedersen/

DEPARTMENT OF MATHEMATICS, WRIGHT STATE UNIVERSITY, DAYTON, OHIO 45435
E-mail address: feng.tian@wright.edu

Contemporary Mathematics
Volume **650**, 2015
http://dx.doi.org/10.1090/conm/650/13033

Approximations in L_p-norms and Besov spaces on compact manifolds

Isaac Z. Pesenson

ABSTRACT. The objective of the paper is to describe Besov spaces on general compact Riemannian manifolds in terms of the best approximation by eigenfunctions of elliptic differential operators.

1. Introduction

Approximation theory and its relations to function spaces on Riemannian manifolds is an old subject which still attracts attention of mathematicians [**2**], [**5**], [**7**], [**9**], [**10**].

The goal of the paper is to give an alternative proof of a theorem in [**6**] (see Theorem 1.1 below) which characterizes functions in Besov spaces on compact Riemannian manifolds by best approximations by eigenfunctions of elliptic differential operators. This theorem was an important ingredient of a construction which led to a descriptions of Besov spaces in terms of bandlimited localized frames (see [**6**] for details).

Let $\mathbf{M}$, dim $\mathbf{M} = n$, be a connected compact Riemannian manifold without boundary and the corresponding space $L_p(\mathbf{M}), 1 \leq p \leq \infty$, is constructed by using Riemannian measure. To define Sobolev spaces, we fix a covering $\{B(y_\nu, r_0)\}$ of $\mathbf{M}$ of finite multiplicity by balls $B(y_\nu, r_0)$ centered at $y_\nu \in \mathbf{M}$ of radius $r_0 < \rho_{\mathbf{M}}$, where $\rho_{\mathbf{M}}$ is the injectivity radius of the manifold. For a fixed partition of unity $\Psi = \{\psi_\nu\}$ subordinate to this covering the Sobolev spaces $W_p^k(\mathbf{M}), k \in \mathbb{N}, 1 \leq p < \infty$, are introduced as the completion of $C^\infty(\mathbf{M})$ with respect to the norm

$$(1.1) \qquad \|f\|_{W_p^k(\mathbf{M})} = \left(\sum_\nu \|\psi_\nu f\|_{W_p^k(\mathbf{R}^n)}^p \right)^{1/p}.$$

Similarly,

$$(1.2) \qquad \|f\|_{B_{p,q}^\alpha(\mathbf{M})} = \left(\sum_\nu \|\psi_\nu f\|_{B_{p,q}^\alpha(\mathbf{R}^n)}^p \right)^{1/p},$$

where $B_{p,q}^\alpha(\mathbf{R}^n)$ is the Besov space $B_{p,q}^\alpha(\mathbf{M}), \alpha > 0, 1 \leq p < \infty, 0 \leq q \leq \infty$. It is known [**13**] that such defined spaces are independent on the choice of a partition of unity.

2010 *Mathematics Subject Classification.* Primary 43A85, 42C40, 41A17, 41A10.

Let L be a second order elliptic differential operator on $\mathbf{M}$ with smooth coefficients. By duality such an operator can be extended to the space of distributions on
$\mathbf{M}$. This extension, with domain consisting of all $f \in L_2(\mathbf{M})$ for which $Lf \in L_2(\mathbf{M})$
defines an operator in the space $L_2(\mathbf{M})$. We will assume that this operator is self-
adjoint and non-negative.

One can consider the positive square root $L^{1/2}$ and by duality extend it to the
space of distributions on $\mathbf{M}$. The corresponding operator $L^{k/2}$, $k \in \mathbb{N}$, in a space
$L_p(\mathbf{M}), 1 \leq p \leq \infty$, is defined on the set of all distributions $f \in L_p(\mathbf{M})$ for which
$L^{k/2}f \in L_p(\mathbf{M})$. It is known that the domain of the operator $L^{k/2}$ is exactly the
Sobolev space $W_p^k(\mathbf{M})$.

In every space $L_p(\mathbf{M})$, $1 \leq p \leq \infty$, such defined operators (which will all be
denoted as L) have the same spectrum $0 = \lambda_0 < \lambda_1 \leq \lambda_2 \leq \ldots$ and the same set of
eigenfunctions. Let $u_0, u_1, u_2, \ldots$ be a corresponding complete set of eigenfunctions
which are orthonormal in the space $L_p(\mathbf{M})$.

The notation $\mathbf{E}_\omega(L)$, $\omega > 0$, will be used for the span of all eigenfunctions of
L, whose corresponding eigenvalues are not greater than ω.

For $1 \leq p \leq \infty$, if $f \in L_p(\mathbf{M})$, we let

$$(1.3) \qquad \mathcal{E}(f, \omega, p) = \inf_{g \in \mathbf{E}_\omega(L)} \|f - g\|_{L_p(\mathbf{M})}.$$

The following theorem was proved in [6].

THEOREM 1.1. *If $\alpha > 0$, $1 \leq p \leq \infty$, and $0 < q \leq \infty$ then $f \in B_{p,q}^\alpha(\mathbf{M})$ if and
only if $f \in L_p(\mathbf{M})$ and*

$$(1.4) \qquad \|f\|_{\mathcal{A}_{p,q}^\alpha(\mathbf{M})} := \|f\|_{L_p(\mathbf{M})} + \left(\sum_{j=0}^\infty (2^{\alpha j} \mathcal{E}(f, 2^{2j}, p))^q \right)^{1/q} < \infty.$$

Moreover,

$$(1.5) \qquad \|f\|_{\mathcal{A}_{p,q}^\alpha(\mathbf{M})} \sim \|f\|_{B_{p,q}^\alpha(\mathbf{M})}.$$

Our objective is to give a proof of this Theorem which is different from the
proof presented in [6]. The new proof relies on powerful tools of the theory of
interpolation of linear operators.

2. Kernels and Littlewood-Paley decomposition on compact Riemannian manifolds

Assuming $F \in C_0^\infty(\mathbf{R})$ and using the spectral theorem, one can define the
bounded operator $F(t^2 L)$ on $L_2(\mathbf{M})$. In fact, for $f \in L_2(\mathbf{M})$,

$$(2.1) \qquad [F(t^2 L)f](x) = \int K_t^F(x, y) f(y) dy,$$

where

$$(2.2) \qquad K_t^F(x, y) = \sum_l F(t^2 \lambda_l) u_l(x) u_l(y).$$

We call K_t^F the kernel of $F(t^2 L)$. $F(t^2 L)$ maps $C^\infty(\mathbf{M})$ to itself continuously, and
may thus be extended to be a map on distributions. In particular we may apply
$F(t^2 L)$ to any $f \in L_p(\mathbf{M}) \subseteq L_1(\mathbf{M})$ (where $1 \leq p \leq \infty$), and by Fubini's theorem
$F(t^2 L)f$ is still given by (2.1).

The following Theorem about K_t^F was proved in [**6**] for general elliptic second order differential self-adjoint positive operators.

THEOREM 2.1. *Assume $F \in C_0^\infty(\mathbf{R})$ and let $K_t^F(x, y)$ be the kernel of $F(t^2 L)$. Then for any $n \in \mathbb{N}$ there exists $C(F, N) > 0$ such that*

$$(2.3) \qquad |K_t^F(x, y)| \leq \frac{C(F, N)}{t^n \left[1 + \frac{d(x,y)}{t}\right]^N}, \quad n = \dim \mathbf{M},$$

for $0 < t \leq 1$ and all $x, y \in \mathbf{M}$. The constant $C(F, N)$ depends on the norm of F in the space $C^N(\mathbf{R})$.

This estimate has the following important implication.

COROLLARY 2.2. *Consider $1 \leq \alpha \leq \infty$, with conjugate index α'. In the situation of Theorem 2.3, there is a constant $C > 0$ such that*

$$(2.4) \qquad \left(\int |K_t^F(x, y)|^\alpha dy\right)^{1/\alpha} \leq Ct^{-n/\alpha'} \qquad \qquad \text{for all } x.$$

PROOF. First, we note that if $N > n$, $x \in \mathbf{M}$ and $t > 0$, then

$$(2.5) \qquad \int_\mathbf{M} \frac{1}{[1 + (d(x,y)/t)]^N} dy \leq Ct^n, \quad n = \dim \mathbf{M},$$

with C independent of x or t. Indeed, there exist $c_1, c_2 > 0$ such that for all $x \in M$ and all sufficiently small $r \leq \delta$ one has

$$c_1 r^n \leq |B(x, r)| \leq c_2 r^n,$$

and if $r > \delta$

$$c_3 \delta^n \leq |B(x, r)| \leq |\mathbf{M}| \leq c_4 r^n.$$

For fixed x, t let $A_j = B(x, 2^j t) \setminus B(x, 2^{j-1} t)$. Then $|A_j| \leq c_4 2^{nj} t^n$ and for every A_j one has

$$\int_{A_j} \frac{1}{[1 + (d(x,y)/t)]^N} dy \leq c_4 2^{(n-N)j} t^n.$$

In other words,

$$\int_\mathbf{M} \frac{1}{[1 + (d(x,y)/t)]^N} dy = \sum_j \int_{A_j} \frac{1}{[1 + (d(x,y)/t)]^N} dy \leq c_4 \sum_j 2^{(n-N)j} t^n.$$

Using this estimate and (2.3) one obtains (2.5).

We note that if $d(x, y) \geq t$, then $(d(x, y)/t)^{-N} \leq 2^N (1 + d(x, y)/t)^{-N}$. This observation implies the following fact:

for any $N > n$ there exists C_N such that for all $x \in \mathbf{M}$, $t > 0$

$$\int_{d(x,y) \geq t} d(x, y)^{-N} dy \leq C_N t^{n-N}.$$

This inequality, in turn, implies that our Corollary holds for $\alpha = \infty$.

This completes the proof. $\qquad\qquad\qquad\qquad\qquad\qquad\qquad\qquad\qquad\qquad$ □

THEOREM 2.3. *If $F \in C_0^\infty(\mathbf{R})$ and $(1/q)+1 = (1/p)+(1/\alpha)$ then for the same constant C as in (2.4) one has*

$$\|F(t^2 L)\|_{L_p(\mathbf{M}) \to L_q(\mathbf{M})} \le C t^{-n/\alpha'}, \quad n = dim \, \mathbf{M},$$

for all $0 < t \le 1$. In particular,

$$\|F(t^2 L)\|_{L_p(\mathbf{M}) \to L_p(\mathbf{M})} \le C, \quad n = dim \, \mathbf{M},$$

for all $0 < t \le 1$.

PROOF. The proof follows from Corollary 2.2 and the following Young inequalities. $\qquad\square$

LEMMA 2.4. *Let $\mathcal{K}(x,y)$ be a measurable function on $\mathbf{M} \times \mathbf{M}$. Suppose that $1 \le p, \alpha \le \infty$, and that $(1/q)+1 = (1/p)+(1/\alpha)$. If there exists a $C > 0$ such that*

$$(2.6) \qquad \left(\int_{\mathbf{M}} |\mathcal{K}(x,y)|^\alpha dy \right)^{1/\alpha} \le C \qquad \text{for all } x \in \mathbf{M},$$

and

$$(2.7) \qquad \left(\int_{\mathbf{M}} |\mathcal{K}(x,y)|^\alpha dx \right)^{1/\alpha} \le C \qquad \text{for all } y \in \mathbf{M},$$

then for the same constant C for all $f \in L_p(\mathbf{M})$ one has the inequality

$$\left\| \int_{\mathbf{M}} \mathcal{K}(x,y)f(y)dy \right\|_{L_q(\mathbf{M})} \le C\|f\|_{L_p(\mathbf{M})}.$$

PROOF. Let $\beta = q/\alpha \ge 1$, so that $\beta' = p'/\alpha$. For any x, we have

$$
\begin{aligned}
|(\mathcal{K}f)(x)| &\le \int |\mathcal{K}(x,y)|^{1/\beta'}|\mathcal{K}(x,y)|^{1/\beta}f(y)|dy \\
&\le \left(\int |\mathcal{K}(x,y)|^{p'/\beta'} dy \right)^{1/p'} \left(\int |\mathcal{K}(x,y)|^{p/\beta}|f(y)|^p dy \right)^{1/p} \\
&\le c^{1/\beta'} \left(\int |\mathcal{K}(x,y)|^{p/\beta}|f(y)|^p dy \right)^{1/p}
\end{aligned}
$$

since $p'/\beta' = \alpha$, $\alpha/p' = 1/\beta'$. Thus

$$
\begin{aligned}
\|\mathcal{K}f\|_q^p &\le c^{p/\beta'} \left(\int \left(\int |\mathcal{K}(x,y)|^{p/\beta}|f(y)|^p dy \right)^{q/p} dx \right)^{p/q} \\
&\le c^{p/\beta'} \int \left(\int |\mathcal{K}(x,y)|^{pq/\beta p}|f(y)|^{pq/p}dx \right)^{p/q} dy \\
&= c^{p/\beta'} \int \left(\int |\mathcal{K}(x,y)|^\alpha dx \right)^{p/q} |f(y)|^p dy \\
&\le c^{p/\beta'} c^{p/\beta}\|f\|_p^p
\end{aligned}
$$

as desired. (In the second line, we have used Minkowski's inequality for integrals.) $\qquad\square$

3. Interpolation and Approximation spaces

The goal of this section is to remind the reader of certain connections between interpolation spaces and approximation spaces which will be used later. The general theory of interpolation spaces can be found in [1], [3], [8]. The notion of Approximation spaces and their relations to Interpolations spaces can be found in [4], [12], and in [1], Ch. 3 and 7.

It is important to realize that relations between Interpolation and Approximation spaces cannot be described in the language of normed spaces. One has to use the language of quasi-normed linear spaces to treat interpolation and approximation spaces simultaneously.

Let E be a linear space. A quasi-norm $\|\cdot\|_E$ on E is a real-valued function on E such that for any $f, f_1, f_2 \in E$ the following holds true

(1) $\|f\|_E \geq 0$;
(2) $\|f\|_E = 0 \iff f = 0$;
(3) $\|-f\|_E = \|f\|_E$;
(4) $\|f_1 + f_2\|_E \leq C_E(\|f_1\|_E + \|f_2\|_E), C_E > 1$.

We say that two quasi-normed linear spaces E and F form a pair, if they are linear subspaces of a linear space $\mathcal{A}$ and the conditions

$$\|f_k - g\|_E \to 0, \quad \|f_k - h\|_F \to 0, \ k \to \infty, \ f_k, \, g, \, h \in \mathcal{A},$$

imply equality $g = h$. For a such pair E, F one can construct a new quasi-normed linear space $E \bigcap F$ with quasi-norm

$$\|f\|_{E \bigcap F} = \max\left(\|f\|_E, \|f\|_F\right)$$

and another one $E + F$ with the quasi-norm

$$\|f\|_{E+F} = \inf_{f=f_0+f_1, f_0 \in E, f_1 \in F} \left(\|f_0\|_E + \|f_1\|_F\right).$$

All quasi-normed spaces H for which $E \bigcap F \subset H \subset E + F$ are called intermediate between E and F. A vector space homomorphism $T : E \to F$ is called bounded if

$$\|T\| = \sup_{f \in E, f \neq 0} \|Tf\|_F / \|f\|_E < \infty.$$

One says that an intermediate quasi-normed linear space H interpolates between E and F if every bounded homomorphism $T : E + F \to E + F$ which is also bounded when restricted to E and F is bounded homomorphism of H into H.

On $E + F$ one considers the so-called Peetre's K-functional

$$(3.1) \qquad K(f,t) = K(f,t,E,F) = \inf_{f=f_0+f_1, f_0 \in E, f_1 \in F} \left(\|f_0\|_E + t\|f_1\|_F\right).$$

The quasi-normed linear space $(E,F)^K_{\theta,q}, 0 < \theta < 1, 0 < q \leq \infty$, or $0 \leq \theta \leq 1, q = \infty$, is introduced as a set of elements f in $E + F$ for which

$$(3.2) \qquad \|f\|_{\theta,q} = \left(\int_0^\infty \left(t^{-\theta} K(f,t)\right)^q \frac{dt}{t}\right)^{1/q}.$$

It turns out that $(E,F)^K_{\theta,q}, 0 < \theta < 1, 0 \leq q \leq \infty$, or $0 \leq \theta \leq 1, q = \infty$, with the quasi-norm (3.2) interpolates between E and F. The following Reiteration Theorem is one of the main results of the theory (see [1], [3], [8], [12]).

THEOREM 3.1. *Suppose that E_0, E_1 are complete intermediate quasi-normed linear spaces for the pair E, F. If $E_i \in \mathcal{K}(\theta_i, E, F)$, which means*

$$K(f, t, E, F) \leq Ct^{\theta_i} \|f\|_{E_i}, i = 0, 1,$$

where $0 \leq \theta_i \leq 1, \theta_0 \neq \theta_1$, then

$$(E_0, E_1)_{\eta, q}^K \subset (E, F)_{\theta, q}^K,$$

where $0 < q < \infty, 0 < \eta < 1, \theta = (1 - \eta)\theta_0 + \eta\theta_1$.

If for the same pair E, F and the same E_0, E_1 one has $E_i \in \mathcal{J}(\theta_i, E, F)$, i. e.

$$\|f\|_{E_i} \leq C\|f\|_E^{1-\theta_i}\|f\|_F^{\theta_i}, i = 0, 1,$$

where $0 \leq \theta_i \leq 1, \theta_0 \neq \theta_1$, then

$$(E, F)_{\theta, q}^K \subset (E_0, E_1)_{\eta, q}^K,$$

where $0 < q < \infty, 0 < \eta < 1, \theta = (1 - \eta)\theta_0 + \eta\theta_1$.

It is important to note that in all cases considered in the present article the space F will be continuously embedded as a subspace into E. In this case (3.1) can be introduced by the formula

$$K(f, t) = \inf_{f_1 \in F} \left(\|f - f_1\|_E + t\|f_1\|_F \right),$$

which implies the inequality

(3.3) $K(f, t) \leq \|f\|_E.$

This inequality can be used to show that the norm (3.2) is equivalent to the norm

(3.4) $\|f\|_{\theta, q} = \|f\|_E + \left(\int_0^\varepsilon \left(t^{-\theta} K(f, t) \right)^q \frac{dt}{t} \right)^{1/q}, \varepsilon > 0,$

for any positive ε.

Let us introduce another functional on $E + F$, where E and F form a pair of quasi-normed linear spaces

$$\mathcal{E}(f, t) = \mathcal{E}(f, t, E, F) = \inf_{g \in F, \|g\|_F \leq t} \|f - g\|_E.$$

DEFINITION 3.2. *The approximation space $\mathcal{E}_{\alpha, q}(E, F), 0 < \alpha < \infty, 0 < q \leq \infty$ is a quasi-normed linear spaces of all $f \in E + F$ with the following quasi-norm*

(3.5) $\left(\int_0^\infty \left(t^\alpha \mathcal{E}(f, t) \right)^q \frac{dt}{t} \right)^{1/q}.$

For a general quasi-normed linear spaces E the notation $(E)^\rho$ is used for a quasi-normed linear spaces whose quasi-norm is $\| \cdot \|^\rho$.

The following Theorem describes relations between interpolation and approximation spaces (see [1], Ch. 7).

THEOREM 3.3. *If $\theta = 1/(\alpha + 1)$ and $r = \theta q$, then*

$$(\mathcal{E}_{\alpha, q}(E, F))^\theta = (E, F)_{\theta, q}^K.$$

The following important result is known as the Power Theorem (see [1], Ch. 7).

THEOREM 3.4. *Suppose that the following relations satisfied:* $\nu = \eta \rho_1 / \rho$, $\rho = (1 - \eta)\rho_0 + \eta \rho_1$, *and* $q = \rho r$ *for* $\rho_0 > 0, \rho_1 > 0$. *Then, if* $0 < \eta < 1, 0 < r \leq \infty$, *the following equality holds true*

$$\left((E)^{\rho_0}, (F)^{\rho_1}\right)_{\eta,r}^K = \left((E,F)_{\nu,q}^K\right)^{\rho}.$$

The Theorem we prove next represents a very abstract version of what is known as Direct and Inverse Approximation Theorems.

THEOREM 3.5. *Suppose that* $\mathcal{T} \subset F \subset E$ *are quasi-normed linear spaces and* E *and* F *are complete. If there exist* $C > 0$ *and* $\beta > 0$ *such that for any* $f \in F$ *the following Jackson-type inequality is satisfied*

$$(3.6) \qquad t^{\beta} \mathcal{E}(t, f, \mathcal{T}, E) \leq C \|f\|_F, t > 0,$$

then the following embedding holds true

$$(3.7) \qquad (E, F)_{\theta,q}^K \subset \mathcal{E}_{\theta\beta,q}(E, \mathcal{T}), \ 0 < \theta < 1, \ 0 < q \leq \infty.$$

If there exist $C > 0$ *and* $\beta > 0$ *such that for any* $f \in \mathcal{T}$ *the following Bernstein-type inequality holds*

$$(3.8) \qquad \|f\|_F \leq C \|f\|_{\mathcal{T}}^{\beta} \|f\|_E,$$

then

$$(3.9) \qquad \mathcal{E}_{\theta\beta,q}(E, \mathcal{T}) \subset (F, F)_{\theta,q'}^K, \ 0 < \theta < 1, \ 0 < q \leq \infty.$$

PROOF. It is known ([1], Ch.7) that for any $s > 0$, for

$$(3.10) \quad t = K_{\infty}(f, s) = K_{\infty}(f, s, \mathcal{T}, E) = \inf_{f = f_1 + f_2, f_1 \in \mathcal{T}, f_2 \in E} \max(\|f_1\|_{\mathcal{T}}, s \|f_2\|_E)$$

the following inequality holds

$$(3.11) \qquad s^{-1} K_{\infty}(f, s) \leq \lim_{\tau \to t - 0} \inf \mathcal{E}(f, \tau, E, \mathcal{T}).$$

Since

$$(3.12) \qquad K_{\infty}(f, s) \leq K(f, s) \leq 2K_{\infty}(f, s),$$

the Jackson-type inequality (3.6) and the inequality (3.11) imply

$$(3.13) \qquad s^{-1} K(f, s, \mathcal{T}, E) \leq C t^{-\beta} \|f\|_F.$$

The equality (3.10), and inequality (3.12) imply the estimate

$$(3.14) \qquad t^{-\beta} \leq 2^{\beta} \left(K(f, s, \mathcal{T}, E)\right)^{-\beta}$$

which along with the previous inequality gives the estimate

$$K^{1+\beta}(f, s, \mathcal{T}, E) \leq C s \|f\|_F$$

which in turn imply the inequality

$$(3.15) \qquad K(f, s, \mathcal{T}, E) \leq C s^{\frac{1}{1+\beta}} \|f\|_F^{\frac{1}{1+\beta}}.$$

At the same time one has

$$(3.16) \qquad K(f, s, \mathcal{T}, E) = \inf_{f = f_0 + f_1, f_0 \in \mathcal{T}, f_1 \in E} (\|f_0\|_{\mathcal{T}} + s \|f_1\|_E) \leq s \|f\|_E,$$

for every f in E. The inequality (3.15) means that the quasi-normed linear space $(F)^{\frac{1}{1+\beta}}$ belongs to the class $\mathcal{K}(\frac{1}{1+\beta}, \mathcal{T}, E)$ and (3.16) means that the quasi-normed linear space E belongs to the class $\mathcal{K}(1, \mathcal{T}, E)$. This fact allows us to use the Reiteration Theorem to obtain the embedding

$$(3.17) \qquad \left((F)^{\frac{1}{1+\beta}}, E\right)^K_{\frac{1-\theta}{1+\theta\beta}, q(1+\theta\beta)} \subset (\mathcal{T}, E)^K_{\frac{1}{1+\theta\beta}, q(1+\theta\beta)}$$

for every $0 < \theta < 1, 1 < q < \infty$. But the space on the left is the space

$$\left(E, (F)^{\frac{1}{1+\beta}}\right)^K_{\frac{\theta(1+\beta)}{1+\theta\beta}, q(1+\theta\beta)},$$

which according to the Power Theorem is the space

$$\left((E, F)^K_{\theta, q}\right)^{\frac{1}{1+\theta\beta}}.$$

All these results along with the equivalence of interpolation and approximation spaces give the embedding

$$(E, F)^K_{\theta, q} \subset \left((\mathcal{T}, E)^K_{\frac{1}{1+\theta\beta}, q(1+\theta\beta)}\right)^{1+\theta\beta} = \mathcal{E}_{\theta\beta, q}(E, \mathcal{T}),$$

which proves the embedding (3.7). Conversely, if the Bernstein-type inequality (3.18) holds then one has the inequality

$$(3.18) \qquad \|f\|_F^{\frac{1}{1+\beta}} \le C \|f\|_{\mathcal{T}}^{\frac{\beta}{1+\beta}} \|f\|_E^{\frac{1}{1+\beta}}.$$

Along with obvious equality $\|f\|_E = \|f\|_{\mathcal{T}}^0 \|f\|_E$ and the Iteration Theorem one obtains the embedding

$$(\mathcal{T}, E)^K_{\frac{1}{1+\theta\beta}, q(1+\theta\beta)} \subset \left((F)^{\frac{1}{1+\beta}}, E\right)^K_{\frac{1-\theta}{1+\theta\beta}, q(1+\theta\beta)}.$$

To finish the proof of the theorem one can use the same arguments as above. Theorem is proven. □

4. Approximation in spaces $L_p(\mathbf{M})$, $1 \le p \le \infty$

In this section we are going to prove Theorem 1.1 by applying relations between Interpolation and approximation spaces described in Section 3.

For $1 \le p \le \infty$, if $f \in L_p(\mathbf{M})$, we let

$$(4.1) \qquad \mathcal{E}(f, \omega, p) = \inf_{g \in \mathbf{E}_\omega(L)} \|f - g\|_{L_p(\mathbf{M})}.$$

4.1. The Jackson inequality.

LEMMA 4.1. *For every $k \in \mathbb{N}$ there exists a constant $C(k)$ such that for any $\omega > 1$*

$$\mathcal{E}(f, \omega, p) \le C(k)\omega^{-k}\|L^{k/2}f\|_{L_p(\mathbf{M})}, \quad f \in W_p^k(\mathbf{M}).$$

PROOF. Let h be a C^∞ function on $[0, \infty)$ which equals 1 on $[0, 1]$, and which is supported in $[0, 4]$. Define, for $\lambda > 0$,

$$F(\lambda) = h(\lambda/4) - h(\lambda)$$

so that F is supported in $[1, 16]$. For $j \ge 1$, we set

$$F_j(\lambda) = F(\lambda/4^{j-1}).$$

We also set $F_0 = h$, so that $\sum_{j=0}^\infty F_j \equiv 1$. For $\lambda > 0$ we define

$$\Psi(\lambda) = F(\lambda)/\lambda^{k/2}$$

so that Ψ is supported in $[1, 16]$. For $j \geq 1$, we set

$$\Psi_j(\lambda) = \Psi(\lambda/4^{j-1}),$$

so that

$$F_j(\lambda) = 2^{-(j-1)k}\Psi_j(\lambda)\lambda^{k/2}.$$

Now for a given ω we change the variable λ to the variable $4\lambda/\omega$. Clearly, the support of $h(4\lambda/\omega)$ is the interval $[0, \omega]$ and we have the following relation

$$F_j(4\lambda/\omega) = \omega^{-k/2}2^{-(j-2)k}\Psi_j(4\lambda/\omega)\lambda^{k/2}.$$

It implies that if f is a distribution on $\mathbf{M}$ then

$$F_j\left(\frac{4}{\omega}L\right)f = \omega^{-k/2}2^{-(j-2)k}\Psi_j\left(\frac{4}{\omega}L\right)(L^{k/2}f),$$

in the sense of distributions. If now $f \in W_p^k(\mathbf{M})$, so that $L^{k/2}f \in L_p(\mathbf{M})$, we see that for $\alpha' = \infty$ according to Theorem 2.3

$$\left\|\Psi_j\left(\frac{4}{\omega}L\right)(L^{k/2}f)\right\|_{L_p(\mathbf{M})} \leq C'(k)\|L^{k/2}f\|_{L_p(\mathbf{M})},$$

which implies the inequality

$$\mathcal{E}(f, \omega, p) \leq \left\|f - h\left(\frac{4}{\omega}L\right)f\right\|_{L_p(\mathbf{M})} \leq \sum_{j \geq 1}\left\|F_j\left(\frac{4}{\omega}L\right)f\right\|_{L_p(\mathbf{M})} \leq$$

$$C'(k)\omega^{-k/2}\sum_{j \geq 1}2^{-(j-2)k}\|L^{k/2}f\|_{L_p(\mathbf{M})} \leq C(k)\omega^{-k/2}\|L^{k/2}f\|_{L_p(\mathbf{M})}.$$

The proof is complete. $\qquad\qquad\qquad\qquad\qquad\qquad\qquad\qquad\qquad\qquad\qquad$ $\square$

4.2. The Bernstein inequality.

LEMMA 4.2. *There exists a constant $c(k)$ such that for all $f \in \mathbf{E}_\omega(L)$*

$$(4.2) \qquad \|L^k f\|_{L_p(\mathbf{M})} \leq C(k)\omega^{2k}\|f\|_{L_p(\mathbf{M})}.$$

PROOF. Consider a function $h \in C_0^\infty(\mathbf{R}_+)$ such that $h(\lambda) = 1$ for $\lambda \in [0, 1]$. For a fixed $\omega > 0$ the support of $h(\lambda/\omega)$ is $[0, \omega]$ and it shows that for any $f \in \mathbf{E}_\omega(L)$ one has the equality $h(\omega^{-1}L)f = f$.

According to Theorems 2.1 and 2.3 the operator $H(L) = (\omega^{-2}L)^k h(\omega^{-2}L)$ is bounded from $L_p(\mathbf{M})$ to $L_q(\mathbf{M})$. Thus for every $f \in \mathbf{E}_\omega(L)$ we have

$$\|L^k f\|_{L_p(\mathbf{M})} = \omega^{2k}\|(\omega^{-2}L)^k h(\omega^{-2}L)f\|_{L_p(\mathbf{M})} =$$

$$\omega^{2k}\|H(L)f\|_{L_p(\mathbf{M})} \leq C(k)\omega^{2k}\|f\|_{L_p(\mathbf{M})}.$$

$\qquad\qquad\qquad\qquad\qquad\qquad\qquad\qquad\qquad\qquad\qquad\qquad\qquad\qquad$ $\square$

REMARK 4.3. In the inequality (4.2) the constant depends on the exponent k. One can show [11] that in the case of a compact homogeneous manifold a similar inequality holds with a constant that depends just on the manifold.

4.3. Besov spaces and approximations. It is known [13] that the Besov space $B^\alpha_{p,q}(\mathbf{M}), k \in \mathbb{N}, 1 \leq p < \infty, 0 \leq q \leq \infty$, which was defined in (1.2), is the interpolation space

$$B^\alpha_{p,q}(\mathbf{M}) = (L_p(\mathbf{M}), W^r_p(\mathbf{M}))^K_{\alpha/r,q},$$

where K is Peetre's interpolation functor.

Let us compare the situation on manifolds with the abstract conditions of the Theorem 3.5. We treat linear normed spaces $W^r_p(\mathbf{M})$ and $L_p(\mathbf{M})$ as the spaces E and F respectively. We identify $\mathcal{T}$ with the linear space $\mathbf{E}_\omega(L)$ which is equipped with the quasi-norm

$$\|f\|_{\mathcal{T}} = \inf_\omega \{\omega : f \in \mathbf{E}_\omega(L)\}, \quad f \in \mathbf{E}_\omega(L).$$

Thus, Lemmas 4.1, 4.2 and Theorem 3.5 imply the following result.

THEOREM 4.4. *If $\alpha > 0$, $1 \leq p \leq \infty$, and $0 < q \leq \infty$ then $f \in B^\alpha_{p,q}(\mathbf{M})$ if and only if $f \in L_p(\mathbf{M})$ and*

$$(4.3) \qquad \|f\|_{\mathcal{A}^\alpha_{p,q}(\mathbf{M})} := \|f\|_{L_p(\mathbf{M})} + \left(\int_0^\infty (t^\alpha \mathcal{E}(f,t,p))^q \frac{dt}{t} \right)^{1/q} < \infty.$$

Moreover,

$$(4.4) \qquad \|f\|_{\mathcal{A}^\alpha_{p,q}(\mathbf{M})} \sim \|f\|_{B^\alpha_{p,q}(\mathbf{M})}.$$

By discretizing the integral term (see [1]) we obtain Theorem 1.1.

References

[1] J. Bergh, J. Löfström, *Interpolation spaces*, Springer-Verlag, 1976.
[2] Gavin Brown and Feng Dai, *Approximation of smooth functions on compact two-point homogeneous spaces*, J. Funct. Anal. **220** (2005), no. 2, 401–423, DOI 10.1016/j.jfa.2004.10.005. MR2119285 (2005m:41054)
[3] Paul L. Butzer and Hubert Berens, *Semi-groups of operators and approximation*, Die Grundlehren der mathematischen Wissenschaften, Band 145, Springer-Verlag New York Inc., New York, 1967. MR0230022 (37 #5588)
[4] P. L. Butzer and K. Scherer, *Jackson and Bernstein-type inequalities for families of commutative operators in Banach spaces*, J. Approximation Theory **5** (1972), 308–342. Collection of articles dedicated to J. L. Walsh on his 75th birthday, III. MR0346396 (49 #11121)
[5] F. Filbir and H. N. Mhaskar, *A quadrature formula for diffusion polynomials corresponding to a generalized heat kernel*, J. Fourier Anal. Appl. **16** (2010), no. 5, 629–657, DOI 10.1007/s00041-010-9119-4. MR2673702 (2011j:41007)
[6] Daryl Geller and Isaac Z. Pesenson, *Band-limited localized Parseval frames and Besov spaces on compact homogeneous manifolds*, J. Geom. Anal. **21** (2011), no. 2, 334–371, DOI 10.1007/s12220-010-9150-3. MR2772076 (2012c:43013)
[7] Daryl Geller and Isaac Z. Pesenson, *Kolmogorov and linear widths of balls in Sobolev spaces on compact manifolds*, Math. Scand. **115** (2014), no. 1, 96–122. MR3250051
[8] S. G. Kreĭn, Yu. Ī. Petunīn, and E. M. Semënov, *Interpolation of linear operators*, Translations of Mathematical Monographs, vol. 54, American Mathematical Society, Providence, R.I., 1982. Translated from the Russian by J. Szűcs. MR649411 (84j:46103)
[9] Erlan Nursultanov, Michael Ruzhansky, Sergey Tikhonov, *Nikolskii inequality and Besov, Triebel-Lizorkin, Wiener and Beurling spaces on compact homogeneous manifolds*, arXiv:1403.3430
[10] I. Z. Pesenson, *Best approximations in a space of the representation of a Lie group* (Russian), Dokl. Akad. Nauk SSSR **302** (1988), no. 5, 1055–1058; English transl., Soviet Math. Dokl. **38** (1989), no. 2, 384–388. MR981052 (90c:41071)
[11] Isaac Pesenson, *Bernstein-Nikolskii inequalities and Riesz interpolation formula on compact homogeneous manifolds*, J. Approx. Theory **150** (2008), no. 2, 175–198, DOI 10.1016/j.jat.2007.06.001. MR2388855 (2009g:43005)

[12] J. Peetre and G. Sparr, *Interpolation of normed abelian groups*, Ann. Mat. Pura Appl. (4) **92** (1972), 217–262. MR0322529 (48 #891)

[13] Hans Triebel, *Theory of function spaces. II*, Monographs in Mathematics, vol. 84, Birkhäuser Verlag, Basel, 1992. MR1163193 (93f:46029)

DEPARTMENT OF MATHEMATICS, TEMPLE UNIVERSITY, PHILADELPHIA, PENNSYLVANIA 19122
E-mail address: pesenson@temple.edu

650 **Jens G. Christensen, Susanna Dann, Azita Mayeli, and Gestur Ólafsson, Editors,** Trends in Harmonic Analysis and Its Applications, 2015

647 **Gary Kennedy, Mirel Caibăr, Ana-Maria Castravet, and Emanuele Macrì, Editors,** Hodge Theory and Classical Algebraic Geometry, 2015

646 **Weiping Li and Shihshu Walter Wei, Editors,** Geometry and Topology of Submanifolds and Currents, 2015

645 **Krzysztof Jarosz, Editor,** Function Spaces in Analysis, 2015

644 **Paul M. N. Feehan, Jian Song, Ben Weinkove, and Richard A. Wentworth, Editors,** Analysis, Complex Geometry, and Mathematical Physics, 2015

643 **Tony Pantev, Carlos Simpson, Bertrand Toën, Michel Vaquié, and Gabriele Vezzosi, Editors,** Stacks and Categories in Geometry, Topology, and Algebra, 2015

642 **Mustapha Lahyane and Edgar Martínez-Moro, Editors,** Algebra for Secure and Reliable Communication Modeling, 2015

641 **Maria Basterra, Kristine Bauer, Kathryn Hess, and Brenda Johnson, Editors,** Women in Topology, 2015

640 **Gregory Eskin, Leonid Friedlander, and John Garnett, Editors,** Spectral Theory and Partial Differential Equations, 2015

639 **C. S. Aravinda, William M. Goldman, Krishnendu Gongopadhyay, Alexander Lubotzky, Mahan Mj, and Anthony Weaver, Editors,** Geometry, Groups and Dynamics, 2015

638 **Javad Mashreghi, Emmanuel Fricain, and William Ross, Editors,** Invariant Subspaces of the Shift Operator, 2015

637 **Stéphane Ballet, Marc Perret, and Alexey Zaytsev, Editors,** Algorithmic Arithmetic, Geometry, and Coding Theory, 2015

636 **Simeon Reich and Alexander J. Zaslavski, Editors,** Infinite Products of Operators and Their Applications, 2015

635 **Christopher W. Curtis, Anton Dzhamay, Willy A. Hereman, and Barbara Prinari, Editors,** Nonlinear Wave Equations, 2015

634 **Steven Dougherty, Alberto Facchini, André Leroy, Edmund Puczyłowski, and Patrick Solé, Editors,** Noncommutative Rings and Their Applications, 2015

633 **Delaram Kahrobaei and Vladimir Shpilrain, Editors,** Algorithmic Problems of Group Theory, Their Complexity, and Applications to Cryptography, 2015

632 **Gohar Kyureghyan, Gary L. Mullen, and Alexander Pott, Editors,** Topics in Finite Fields, 2015

631 **Siddhartha Bhattacharya, Tarun Das, Anish Ghosh, and Riddhi Shah, Editors,** Recent Trends in Ergodic Theory and Dynamical Systems, 2015

630 **Pierre Albin, Dmitry Jakobson, and Frédéric Rochon, Editors,** Geometric and Spectral Analysis, 2014

629 **Milagros Izquierdo, S. Allen Broughton, Antonio F. Costa, and Rubí E. Rodríguez, Editors,** Riemann and Klein Surfaces, Automorphisms, Symmetries and Moduli Spaces, 2014

628 **Anita T. Layton and Sarah D. Olson, Editors,** Biological Fluid Dynamics: Modeling, Computations, and Applications, 2014

627 **Krishnaswami Alladi, Frank Garvan, and Ae Ja Yee, Editors,** Ramanujan 125, 2014

626 **Veronika Furst, Keri A. Kornelson, and Eric S. Weber, Editors,** Operator Methods in Wavelets, Tilings, and Frames, 2014

625 **Alexander Barg and Oleg R. Musin, Editors,** Discrete Geometry and Algebraic Combinatorics, 2014